OREGON FOSSILS

Second Edition

Elizabeth L. Orr and William N. Orr

Oregon State University Press
Corvallis

Front cover photographs: *Oreodont* (top) was a herbivore about the size of a goat that ran in vast herds in eastern Oregon between 40 and 10 million years ago. It is one of the commonest mammals of the John Day formation. *Salenia* (bottom) was a deep water spiny sea urchin about the size of a quarter that was found in the 40 million-year-old Keasey formation where it is exposed west of Portland in the vicinity of Scappoose.

The paper in this book meets the guidelines for permanence and durability of the Committee on Production Guidelines for Book Longevity of the Council on Library Resources and the minimum requirements of the American National Standard for Permanence of Paper for Printed Library Materials Z39.48-1984.

Library of Congress Cataloging-in-Publication Data
Orr, Elizabeth L.
  Fossils of Oregon / Elizabeth L. Orr and William N. Orr. -- 2nd ed.
      p. cm.
  Rev. ed. of: Fossils of Oregon. 1st ed. Dubuque, Iowa : Kendall/Hunt Pub. Co., c1999.
  Includes bibliographical references and index.
  ISBN 978-0-87071-573-0 (alk. paper)
  1. Geology--Oregon. I. Orr, William N., 1939- II. Orr, Elizabeth L. Geology of Oregon. III. Title.
  QE155.O77 2009
  560.9795--dc22
                                                            2009011692

Oregon State University Press
121 The Valley Library
Corvallis OR 97331-4501
541-737-3166 • fax 541-737-3170
http://oregonstate.edu/dept/press

To the sally seekers

Contents

PREFACE

The *Handbook of Oregon Plant and Animal Fossils*, first published in 1981, was rewritten as *Oregon Fossils* in 1999. The purpose of this edition is to summarize new work and discoveries over the past ten years. The authors wanted to cover Oregon's complete fossil record, but that ambitious goal had to be set aside when it became apparent that to record the sheer diversity and number of fossils would be overwhelming. Ultimately, they focused on the most significant discoveries or those that are unique to Oregon's fossil record.

In an undertaking such as this, research and writing are only part of the process. With that in mind, we thank John Beaulieu, Robert McWilliams, Ellen Moore, Guy DiTorris, and Greg Retallack for their help and reviews. For accuracy we depended on editing by Teresa Jesionowski and indexing by Jean Mooney. The efficient staff at Oregon State University Press organized, assembled, and produced the final book. As always, those who work at the Oregon Department of Geology and Mineral Industries are generous with their photographs and other resources. Finally, no matter what venue, publications could not exist without the many talents of librarians.

Elizabeth L. Orr and William N. Orr

INTRODUCTION

Although deciphering Oregon's geologic past began with government-sponsored surveys during the mid-1800s, formal attention to fossils and paleontology did not begin until some 40 years later. These studies were initiated by Thomas Condon, who brought worldwide attention to the state's fossil wealth, and whose book *The Two Islands* was the first to tell Oregon's geologic story. Born in southern Ireland in 1822, Condon was 11 years of age when his family moved to New York. After graduating from the theological seminary at Auburn, he married Cornelia Holt in 1852, and in the fall of that year they sailed for Oregon under the auspices of the Congregational Church. It was prophetic that he accepted a position at The Dalles, because there he was able to pursue his long-time interest in geology at the nearby John Day fossil region. Soon Condon's collection of fossils, his enthusiasm, and his geologic knowledge attracted inquiries and brought distinguished visitors to his home. As his reputation spread, he filled requests for specimens, which were sent to paleontologists elsewhere. As an outgrowth of his own interest in geology and paleontology, he accepted a position as the State Geologist in 1872, resigning in 1876 to accept an appointment as geology and natural history professor at the fledgling University of Oregon in Eugene. For more than 40 years he acquired fossils, which form the basis for the Condon collection at the Museum of Natural and Cultural History. Retiring from teaching at age 82, he died shortly thereafter in 1907.

Thomas Condon by the Condon oak, which stands today next to Villard Hall on the University of Oregon campus. (Photo courtesy Condon collection.)

As enthusiasm to uncover exotic new fossils gathered speed in the late 1800s, great bone-collecting expeditions sent wagon-loads of material to fill museum repositories. Here the specimens were examined, and many new species illustrated. From the 1930s through the 1960s, researchers itemized, catalogued, and published on Oregon's regional plant and animal communities, using the faunas and floras for age-dating strata. With the renaissance of continental drift as plate tectonics, scientists realized that fossils were the best tool for plotting the movements of the crust. Wandering microcontinents, which form an integral part of Oregon's past, could be traced back to their original locations. More recently, paleontologists have moved into the realm of geologic changes on a global scale. The large body of data now available enables them to tie the state's history into worldwide events of extinctions, warming, and volcanic activity.

Thanks to movies such as *Jurassic Park* and NOVA programs, public interest in fossils is at an all-time high, and there are numerous resources to provide collectors with assistance. Literature, talks and classes, displays, field trips, and fossil identification are available. In Portland, the Oregon State Department of Geology and Mineral Industries offers information and expertise, and their Nature of the Northwest store is an invaluable source of books, reports, maps, videos, and CDs. The Oregon Museum of Science and Industry in Portland holds classes and has fossil displays. OMSI also sponsors Camp Hancock in Wheeler County where young students can experience fieldwork in fossil collecting. In Eugene, the Museum of Natural and Cultural History, which houses the Condon collection, displays specimens and offers information and classes, as well as expertise in fossil identification. In eastern Oregon, the John Day National Monument displays fossils in their new center, and the Oregon Paleo Lands Institute in Fossil holds field trips and classes. Many cities and counties have rock, fossil, and mineral clubs that welcome members and inquiries. The Geologic Society of the Oregon Country in Portland has sponsored speakers and field trips for more than 50 years, while the North American Research Group, based at the Rice Museum in Beaverton, has a large group of enthusiastic and knowledgeable participants.

Paleontology examines prehistoric life and ranges from the study of shells, wood, leaves, and bones to single-celled animals or plants. Fossils need not be the body of an organism, shell, or skeleton but may be a mold, an impression, a track, and even a burrow. The word *fossil* means "dug up," and it usually calls up visions of dinosaurs or large animals, but vertebrates constitute only a very small portion of the science.

To be elevated to the status of a fossil, plant or animal remains must have a certain degree of antiquity. Oregon's most ancient organisms are Paleozoic marine corals and microfossils, which lived 400 million years ago, while the youngest are from the Ice Age, inhabiting the state at the end of the last glacial withdrawal, about 11,000 years ago. Even though a creature may have become extinct, it does not automatically become a fossil. In fact, many so-called fossils fail to meet the *fossil* criteria because they are too young. True fossils, then, are remnants which date back at least to the end of the Pleistocene or Ice Ages.

The probability that an animal will be preserved in a given rock is a complex equation that includes the actual numbers or frequency of the organism, its relative preservability as bone, shell, or leaf, and the environment in which it is to be entombed. The optimum conditions for preservation differ for individual groups of organisms; therefore, the various possibilities are dealt with separately for each. In general terms, both invertebrates, which are diverse and numerous, and plants, that contribute thousands of leaves or pollen, are better represented in the fossil record than are vertebrate skeletons, which are more limited and often fail to be preserved. Heat, pressure, and percolating water, the greatest liabilities to preservation, damage or destroy most remains entirely.

Fossils are a nonrenewable resource, and the more rare forms such as the bones of vertebrates should be treated as historic artifacts with the same respect shown to Stone Age tools, flintlock firearms, and antique automobiles. Just as these objects reflect the creative effort of humans to meet a need, an individual fossil represents the product of natural selection (trial and error) by nature over millions of years of environmental changes. Where the fossilized specimens are limited, the site as well as the remains should be conserved. Once an area is mined out or the fossil removed without regard for its origin, the information that the specimen might provide is in danger of being lost.

Knowing where to look for fossils goes a long way toward actually finding them, and the inexperienced person might spend a considerable amount of time looking in the wrong places. Most rocks making up the earth's crust are sedimentary or volcanic, and volcanic rocks, which began as molten lava, have few fossils. Sedimentary rocks, on the other hand, were deposited in an aquatic setting such as in streams, lakes, or the ocean, and these environments are more conducive to preserving material. With vast tracts of Oregon in volcanic terrains, exposures of sedimentary rocks in quarries, roadway and railroad cuts, at building construction sites, in stream valleys, or from beach cliffs are good places to start. It also might be best to start at well-known classic localities, which are accessible and where there is a good possibility of a find. Although museums and educational displays generally show complete and unbroken specimens, most fossils are only partially preserved, as a variety of environmental conditions scatter or break apart the material.

Plotting the frequency and location of fossil occurrences in Oregon results in an interesting, if possibly misleading, picture of distribution. The highest incidence of fossil sites is in the Willamette Valley, followed by those on the western slopes of the Coast Range and along the beach, in the Basin and Range province, and in central Oregon in the John Day River valley. There are large gaps in fossil locales in portions of the central coastal range, in the Cascades, and in the Wallowa Mountains; and there are only a few places in the Klamath province and on the High Lava Plains where remains have been exposed.

To a certain extent, this distribution is the result of human activity and not the consequence of where the most fossils were preserved. Cultural activity in the

Willamette Valley has led to far more fossils being unearthed here than elsewhere. Heavy vegetation and little development within the Coast Range have limited the number of finds in this region. Oregon's most celebrated and richest trove of fossils lies in the John Day area where soils are thin and vegetation sparse. First discovered in the 1880s, this site has been the focus of attention ever since. Fossil Lake is equally well known for its variety of fossils. Discovered in the 1860s, the shiny jet-black bones are visible where they are scattered across the surface.

There are specific rules for collecting. Lawmakers have recognized the need to protect natural resources such as special plants and animals, native terrestrial and aquatic ecosystems, and geologic features. Consequently, fossil collecting is limited or prohibited in areas set aside for parks, in state- and federally owned forests, on federal natural areas, in marine preserves, campgrounds, roadside parks, or day-use areas. State ownership of the ocean shore between low tide and the line of established vegetation excludes large-scale collecting. Occasional pieces of driftwood, shells, agates, or fossils may be taken in limited quantities, but such natural objects may not be sold commercially.

Issues of ownership arise with fossils just as with timber, water, and mineral resources. Private land ownership reigns supreme for fossil collecting since the landowner alone retains the rights to collect, trade, sell, or assign the rights to others. Private property is exempt from state and federal regulations regarding fossils, and permission should always be obtained before collecting and trespassing.

A great deal of Oregon's ancient past is written in its fossil record. Plants and animals are used to reconstruct the succession of life as well as to tell the story of climate variation, sea level changes, mountain building, and even moving continents. An understanding of ancient environments is critical when viewing the current trends of global warming or other perturbations in the earth's domain. One of the most fascinating aspects of ancient global climate modifications is that fossils may record the events. With such a remarkably comprehensive and high quality Tertiary fossil record, Oregon displays virtually all of the processes and consequences of climate alterations. In many ways Oregon may be the Rosetta Stone to a complete understanding of Tertiary climate and global environmental changes.

PLANTS

Oregon's fossil record stands out for the quantity and quality of material it has contributed to the knowledge of plant life, preserved as leaves, seeds, fruits, flowers, wood, pollen grains, and spores. Floral remains from the Paleozoic and Mesozoic are scarce and known only from the Klamath and Blue mountains, but despite this they play a key role in deciphering the complex story of plate tectonics and ancient climates during these eras, when they grew in habitats well to the west. In contrast to those of older time periods, Oregon's Tertiary plants were home grown, displaying an impressive record of forest, floodplain, delta, and lake settings. The diversity and abundance of floral communities both east and west of the Cascades reflect a wide variety of changing physiographies and climates that tell the story of the state's geologic past.

A Sunday outing to look for fossils in southwest Oregon. (Photo courtesy Grants Pass Art Studio and Oregon Department of Geology and Mineral Industries.)

Analyzing floras to age-date or define ancient environments presents special problems. Quite often, when fossil plants are used to determine the geologic age of a rock, it has been shown that environments, not time, are being correlated. Stated another way, two floras that are identical certainly represent similar settings, but they may belong to completely different time intervals. An inventory of the many climate events in a given geographic area should be made before dating strata of unknown age. Another of the many vexing difficulties that arise when comparing multiple fossil floras is determining whether different plant localities are part of the same community. Such is the case with the Rockville flora found in the Sucker Creek Formation near the Rockville post office in Malheur County. Although sometimes considered a separate flora, angiosperms from Rockville are similar enough to those in the nearby Succor Creek locality that both are considered as part of the same floral horizon.

Similar misinterpretations can arise when making determinations of geologic dates based on plant types alone. For example, although the Bridge Creek, dated at 33.6 million years in the past, is among the oldest of the Oregon Oligocene floras, it has long been considered much younger because of its more temperate floral characteristics. The overriding assumption that tropical vegetation was invariably replaced by temperate plants automatically made the temperate floras seem younger. But the temperate nature of the Bridge Creek is actually due to the fact that the plants grew far inland. This contrasts to western Oregon plant communities, which appear more tropical in complexion because they inhabited a region influenced by the nearby ocean.

Paleoenvironments and Plants

The most fascinating and useful aspect of fossilized plant remains is their role in unraveling prehistoric climates. Environmental settings of past ages—from alpine to low elevations, from dry to wet, from tropical to cool climates—can be traced through careful examination of plant assemblages. A warming trend in the early Eocene was followed by a profound climate event at the end of that epoch, when the early Oligocene saw a drop in mean annual temperature of 10 to 11 degrees Fahrenheit. These changes brought on a marked decline in tropical vegetation. Plants recovered in diversity during the Oligocene, whereas marine invertebrates continued to decline regionally as the ocean retreated. A warm humid episode again took place during the later Oligocene and in the Miocene, 25 million years ago, but overall temperatures for the last 50 million years show a cooling trend as floras correspondingly became more temperate.

The composition of plant communities in a region is a complex by-product of interrelated factors including, in part, rainfall, temperature, sunlight, latitude, soils, natural catastrophes, and successions. It is only by understanding the multitude of characteristics that go into producing a plant assemblage that former habitats can be deciphered. As early as 1916, leaf shapes were used to estimate paleoenvironmental changes. Leaves with entire or smooth margins were regarded as reflecting dry lowland tropical areas, whereas those having serrated margins were thought to be characteristic of higher cooler latitudes. A comparison of smooth and serrate leaf margins not only could illustrate the different types of tropical and temperate communities, but it also could aid in determining latitudes and climates.

Interpreting paleoenvironments using pollen and spores can be more definitive than by analyzing leaves, wood, or seeds. Since pollen grains tend to be extremely abundant, they usually provide a better picture of the local physiography and climate than do leaves. Fossil pollen and spores are rarely attached to the parent plant, and in most cases they are treated and classified as separate entities. For example, an individual pollen grain may be identified as being from an oak tree, but the association of a particular species of oak with the corresponding pollen

Ralph Chaney, a professor at the University of California, Berkeley, for close to 50 years, was one of the first to apply the technique of relating leaf characteristics to prehistoric environments. Chaney, who published widely on the floras and paleoecology, suggested that rainfall played an important role in leaf shape and reasoned that the contrast between smooth and serrate leaf margins is largely determined by available moisture. Since leaves form drip points for water to run off the surface, he concluded that leaves with serrate margins would dominate in areas of high rainfall. Chaney's thorough knowledge of Tertiary plants enabled him to reconstruct ancient topographies and document environmental changes based on floral sequences. The premier western paleobotanist for almost half a century, Chaney became interested in geology and the Northwest while at the University of Chicago, ultimately finishing a Ph.D. in 1918 on the Eagle Creek Flora of the Columbia River Gorge. In the photograph he is standing next to a large slab of poplar leaves from exposures at Vanora Grade near Madras. (Photo courtesy University of Oregon Archives.)

may be difficult. Often pollen for a given species is so distinctive that its presence in a fossil assemblage may confirm the identification of a questionable leaf. Unfortunately comparisons between pollen and leaves from the same strata are rare.

Plant communities go through slow successional changes over time from the pioneer phase of grasses, to an intermediate stage of hardwoods, followed by a climax community of conifers. Catastrophes such as fires, floods, volcanic eruptions, plant disease, insects, or climate changes can interrupt, slow, or even stop and reverse the natural succession. Once paleobotanists have examined the composition of the vegetation and documented regional successions in the fossil record, sudden or dramatic changes in the normal patterns can be recognized, and environmental adjustments made accordingly.

Plant Preservation

Unlike the bones of mammals and invertebrate shells, the remains of leaves, wood, and pollen require special conditions for fossilization. While mineral-rich bone, teeth, and shells need only to be buried to earn a place in the geologic record, vegetable and woody matter is highly susceptible to decay, bacteria, oxidization, and insects. A critical factor to successful preservation of the delicate soft tissue of plants is rapid burial in an environment free of oxygen, such as a swamp or lakebed. Even the most fragile thin leaves fossilize by being pressed flat within

Daniel Axelrod worked with regional Tertiary plant origins and distributions, reaching paleoclimate determinations by comparing fossil plant species with those from modern settings. In addition to climates, Axelrod demonstrated that plants provided clues to prehistoric landscapes and physiography. By recording modern plants from different altitudes—from hardwoods in lowlands to mountain conifers at higher elevations—he estimated paleo-altitudes for individual fossil floras. Originally from Brooklyn, Axelrod, known as "The Axe" by his students, received his Ph.D. from Ralph Chaney at the University of California, Berkeley, then went on to work in the University of California system. He taught for over 30 years, dying in 1998. Axelrod is pictured standing in Mel Ashwill's yard in Madras, Oregon. (Photo courtesy M. Ashwill.)

the very fine particles of clay and mud of a quiet lake floor. Coarse-grained rocks represent areas of rapid deposition where swiftly flowing streams tended to destroy leaves. Similarly, sand or gravel that allows the passage of water through the sediments after burial contributes to the destruction of plant tissue by solution and oxidation.

Other considerations in preservation are dispersal and the chemical makeup of the plant material. Because leaves and wood float, they can be carried with a minimum of abrasion to favorable burial sites. Large, thick tropical leaves are more easily moved by streams and are thus more readily preserved than the small, thin ones typical of temperate environments. Dispersal of gymnospermous needles by wind, moreover, is not nearly as thorough as it is for deciduous leaves. Pollen may be airborne or carried by water for weeks before it settles. Additionally the chemistry of plant tissue itself contributes to fossilization, since pollen, wood, leaves, and fruit contain a variety of organic compounds that may retard decay.

Leaves, wood, and pollen may be fossilized in a variety of ways. Leaves are most frequently preserved as a thin carbon layer. This type of fossilization is called "distillation." In this process the volatile organic material is removed from the leaf without disturbing the fine details. What remains after distillation is a delicate carbon film displaying the leaf outline, veins, and cell structures. Like leaves, wood may undergo distillation by the slow removal of most other elements except the carbon. What appears to be charcoal in sediments is usually distilled wood. Burning wood to make charcoal fuses the walls to create a distinctive trellis-like shape that can be seen in thin section. Very often wood is permineralized, which entails the infilling of all the vacant pore spaces by minerals such as calcite or silica. Water percolating through porous rock is normally rich in dissolved minerals,

Jack Wolfe, a paleobotanist at the University of Arizona, Tucson, has challenged much of the early thinking on the relationship between plant physiognomy, paleolatitudes, and paleoclimates. After a career of studying major floristic patterns and climates in the western Tertiary for the U.S. Geological Survey, he pointed out that there is no exact match between Tertiary and modern forests and that care must be taken before drawing definitive conclusions based on plant habitats alone, the so-called nearest living relative method. Furthermore, the older strategy of reconstructing ancient forests by a count of leaf types alone might be flawed because of the many variables involved in preservation. Wolfe's own synthesis for determining environments draws data from close to 30 leaf characteristics such as size, lobed or serrate margin, and a ratio of length to width, a system he called Climate-Leaf Multivariate Program (CLAMP). Applying CLAMP to regional floras, he deduced mean annual temperature ranges for many mid-latitude Tertiary floras of western North America and showed that

climate change, not elevation, is the most important factor in floral composition at a given site. He also refined his calculations to adjust for differences between the elevation of the present-day landscape and that of the Tertiary. As a high school student, Wolfe worked for the Oregon Museum of Science and Industry (OMSI) at Camp Hancock and went on to Harvard to study with the premier paleobotanists Richard Scott and Elso Barghoorn. His Ph.D. dissertation in 1960 treated early Miocene floras of northwest Oregon. In 2001 Wolfe received a medal from the Paleontological Society for his outstanding work, dying in a falling accident from an outcrop in the Sierras four years later. (Photo courtesy S. Manchester.)

augmenting permineralization. Fossil wood in this condition may appear fresh and unaltered, but it is strikingly heavy because of saturation by minerals. The unaltered appearance of permineralized wood has been dramatically displayed when individuals, thinking such wood is modern, have actually attempted to run it through a saw, destroying the blade in the process. In this condition, wood can occasionally be polished on a lapidary wheel, but frequently the framework of the woody tissue makes the specimen crumbly. Wood that lacks any of the original tissue has undergone complete replacement. During the replacement process silica, in the form of calcedony, opal, jasper, or agate, has been exchanged for the wood tissue. This slow alteration permits the preservation of delicate features such as annual rings and details of individual cells. Like permineralization, replaced or petrified specimens are noticeably heavier than ordinary wood, but they will usually take a nice polish.

Pollen and spores are rarely altered by fossilization. That is, pollen undergoes little substantive change other than being flattened and darkened with burial. While the waxy, resinous walls of these tiny grains are especially resistant to decay, they may be oxidized unless rapidly covered.

In spite of the multiple difficulties that stand in the way of preservation, the plant fossil record is remarkably good. Plants are generally better represented than mammals, and several factors contribute to this. A mammal skeleton will produce around 100 separate bones as potential fossil material, whereas a typical deciduous or hardwood tree will annually contribute tens of thousands of leaves and possibly one hundred times that in its lifetime. The sum total of leaves from a needle tree exceeds that. The amount of pollen produced by a plant in either a lifetime or on an annual basis is estimated in the millions. This overwhelming statistical edge guarantees a prominent place for plants in the fossil record.

Paleozoic and Mesozoic Landscapes

The seas that covered much of the Northwest during the Paleozoic, beginning over 400 million years ago and continuing into the Mesozoic, were dotted by offshore volcanic island chains and submerged oceanic plateaus. These landmasses, or terranes, made their way toward the North American continent, driven by the engine of plate tectonics. Lodging against the West Coast, today their fragments, with entombed fossilized plant remains, are found in the Blue Mountains and Klamaths of Oregon.

Oregon's oldest fossil plant community, going back 300 million years, is found in Pennsylvanian age rocks of the Spotted Ridge Formation in the Crooked River valley. These 1,000-foot-thick, fine-grained mudstones lie between two marine formations, the older Coffee Creek of the Mississippian Period and the younger Permian Coyote Butte. While the Coffee Creek and Coyote Butte contain invertebrate fossils, the Spotted Ridge is dominated by plants with just a few shallow-water shells. All of these rock formations are part of the Grindstone terrane.

In the Midwest and northeast United States, floras from the Pennsylvanian Period, comparable to those of the Spotted Ridge, are commonly found associated with

Imaginary view of a forest 250 million years ago. (From Pouchet 1882.)

coal layers. The lack of coal swamps in Oregon may be evidence that forests grew at higher elevations. Spotted Ridge plant material is so limited that speculations on the local paleoecology for this interval involve a certain amount of guesswork, although the environment was probably not unlike the warm, humid climates developing elsewhere in North America. An overriding consideration is that the flora is part of an exotic terrane not yet traced back to its point of origin somewhere along the Pacific Rim.

The Spotted Ridge locality was discovered in 1937, but the specimens proved too fragmentary even for generic determinations to be made. A later, more thorough excavation at the site produced several species of ferns and fern-like foliage. In 1956 Charles Read and Sergius Mamay, specialists in Paleozoic and Mesozoic floras, fully described and illustrated the Spotted Ridge collection, naming three new species, the horsetails *Mesocalamites* and *Phyllotheca* along with the fern-like *Dicranophyllum*. At the most productive outcrop at Mills Ranch in Crook County, seed ferns, giant horsetails with whorled leaves, exceptionally tall *Cordaites* trees bearing long, strap-like leaves, and the scale tree, *Lepidodendron*, an ancient relative of modern clubmosses, can be found. Distinctive diamond-shaped leaf scars that resemble scales covered the *Lepidodendron* trunk. *Asterophyllites, Mesocalamites,* and *Phyllotheca*, all three ancestral joint grass, *Pecopteris* (fern-like), *Cordaianthus* (a cone of a conifer relative *Cordaites*), *Schizopteris* (fern), *Dicranophyllum* (fern-like), and *Stigmaria* (the root system of a *Lepidodendron* scale tree) characterize this flora.

No plants of Permian age have been found in Oregon, and the following Triassic interval was also virtually barren except for unique marine dasycladacean algae, *Diplopora oregonensis,* from the Hurwal Formation in Wallowa County. Recently discovered by George Stanley of the University of Montana, the

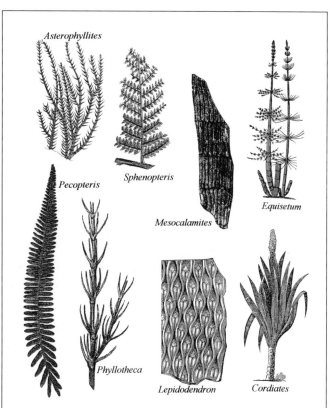

Pennsylvanian age plants in Oregon are the oldest in the state. (After Le Conte 1892; Lesquereux 1879; Seward 1898.)

200-million-year-old plant is the oldest green algae known from the United States. Even though common in similar tropical sediments of central Europe, the algae are unknown from other Triassic rocks of the eastern and western Pacific. Shales and siltstones from these strata are part of the Wallowa volcanic archipelago, terrane rocks that were accreted as a belt to the edge of the continent along northeast Oregon, Washington, Idaho, and British Columbia. Draped over volcanic rocks that made up an island chain, Hurwal sands and shales represent a shallow ocean basin with lagoons in which limestone reefs developed.

Like the Pennsylvanian floras, those from the Jurassic Period, between 145 to 208 million years ago, are also preserved in fragments of exotic terranes. While the vast plant system of the Mesozoic was originally thought to reflect tropical conditions worldwide, the far-reaching nature of Jurassic floras may more likely be the effect of a large continent broken up and dispersed by tectonic processes.

Ferns, cycads, conifers, and ginkgos were common during the Jurassic. Most are now extinct, but descendants of cycads grow in the tropics today, and the fan-shaped leaves of the ginkgo, whose modern equivalents are less deeply lobed, are found throughout the world. Preserved needles of a pine tree suggest that these early conifers may have had some resemblance to those of the present time, but it is doubtful that any true members of *Pinus* existed during this period. Although this Jurassic vegetation looks unusual, it is by no means as distinctive and archaic in appearance as that of the earlier Pennsylvanian.

In Oregon, Jurassic plants have been recorded only in Douglas County and across the state in rocks of the Snake River Canyon. Interest in Jurassic plants was high during the 1800s when attention was focused on southern Oregon. A sizable collection was made by Joseph Diller and others of the U.S. Geological Survey at the abandoned Nichols railroad station and at Buck Mountain. The finest specimens, however, came from the distinctive black slates in the bed of Thompson Creek, described by Frank Knowlton in 1910. Sediments of the Riddle Formation, in which these plants are found, are part of the Snow Camp terrane. The package of conglomerates, sandstones, and siltstones that make up the Snow Camp is believed to have been moved by faulting processes as much as 200 miles northward from the vicinity of the Sacramento Valley in California. Associated with the plants, invertebrate shells of *Buchia piochii* (*Aucella*), a mussel-like Jurassic clam, occur in fragmented condition. In some regions of the Klamaths, reef-like mounds of *Buchia* are indicative of nearshore wave-swept settings.

In 1991, the focus on Jurassic floras switched to the opposite side of the state with the discovery of similar plants in exposures along the Snake River Canyon. Petrified wood, as well as leaf and seed impressions, in the Coon Hollow Formation are part of the Wallowa terrane. At Weber State University, Sidney Ash reported that the assemblage was dominated by ferns, conifers, and ginkgos, while broad-leafed plants were lacking. Ferns include *Adiantites, Cladophlebis,* and *Dicksonia.* The seed fern *Sagenopteris* is rare, but several conifers—*Brachyphyllum, Mesembrioxylon, Pagiophyllum,*and*Podozamites*—resembling modern junipers are plentiful. *Neocalamites* (horsetail) is poorly preserved, and the *Ginkgo* leaves are small and moderate in number. All of the wood recovered to date is similar to that of the conifer *Mesembrioxylon.* Tree trunks up to a 18 inches in diameter and 6 feet long were dispersed and broken, suggesting transport by streams before burial. A fern species *Phlebopteris tracyi* with fronds up to 18 inches in length and the quillwort *Isoetites rolandii,* both newly discovered in the Coon Hollow Formation, have not been found in the Riddle flora from the Klamath Mountains.

Coon Hollow plants, deposited in fine-grained sandstones, mudstones, and conglomerates, reflect local nonmarine conditions in the Blue Mountains volcanic archipelago. Swampy lowlands supported ferns and quillworts, while drier hills were covered with conifers. In the moist temperate climate, turbid streams washed out onto nearby deltas and floodplains before depositing their sediment load. As the ocean encroached upon the land, plant communities were replaced by marine invertebrates, which are prominent in the upper layers.

> **A selection of Jurassic and Cretaceous plants:**
> Bryophytes: *Marchantites*
> Lycopods: *Isoetites*
> Joint grass: *Equisetum, Neocalamites*
> Ferns: *Adiantites, Cladophlebis, Coniopteris, Danaeopsis, Dicksonia, Hausmannia, Phlebopteris, Ruffordia, Sagenopteris, Scleropteris, Taeniopteris, Thyrsopteris*
> Maidenhair: *Ginkgo*
> Cycads: *Ctenis, Ctenophyllum, Nilssonia, Pterophyllum, Ptilozamites, Williamsonia*
> Conifers: *Araucarites, Brachyphyllum, Cyclopitys, Mesembrioxylon, Pagiophyllum, Pinus, Podozamites, Sequoia, Taxites, Yuccites*

By the Cretaceous Period, more than 100 million years ago, most of the major exotic terranes that form the foundation of the Pacific Northwest had already been accreted to the West Coast. During this period much of what is now Oregon was covered by a shallow seaway with a shoreline extending from the Klamath Mountains, into the eastern part of the state, and up into Washington. A tropical rainy climate characterizes the Cretaceous interval.

Cretaceous floras are rare in Oregon, but, in contrast to those of the earlier Pennsylvanian and Jurassic, these were probably native to the region. Like those of the previous periods, rocks of the Cretaceous are found only in the southwest and northeast regions of the state where the land, perhaps as small islands, dotted a shallow seaway. In Curry County, floral remains along Elk River occur in the Rocky Point Formation, a 1,000-foot-thick layer of sandstone and siltstone above the Humbug Mountain Conglomerate. Both are part of the Elk terrane, a tectonic fragment transported northward from California by faulting. A small number of

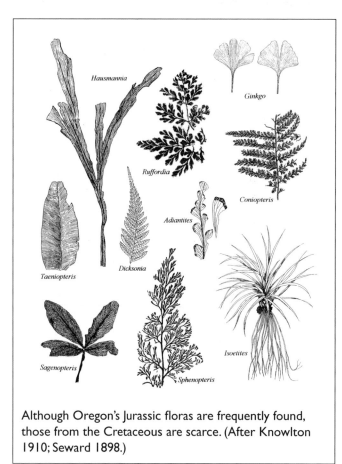

Hausmannia

Ginkgo

Ruffordia

Coniopteris

Adiantites

Taeniopteris

Dicksonia

Isoetites

Sagenopteris

Sphenopteris

Although Oregon's Jurassic floras are frequently found, those from the Cretaceous are scarce. (After Knowlton 1910; Seward 1898.)

worn plants and shells, scattered throughout the unit, include the ferns *Dicksonia,* the most common, as well as *Cladophlebis* and *Thyrsopteris,* the cycads *Ctenis* and *Ctenophyllum,* and the conifers *Podozamites* and *Taxites,* along with the clam *Buchia crassicollis.*

In the Blue Mountains, rare Cretaceous plant remains from near Mitchell in Wheeler County consist of a palm seed and fern spores. In 1958 a palm seed, encased in a concretion with an ammonite, was described by paleobotanist Roland Brown of the U.S. Geological Survey. Brown wrote that the seed belonged to the genus *Attalea,* and "so far as I am aware this specimen is the first plant record from the late Cretaceous of north central Oregon." The oldest microfloras in the state, dated from Cretaceous to early Tertiary, include spore and pollen (palynomorphs) recovered from Texaco Oil Company borehole samples. Four exploratory petroleum wells reveal the presence of a wide tract of Cretaceous sedimentary rocks lying beneath the Tertiary volcanic cover. Drilled to depths of almost a mile into thick marine and nonmarine sediments, the cores yielded 18 fossil spore and pollen genera, dominated by ferns.

One of the most unusual Cretaceous plants is the tree-fern *Tempskya,* found on Lightning Creek north of Sumpter in Baker County as well as at other locales across the western United States. This curious fossil could be related to any of several fern families, but its exact affinity has yet to be determined. *Tempskya* specimens are primarily trunk-like structures, formed from a ropy mass of stems and roots of a climbing plant. In cross-section the stems display distinctive crescent-shaped irregular bodies arranged in a radial pattern around a central core. The consensus seems to be that the false stem grew upright and not horizontally, tapering upward into a conical shape. The remains are typically oxidized to a yellow color and occur as single, extremely hard, silicified, and rounded buds averaging 30 pounds.

The heaviest found was 135 pounds. The association of land plants such as *Tempskya* with ammonites suggests the presence of islands dispersed along a Cretaceous shoreline.

The Cenozoic Era

Cycles of cooling and warming mark the Tertiary of Oregon. Eocene climates were tropical, but temperatures cooled as environments became more temperate in the Oligocene. Following a brief warming trend during the middle Miocene, conditions steadily moderated. These changes are reflected by fossil plant life both east and west of the Cascade Mountains.

The Paleocene Landscape

Oregon Paleocene floras are comparatively rare and appear in only two areas on opposite sides of the state. Microfossil floras (pollen) from Curry

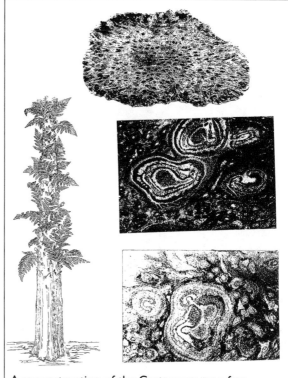

A reconstruction of the Cretaceous tree-fern *Tempskya* with three cross-sections of the stem. (After Ash and Read 1976; Reed and Brown 1937.)

County in the Coast Range and megafossils (leaves) from Umatilla County in the northeast region make up the total accumulation. As offshore volcanic islands constructed a platform of lavas and sediments that became the underpinnings of the Coast Range, sedimentary layers of the Siletz River Volcanics (Roseburg Formation) preserved small numbers of fossil spores and pollen. This distinctive tropical Paleocene to Eocene microflora, located on the West Fork of Floras Creek in Curry County, includes ferns and conifers as part of a brackish environment adjacent to a freshwater swamp.

A limited accumulation of Paleocene leaves from eastern Oregon indicates the presence of a prehistoric lake near Denning Spring in Umatilla County. Strata are similar to those of the Eocene Clarno Formation, but few of the Denning Spring species occur in the Clarno, and only *Hydrangea* and the fern *Dryopteris* are common to both. The water fern *Hydromystria,* found in Paleocene floras through the Rocky Mountains, is unique to the Denning Spring. These striking floral differences indicate that the vegetation grew in a cooler environment than that of the more tropical Clarno. Leaves of *Evodia* (rue family) were the most abundant,

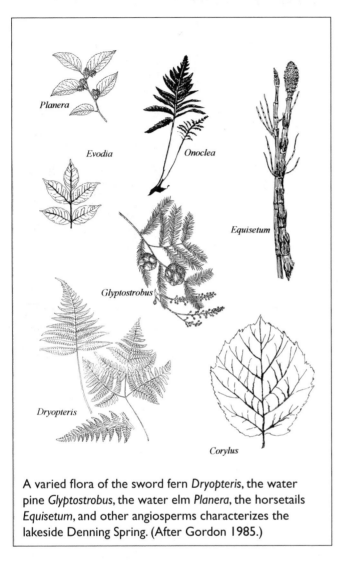

Planera

Evodia

Onoclea

Equisetum

Glyptostrobus

Dryopteris

Corylus

A varied flora of the sword fern *Dryopteris*, the water pine *Glyptostrobus*, the water elm *Planera*, the horsetails *Equisetum*, and other angiosperms characterizes the lakeside Denning Spring. (After Gordon 1985.)

while *Corylus* (hazelnut), *Planera* (water elm), and *Litseaphyllum* (laurel family) were frequent. *Equisetum* (horsetail), six genera of ferns (including *Allantiodiopsis, Onoclea, Woodwardia*), *Betula* (birch), and *Alnus* (alder), along with the conifers *Taxodium* (bald cypress) and *Glyptostrobus* (water pine), make up the remainder of the selection. The sedimentary layer, in which the Denning Spring plants are preserved, is unnamed.

The Eocene Landscape

In Oregon, the Eocene Epoch, from 54 to 33.7 million years ago, was subtropical to tropical. Shorelines lay much further to the east than at present, and the landscape was dominated by volcanism. Clouds of ash from large stratovolcanoes fell over the low-lying coastal plain, while in the eastern part of the state outpourings of ash from local vents mixed with water forming lahars or mudflows that entombed fossils of every description.

Nothing reflects the Eocene environment of Oregon better than the variety of fossil remains from the central part of the state, and the most spectacular of these are the leaves, seeds, and nuts found in mudflows of the Clarno Formation. From the time of discovery in the 1800s to the present day, this assemblage has held a fascination for researchers, who have scrutinized the many details of the flora. In 1878 Leo Lesquereux, who worked for both Harvard and the Smithsonian, listed species, as did John Newberry of Columbia University in 1883. Other researchers who contributed to the knowledge of the Clarno are Frank Knowlton in 1902, Chester Arnold from the 1930s to the 1960s, Richard Scott in the 1950s and 1960s, Thomas Bones around the same time, and Steve Manchester and Greg Retallack beginning in the 1980s.

While the Clarno is as famous for its vertebrate fossils as for its plants, the bones and plants are confined to different levels and environments. The seeds and leaves can be extracted intact, but bones tend to be broken and likely to crumble when removed. Interesting parallels have been drawn between Clarno sedimentation and similar processes resulting from volcanic activity at present-day Paracutin, Mexico. Richard Scott concluded that the Clarno beds were not the product of ash falls, but that both the Clarno and Paracutin regions experienced very rapid deposition of ash washed in with nearby volcanic debris. Much of the Clarno Formation is now thought to have originated as lahars or thick viscous muds of fine ash mobilized by water to move as a liquid. Plants, entombed in this layer, formerly grew along a stream bank. Falling directly into the water, they were deposited downstream at a spot where the current slowed. In many locations, such as at Hancock Canyon, the leaves are fragmented, folded, and bunched within thin layers. Similar to the katsura (*Cercidiphyllum japonicum*) of China and Japan, the famous Hancock Tree, named for Lon Hancock, is one of the numerous large permineralized stumps found within the lahars.

A mean annual temperature of 75–80 degrees Fahrenheit and rainfall around 35 to 80 inches at the lower elevations was indicative of the humid frost-free Clarno

Accompanying Geologic Survey parties to the western regions, paleobotanist Frank Knowlton was frequently in the John Day basin during the late 1800s when he worked for the Smithsonian Institution. A prolific writer, Knowlton's publications on Oregon dealt with Tertiary plants of the Cascades, the Willamette Valley, and the John Day Basin, as well as those from the Jurassic Period. In several articles in 1901, he provided a floral list for the John Day region, but his *Fossil Flora of the John Day Basin* in 1902 was the most comprehensive of the early accounts. (Photo courtesy Archives, U.S. Geological Survey, Denver.)

In 1949 Richard Scott, from the Museum of Paleontology at the University of Michigan, was among a party that collected 400 Clarno leaves, nuts, and seeds. Scott went on to study at the University of London, where he was able to compare these specimens with those of the famous London Clay. The most interesting were the large seeds, pitted on the outside, which Scott called "peach pits." He subsequently named the new species *Palaeophytocrene hancockii*, the first fossil of this genus from North America. Modern relatives of this plant live in the Malay Peninsula where they form extensive jungle vines. From 1955 onward Scott worked on Mesozoic and Tertiary woods, fruits, and pollen for the U.S. Geological Survey. (Photo courtesy Archives, U.S. Geological Survey, Denver.)

A professor at the University of Michigan, Chester Arnold published extensively on fossil plants from the Clarno Formation. He reviewed the many Clarno fern families and genera noting that because they preserve well and frequently have the reproductive parts attached to the leaves, ferns are often better known than the deciduous plants where the seeds and pollen separate from the parent. Although he collected and worked with Tertiary plant communities in Malheur County, his research elsewhere centered on Mesozoic floras. Arnold died in 1977. (Photo courtesy University of Michigan, Bentley Historical Library.)

After years of mining the Clarno Nut Beds, Thomas Bones, a private collector and a printer by profession, estimated that he had removed over 10 tons of seed-bearing rock by packing out 50 pounds at a time in his car. An astonishing 800 pounds had been accumulated by the mid 1940s. Producing sharply detailed photographs of his specimens, Bones then donated the fossils to several regional and national museums. In the picture Bones is holding a chunk of the Clarno Formation impregnated with seeds and nuts. (Photo courtesy E. Baldwin.)

climate. In this setting, a mixed temperate to tropical forest covered the slopes, while more temperate vegetation lived at the higher elevations.

Although ash layers at different locales in the Clarno contain leaves and wood, remains from an area known as the Nut Beds, east of the town of Clarno in Wheeler County, are unique because of the permineralized impressions, molds, and casts of fruits, nuts, and seeds. *The Oregonian* of 1931 described the finds as "silicified pecans, walnuts, and dates, perfectly formed." The article continued, "Some of the nuts are perfect, with shell intact. Others . . . show the nut meat inside." The Nut Beds deposit is a discrete area less than a quarter of a mile across and thirty-feet thick where the concentration of vegetative fragments has been dated at 43.8 million years before the present.

Steve Manchester (on the right of the photograph), of the Florida Museum of Natural History, became interested in fossils at an early age when he was a student at the summer camps organized by the Oregon Museum of Science and Industry. A native of Salem, he finished a Ph.D. in 1981 at Indiana University. His range of research topics on the Clarno provides notable details on families such as the Platanaceae (plane tree), the Musaceae (banana), the Sterculiaceae (cocoa), the Staphyleaceae (bladdernut), the Juglandaceae (walnut), on the winged fruit of *Cedrelospermum* (elm family), and on the fruits and seeds

of individual members of the Fagaceae (beech family), as well as on preserved wood. Co-authored with Herbert Meyer, Manchester's *The Oligocene Bridge Creek Flora of the John Day Formation, Oregon*, provides a comprehensive look at this important assemblage. (Photo courtesy S. Manchester.)

Fossil soils in the Clarno Formation are yet another avenue for depicting prehistoric conditions. Called paleosols, these layers have undergone characteristic modification by plants, climate, and chemical changes before being compacted and covered. Greg Retallack, who has worked extensively with paleosols of eastern Oregon, has demonstrated that they not only preserve plants and animals but also provide clues to paleoenvironments. His assessments of the early Tertiary record in the John Day basin, which cover not only soils and plants, but mammals, reptiles, and insects as well, enabled Retallack to paint a broad overview of the paleosetting. A native of Australia, Retallack has been teaching at the University of Oregon since 1981. His interests are highly eclectic, encompassing the evolution of soils through geologic time, the dispersal of Cretaceous angiosperms as well as worldwide Permian extinctions. An enthusiastic field geologist, he is equally at home in the classroom or laboratory. (Photo courtesy G. Retallack.)

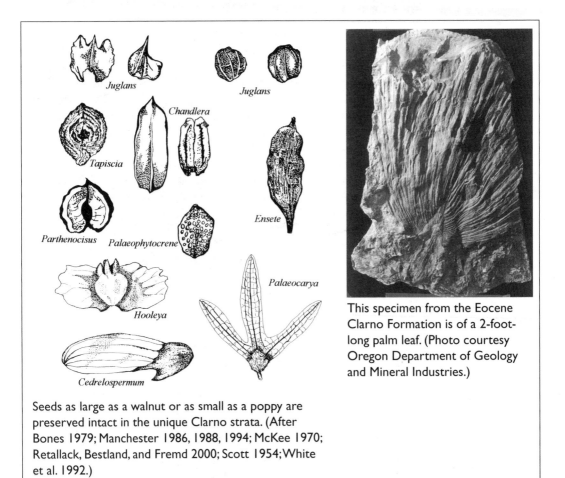

Juglans

Juglans

Chandlera

Tapiscia

Ensete

Parthenocisus   *Palaeophytocrene*

Palaeocarya

Hooleya

Cedrelospermum

This specimen from the Eocene Clarno Formation is of a 2-foot-long palm leaf. (Photo courtesy Oregon Department of Geology and Mineral Industries.)

Seeds as large as a walnut or as small as a poppy are preserved intact in the unique Clarno strata. (After Bones 1979; Manchester 1986, 1988, 1994; McKee 1970; Retallack, Bestland, and Fremd 2000; Scott 1954; White et al. 1992.)

Retallack determined that the intervals containing the Nut Beds and Mammal Quarry, at the uppermost level of the Clarno Formation, are composed of volcanic and stream-deposited mudflows and not lake or still-water sediments. A lack of aquatic plants and fish and the presence of abundant root traces support this conclusion. At other localities near Mitchell and on West Branch Creek in Wheeler County or at Muddy Ranch and along Cherry Creek in Jefferson County, Clarno lake sediments do contain fossilized fish and compressed leaves.

Entombed in the Clarno Nut Beds, over 145 genera and 173 different species are the remnants of a widespread forest which grew in the subtropical climate of eastern Oregon. The bulk of the material includes large thick-leafed plants such as *Juglans* (walnut), *Magnolia,* and *Meliosma* (sabia family). More than 40 percent of the vegetation consists of tropical vines, most of which belong to the moonseed (Menispermaceae) and grape (Vitaceae) families. *Ginkgo*, cycads (*Dioon*), ferns (*Acrostichum, Cyathea, Osmundites*), horsetails (*Equisetum*), and conifers such as *Pinus* and *Taxus* add to the floral diversity. The oldest tropical walnut yet known,

Englehardioxylon nutbedensis, as well as seeds of the temperate *Tapiscia* (Staphyleaceae; bladdernut family), preserved as quartz casts, were recorded here for the first time in North America. More common in the Eocene of Europe, *Tapiscia* was widely distributed in the northern hemisphere as late as 40 million years ago, but its range was reduced drastically by late Eocene cooling.

The tropical nature of the Clarno interval is further demonstrated by a fossilized banana from Wheeler County. The specimen was preserved as an intact mold of fruit and seeds, which had been buried and flattened. The Clarno banana (*Ensete*) was smaller than the domesticated species, which have been selected for size and to eliminate the hard seeds.

As part of the tropical flora, large fossil palm leaves *Sabalites* and wood *Palmoxylon* are found at a number of Clarno locations. The use of form genera *Sabalites* for palm leaves and *Palmoxylon* for palm wood is a taxonomic strategy in paleobotany where the various plant elements such as leaves, wood, and fruit have become separated in the fossil record. Palm wood fragments of the Eocene are almost exclusively found in northeast Oregon where

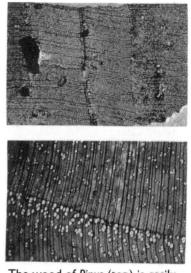

The wood of *Pinus* (top) is easily distinguished by its unmistakable vertical ducts, which are seen as small round holes with a magnifying glass. *Evodia* (bottom) shows a distinct pattern of wood grain interspersed with cell structures. (After Gregory 1976.)

they were limited to tropical settings, whereas later occurrences are restricted to the Miocene Eagle Creek locale along the Columbia River.

Even though wood in the Nut Beds is well preserved and commonly permineralized, individual pieces are water-worn, indicating fluvial transport before burial. Because it is more durable and represents widespread parts of a large watershed, fossil wood may provide a broader picture of a prehistoric forest than vegetative parts do. The paleobotanist Irene Gregory was the first to recognize the value of examining the diagnostic features of fossilized wood. By making thin sections of the tree trunk of an *Acacia* (pea family), as well as the limbs of the shrub *Evodia* (rue family) and the tree *Populus* (poplar family) from the Clarno, she was able to compare anatomical details with those of modern species to aid with identifications and determine climates. Steve Manchester evaluated the xylem or woody tissue of preserved wood, comparing it to that of the modern equivalent, to show that plants readily adjust to the conditions in which they live. Features of the xylem of *Triplochitioxylon oregonensis* (Malvaceae family) display adaptations to the tropical Clarno climate, distinguishing the species from those living today in the arid African savanna.

A selection of plants from the Clarno Formation:

Ferns: *Acrostichum, Anemia, Cyathea, Osmundites*

Cycad: *Dioon*

Joint grass: *Equisetum*

Maidenhair: *Ginkgo*

Conifers: *Diploporus, Pinus, Taxus, Torreya*

Monocotyledons (palm and grass): *Ensete, Graminophyllum, Palmoxylon, Sabal, Sabalites*

Dicotyledons: *Acacia, Actinida, Alangium, Castanopsis, Cedrelospermum, Celtis, Cercis, Chattawaya, Cinnamomum, Cryptocarya, Engelhardioxylon, Evodia, Ficus, Florissantia, Hooleya, Hydrangea, Juglans, Laurocarpum, Lindera, Litsea, Macginitea, Magnolia, Meliosma, Nyssa, Ocotea, Persea, Populus, Prunus, Pterospermum, Quercus, Rhamnus, Sabia, Tapiscia*

Vines: *Ampelocissus, Chandlera, Diploclisia, Iodes, Odontocaryoidea, Palaeophytocrene, Parthenocissus, Tinospora, Vitis*

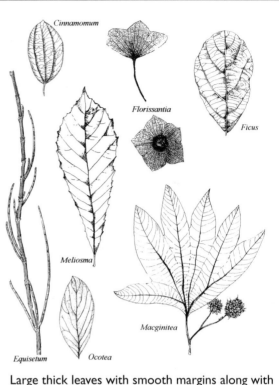

Large thick leaves with smooth margins along with the water-loving *Equisetum* dominated the tropical Eocene of Oregon.

Exposures of the Clarno Formation elsewhere in the John Day basin allow a look at Eocene plants from this stratum. Less than a mile northeast of the Nut Beds, the Hancock Mammal Quarry, better known for its vertebrate remains, includes a flora of compressed casts and molds. Examined by Thomas McKee, a student at Oregon Museum of Science and Industry, many of the specimens were in poor condition, but McKee was able to identify 30 taxa of fruits and seeds. The majority were of *Diploclisia*, a tropical to subtropical vine of the Menispermaceae family, but others such as *Alangium, Ampelocissus, Juglans, Odontocaryoidea,* and *Palaeophytocrene* are similar to those from the Nut Beds.

In western Oregon, Eocene floras are mixed with marine mollusks, signaling the proximity of the beach. From Coos Bay northward, broad marine aprons extended through Douglas, Lane, and Polk counties in the Willamette Valley to Washington and Columbia counties in the far corner of the state. Swamps, deltas, estuaries, and highlands were distributed along the margin of a deep marine trough adjacent to the ancestral Cascade volcanoes, where today basaltic sandstones and siltstones contain subtropical to tropical vegetation reflective of these conditions.

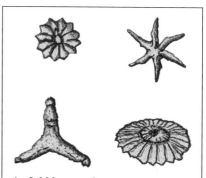

At 3,000 magnification, *Discoasteroides kuepperi, Discoaster lodoensis, Discoaster multiradiatus,* and *Tribrachiatus orthostylus* from the Siletz River Volcanics confirm the transition from Paleocene to Eocene (clockwise from upper left).

The oldest fossils in this sequence are microscopic coccoliths and discoasters, blue-green algae, from the Paleocene-Eocene Siletz River, Lookingglass, and Flournoy formations near Roseburg. Coccolith floras (*Braarudosphaera, Coccolithus, Micrantholithus*) are sparse in the lower strata but more numerous in the upper Lookingglass, providing evidence of shallow water during the early Eocene. Resembling cookies, coccoliths inhabit ocean waters today, whereas snowflake-shaped discoaster floras became extinct in the late Pleistocene.

Leaves from the Coaledo were mentioned as early as the 1800s by Joseph Diller, but it wasn't until 1967 that samples from the Coaledo, Bastendorff, and Tunnel Point formations were examined in detail. Exposed on beach headlands at Cape Arago and Coos Head, these formations contain the pollen of *Ficus* (fig), *Juglans* (walnut), *Laurus* (laurel), *Magnolia, Rhamnus* (buckthorn), and *Sabalites* (palm), substantiating a climate that was considerably warmer and more humid than today. With 50 to 60 inches of rainfall annually and a temperature that rarely fell below freezing, this subtropical environment was characterized by highlands surrounding a coastal basin. In the immediate Coos Bay region, coal seams from the Coaledo Formation were formed from peat in an open embayment adjacent to a low swampy delta. Tropical ferns from a coal layer in the Bateman Formation, found inland near the Umpqua River, show little in common with those of the Coaledo.

Along the Eocene coastline, prevailing warm humid conditions are also indicated by Comstock plants within volcanic ash layers of the Fisher Formation in Douglas and Lane counties. South of Cottage Grove, the Comstock flora, along with similar plants from Hobart Butte, includes subtropical species of *Liquidambar* (sweet gum) that disappeared during the subsequent Oligocene cooling trend. *Cinnamomum dilleri* (camphorwood) makes up nearly one-fourth of the Comstock specimens, with diminishing percentages of *Aralia, Magnolia, Astronium* (cashew family), *Lonchocarpus* (legume family), *Allophylus* (soapberry family), and *Cryptocarya* (laurel family).

Northward in Polk County, scant numbers of 40-million-year-old Eocene plants are scattered along the ancient marine shoreline. At Rickreall, *Cinnamomum* (camphorwood), a fossil fern of the family Polypodiaceae, and wood fragments are found with a dominantly marine invertebrate fauna in the Yamhill Formation.

A selection of Eocene plants from the south and central Coast Range:

Coccoliths and Discoasters:
  Braarudosphaera, Coccolithus, Discoaster, Discoasteroides, Micrantholithus, Pontsphaera, Scyphosphaera
Joint grass: *Equisetum*
Conifer: *Sequoia, Thuja*
Monocotyledons: *Sabalites*
Dicotyledons: *Allophylus, Anona, Aralia, Astronium, Celastrus, Cinnamomum, Cordia, Cryptocarya, Diospyros, Ficus, Juglans, Laurus, Liquidambar, Lonchocarpus, Magnolia, Mallotus, Myrica, Nymphoides, Ocotea, Persea, Platanus, Pterospermum, Quercus, Rhamnus, Trochodendroides*

Coastal Eocene plants. (After Lesquereux 1878; Sudworth 1908.)

A few leaves from the Cowlitz Formation near the town of Timber in Washington County and an equally meager amount from Keasey sediments in Columbia County are associated with invertebrate shells. In the nearshore shallow waters present during Cowlitz time, large thick tropical leaves of the genus *Aralia* are particularly well-preserved, as are *Equisetum* (horsetail) and several ferns. Fine ash layers of the Keasey Formation at Mist yield a flora from a coastal plain where freshwater merged into brackish, then deep marine. Paleobotanist Roland Brown identified leaves of *Quercus* (oak) and *Myrica* (bayberry) as being "tough enough to resist complete destruction while being swept out to burial in the sea." Only one specimen of *Ocotea* (lancewood) and a few fragments of *Thuja* (arborvitae) have been added to this short list.

Little-known plant localities in the southern Cascades near Medford and Ashland can be found in the thick conglomerates, sandstones, and volcanic sedimentary layers of the late Eocene Colestin and early Oligocene Payne Cliffs formations. This subtropical flora includes carbonized logs of *Sequoia*, coal, and fragments of leaves, which were identified by Roland Brown. An extensive braided stream system, flowing northward from the Klamath Mountains, deposited the Payne Cliffs across a flat floodplain.

The Oligocene Landscape

The Oligocene Epoch, spanning the interval from 33.7 to 23.8 million years ago, featured a shallow ocean and wide coastal plain west of a rising chain of Cascade volcanoes. The climate altered dramatically as the steady elevation of the Cascade barrier through the late Oligocene and Miocene epochs brought increasingly drier conditions to eastern Oregon. Moisture from the Pacific Ocean fell as rain and snow on the newly emerging range.

On a global scale, the Eocene-Oligocene boundary is marked by a transition from warm subtropical to a more temperate realm. Ocean water cooled profoundly, and air temperatures dropped correspondingly. This climate evolution was characterized by episodic fluctuations into the late Oligocene and early Miocene. A wide seasonal range of temperatures produced the notable alterations in vegetation that took place during the early Oligocene. Subtropical evergreen plants were replaced by deciduous floras adapted to more moderate conditions and seasonal rainfall.

Fossil remnants of Oligocene forests are found throughout the Willamette Valley and Cascades as well as in the John Day basin and at one unusual location near Yaquina Bay on the coast. In the central region, the remarkable Bridge Creek is famous for the richness, variety, and preservation of its flora, as are the numerous Willamette Valley plant communities such as the Goshen, the Thomas Creek and Bilyeu Creek, the Scio, Lyons, Rujada, Willamette, and Sweet Home.

In eastern Oregon, the Oligocene terrain underwent long serene periods with minimal change that were punctuated by explosive volcanic episodes. Both conditions were responsible for the remarkable flora and fauna in the John Day Formation. Plants of the John Day have been the object of considerable interest since the middle 1800s when they came to the attention of pioneer geologist Thomas Condon, then living at The Dalles. He accompanied soldiers escorting a supply train across the John Day Basin to Harney Valley along the road that followed Bridge Creek for several miles. It was probably on this trip that the first of the cinnamon-colored fossil leaves were found. The largest of the early plant collections from the John Day basin was assembled in the 1870s by naturalist and taxidermist C. D. Voy, a resident of San Francisco, and his specimens were subsequently purchased by the University of California. Aspects of Bridge Creek assemblages have been described by paleobotanists John Newberry in 1882, Leo Lesquereux in 1888, and Frank Knowlton in 1902. Ralph Chaney's reports followed from the 1920s and Roland Brown's from 1935 onward. Chaney, who made major new collections for the University of California, was the first to consider the Bridge Creek flora as a separate ecologic plant community in comparing it to modern assemblages.

Nut Beds   Hancock Canyon   Red Hill

Greg Retallack's trademark drawings reconstruct the fossil soils and vegetation of the middle Big Basin Member of the John Day Formation, some 33 million years in the past. (Courtesy G. Retallack.)

In 1901 John C. Merriam of the University of California divided the John Day Formation into three layers on the basis of color: the older reddish, the middle pale green, and the younger cream to buff sections. Some 71 years later these divisions were formally named as the older Big Basin, the middle Turtle Cove, and the youngest Kimberly and Haystack Valley members by Richard Fisher and John Rensberger of the University of Washington.

Within the formation, plant and animal fossils are unevenly distributed. The Big Basin contains a wealth of plant remains, while mammals are richest in the Turtle Cove and Kimberly intervals, a distribution doubtless reflecting prehistoric environments. While lakes of the Big Basin landscape trapped and preserved leaves, streams and floodplains of the Turtle Cove, Kimberly, and Haystack Valley times were optimum for preservation of bones but lacked the thin, flat laminae (book-page-like layers) optimum for leaf fossilization.

Building on the evidence of paleosols (fossil soils), Greg Retallack has proposed a solution to the problem of two distinct environments represented in the John Day Formation. Noting that the presence of hoofed mammals points to open grassy woodlands, while that of plants signifies a thick forest cover of tall trees, he suggests the mammals and plants occurred in different localities, in different climates, and at different times. In addition, some environments are conducive to preserving vegetation while others clearly favor bone material. Leaves, for example, require low oxygen situations as in the very fine sediments of a lake bottom or swamp, whereas fossilization of mammal remains is enhanced by alkaline carbonate soils.

Examining the succession of paleosols in the Painted Hills area, Retallack further divided the Big Basin Member of the John Day Formation into lower, middle, and upper units based on the color bands exposed there. Deeply weathered paleosol layers, responsible for the colorfully banded beds in the Big Basin Member, reflect the paleoclimate variations that occurred from the Eocene to the Oligocene. The lower iron-rich red paleosols of the warmer Eocene give way to the brown and red colors of the Oligocene as the cooler drier climate prevailed.

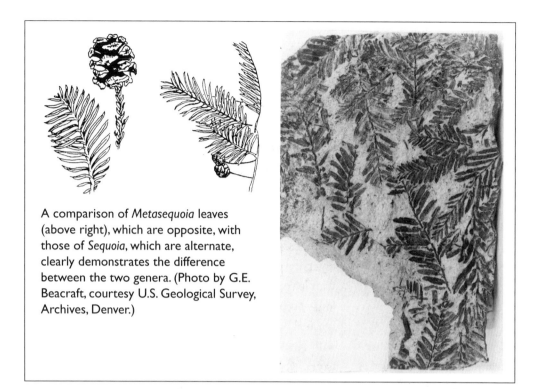

A comparison of *Metasequoia* leaves (above right), which are opposite, with those of *Sequoia*, which are alternate, clearly demonstrates the difference between the two genera. (Photo by G.E. Beacraft, courtesy U.S. Geological Survey, Archives, Denver.)

The lower Big Basin Member has few vertebrates or plants, although both the Whitecap Knoll locality on Iron Mountain in Wheeler County and the Gray Butte in Deschutes County have yielded small plant communities dated around 38 million years ago. This places the floras between the Eocene Clarno and the Oligocene John Day boundary, thus providing evidence of the gradual transition from subtropical to temperate vegetation. As noted by Steve Manchester, the Whitecap Knoll assemblage lacks the broad-leaf evergreen varieties of the Clarno but includes water plants (*Ceratophyllum, Nelumbo*) and deciduous woodland trees (Aceraceae, Fagaceae, Juglandaceae, Platanaceae, Ulmaceae). The best known of the Gray Butte florules, which are found at five separate localities, are the plants at Sumner Spring, occurring in lacustrine (lake) shales. Of the 28 species, which Mel Ashwill examined, the most common dicots are *Alnus, Cruciptera, Macginitea,* and *Quercus,* while the three most frequent conifers are *Picea, Pinus,* and *Sequoia.* Several genera are shared by both the Clarno and Bridge Creek plant communities.

The remarkable Bridge Creek flora is preserved in shales of the middle Big Basin Member, dated at 33.6 to 32.9 million years ago. The climate indicated by the vegetation was moderate with a cool season and about 40 inches of rainfall annually. Forests covered hillsides, and deep valleys were drained by mature

Commonly occurring Bridge Creek plants:

Joint grass: *Equisetum*
Fern: *Polypodium*
Maidenhair: *Ginkgo*
Conifers: *Abies, Cunninghamia, Keteleeria, Metasequoia, Pinus, Sequoia, Torreya*
Monocotyledons: *Nuphar, Typhoides, Zingiberopsis*
Dicotyledons: *Acer, Alnus, Amelanchier, Asterocarpinus, Berberis, Betula, Carya, Cercidiphyllum, Cedrela, Cercis, Cinnamomophyllum, Cladrastis, Comptonia, Cornus, Crataegus, Fagus, Florissantia, Hydrangea, Juglans, Liquidambar, Mahonia, Ostrya, Paracarpinus, Pflakeria, Platanus, Pterocarya, Pyrus, Quercus, Rhus, Ribes, Rosa, Rubus, Tilia, Ulmus, Vitis*

stream systems interspersed with perennial lakes and wetlands. Leaves, fruit, and seed impressions indicate a lakeside habitat was once typical of the John Day and Crooked River basins.

Ninety-one genera and 110 species from the Bridge Creek flora have been identified from leaves, and 52 genera and 58 species from seeds, representing 34 families of angiosperms, three of conifers, three of ferns, and one of horsetails. *Metasequoia occidentalis* (dawn redwood), in conjunction with *Acer* (maple), *Alnus* (alder), *Quercus* (oak), *Juglans* (walnut), and *Ulmus* (elm), essentially characterizes the flora, which has a low diversity of grasses and broad-leafed vegetation. Along with leaves, permineralized wood is also plentiful. Large logs, probably of *Metasequoia*, were burned black and crushed in John Day ash flows before the cells were filled with silica. Although animal remains are rare in Bridge Creek shales, occasional salamanders, frogs, birds, and even bats turn up at leaf localities.

The name "dawn redwood" is somewhat misleading since the plant is neither ancestral to cypress nor redwood. The distinctive feature of *Metasequoia* is its "oppositeness"—branches, needles, and cone scales are all distichous or opposite each other in two symmetrical rows, easily distinguishing *Metasequoia* from *Sequoia* (redwood) or *Taxodium* (bald cypress). This unique quality led to the discovery of living *Metasequoia* trees, long thought to be extinct. A Chinese forester, Tsang Wang, encountered a grove of modern *Metasequoia* in Szechuan Province and brought these trees to the attention of paleobotanists. When this stunning news became known, expeditions were organized in 1947 and 1948 to confirm their existence. Ralph Chaney was among those who traveled to the remote Valley of the Tiger in central China. Accompanied by armed guards, Chaney and his entourage were forced to elude bandits and rely on local villagers for assistance and guidance. Taking photographs and measuring the dawn redwoods, which were up to 100 feet tall, Chaney returned to Berkeley with cones and seeds. Once the *Metasequoia* seeds were planted, the seedlings and trees were reintroduced throughout the Northwest. In May 2005, the Oregon Legislature formally adopted *Metasequoia* as the state fossil.

No lake deposits are evident in the upper Big Basin, but root traces in the paleosols indicate a mixed grassy woodland. Similar conditions continued into the late Oligocene Turtle Cove interval, where abundant grasses supported a varied population of mammals.

Along what would later become the eastern margin of the Willamette Valley, a diversity of dunes, swamps, marshes, river estuaries, and sporadic rocky headlands marked the Oligocene coastline, which stretched only as far south as Salem. Repeated lava flows and clouds of ash from Little Butte fissures and cones throughout the Western Cascades were largely responsible for alterations to the topography and preservation of the stream and lakeside vegetation. Influenced by the nearby coast, floras were temperate to tropical in nature, precipitation of 50 to 60 inches was evenly distributed throughout the year, and moderate temperatures ranged from 30 to 40 degrees Fahrenheit in winter and 70 to 80 degrees in summer. Topography was probably irregular and well drained.

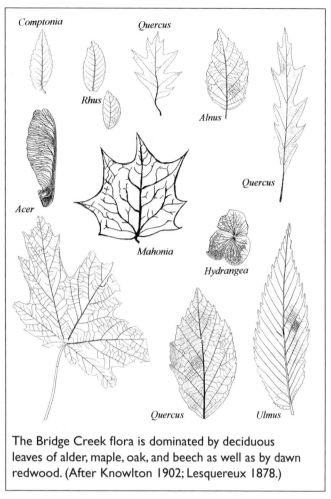

The Bridge Creek flora is dominated by deciduous leaves of alder, maple, oak, and beech as well as by dawn redwood. (After Knowlton 1902; Lesquereux 1878.)

The well-known Goshen assemblage, just south of Eugene, is newly rediscovered whenever construction takes place on the Pacific Highway, and large slabs of siltstones containing the leaves are frequently stacked on the nearby slope. The 2007 Oregon Department of Transportation excavations cut back and removed over 50 feet of the outcrop. In 1876 Thomas Condon made the initial collections, and, after that, the paleobotanists Ralph Chaney and Ethel Sanborn quarried the site in the 1920s. Numerous collectors have visited the location ever since. Striking leaf outlines, preserved as dark brown imprints in Fisher Formation sediments, grew in along stream banks and near lakes. Impressions of unique salamanders can be collected within the same shales as the plants. The diverse flora of 48 species is dominated by species with large leaves, entire margins, and

A selection of Oligocene floras:
  Goshen:
Monocotyledons: *Smilax*
Dicotyledons: *Allophylus, Anona, Aristolochia,*
  Chrysobalanus, Cordia, Cupania, Diospyros,
  Ficus, Ilex, Laurophyllum, Lucuma, Magnolia,
  Meliosma, Nectandra, Ocotea, Platanus,
  Quercus, Siparuna, Symplocos, Tetracera
  Willamette:
Conifers: *Chamaecyparis, Cunninghamia,*
  Pinus, Sequoia
Monocotyledons: *Smilax*
Dicotyledons: *Alnus, Betula, Carpinus,*
  Castanea, Cornus, Engelhardtia, Ficus,
  Hydrangea, Meliosma, Odostemon, Platanus,
  Quercus, Rhus

drip points. *Meliosma goshenensis* (Meliosmaceae family), *Allophylus* (soapberry), *Nectandra* (laurel family), *Ficus* (fig), and *Tetracera* (liana vine) make up one-half of the total. The narrow leaves of *Meliosma* (Meliosmaceae family) and *Lucuma* (egg-fruit) and the heart-shaped leaves of *Aristolochia* (birthwort) are tropical components of the Goshen flora, whereas the twining *Smilax,* in the lily family, grows in temperate areas. The Goshen is significant because it is dated at 33.4 millions of years before the present (mybp), which falls at the earliest Oligocene boundary, and its cool-temperate components lack many species associated with the earlier, warmer Eocene. While slightly younger, the nearby Coburg plant community is considered to be very similar to that of the Goshen.

Also near Goshen on U.S. Highway 99, the Willamette flora was found when the highway department excavated for fill. Occurring in the upper levels of the Little Butte Volcanics, these plants can be distinguished from those of the nearby Goshen because each occupies a separate stratigraphic level and because there are floral differences, which show marked climate variations between the two. In contrast to the more tropical large leaves of the Goshen, the warm-temperate Willamette flora, dated at 30.1 million years ago, is dominated by broad-leafed *Alnus* (alder) and *Quercus* (oak) along with the conifers *Chamaecyparis, Cunninghamia, Keteleeria, Metasequoia,* and *Sequoia.* The unbroken condition of the leaves indicates that they accumulated gradually in a calm-water lake.

The Scio, Lyons, Thomas Creek, and Bilyeu Creek floras in Linn County resemble the Goshen vegetation with large-leafed plants growing at low elevations. All four occur within 20 miles of each other in exposures of the Little Butte Volcanics. The Scio and Bilyeu Creek floras belonged to streamside communities, whereas the Lyons and Thomas Creek are considered to have developed in a lacustrine (lake) setting.

The Scio is largely composed of modern-looking trees and shrubs, many of which presently typify western Oregon. Some exceptions are *Engelhardtia, Grewia, Phoebe,* and *Tetracera* that are native to China and Malaysia. Leaves of *Prunus franklinensis* make up 75 percent of the flora, followed by those of *Amelanchier,*

Equisetum, Vaccinium, Platanus, Sequoia, Grewia, Tilia, Engelhardtia, Phoebe, and *Vitis* in lesser numbers. The warm-temperate floral composition and absence of inland types such as *Quercus* (oak) and *Acer* (maple) imply coastal proximity. The Lyons flora, with few broad-leafed evergreen representatives, grew around a quiet water body. The mixture of temperate *Rosa* (rose), *Pterocarya* (wingnut), *Alnus* (alder), and *Tilia* (linden), as well as *Metasequoia* (dawn redwood) and *Sequoia* in combination with the subtropical *Alangium* (alangium family), *Holmskioldia* (verbena family), and *Meliosma* (meliosma family) can be attributed to marine influences moderating the climate.

Two separate fossil plant localities at Thomas Creek and Bilyeu Creeks are within a few miles of each other near the North Santiam River. Together they include an array of more than 20 families and 43 species and

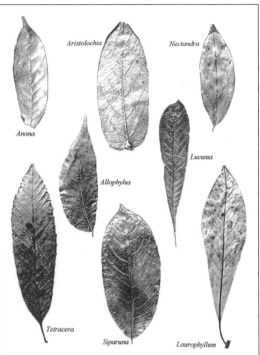

Localities for the Goshen and Willamette are only a short distance apart, but the floras represent different climates. (After Chaney and Sanborn 1933; Lewis 1950; Myers, Kester, and Retallack, 2002.)

A professor of botany and paleobotany at both University of Oregon and Oregon State University, Ethel Sanborn was one of the early contributors to our knowledge of botany in the state. Originally from South Dakota, she took a Ph.D. at Stanford University in 1928, before accepting a position in the geology department in Eugene. Frequently working with Ralph Chaney, Sanborn completed monographs on fossil plant communities from Goshen, Scio, and Comstock in western Oregon. Sanborn is pictured with an unknown person. (Photo courtesy Archives, Oregon State University.)

A selection of Oligocene floras:

Rujada:

Joint grass: *Equisetum*

Conifers: *Abies, Cunninghamia, Keteleeria, Metasequoia, Pinus, Pseudotsuga, Sequoia, Tsuga*

Dicotyledons: *Alnus, Anona, Betula, Castanea, Crataegus, Engelhardtia, Exbucklandia, Fraxinus, Halesia, Juglans, Mahonia, Pinus, Platanus, Populus, Pterocarya, Pyrus, Quercus, Rhus, Salix, Viburnum, Zelkova*

Sweet Home:

Conifers: *Sequoia, Taxodium*

Dicotyledons: *Actinidia, Allophylus, Cedrela, Cinnamomum, Chrysobalanus, Ficus, Juglans, Meliosma, Nectandra, Ocotea, Platanus, Populus, Prunus, Quercus, Reptonia, Schima, Sterculia, Tetracera*

Scio:

Joint grass: *Equisetum*

Maidenhair: *Ginkgo*

Conifers: *Metasequoia*

Dicotyledons: *Alangium, Amelanchier, Engelhardtia, Fraxinus, Grewia, Juglans, Phoebe, Platanus, Populus, Prunus, Salix, Tetracera, Tilia, Vaccinium, Vitis*

Lyons:

Maidenhair: *Ginkgo*

Conifer: *Abies, Chamaecyparis, Cunninghamia, Metasequoia, Sequoia*

Dicotyledons: *Acer, Alangium, Alnus, Exbucklandia, Fraxinus, Holmskioldia, Hydrangea, Meliosma, Nymphoides, Nyssa, Platanus, Populus, Pterocarya, Rosa, Tilia*

Thomas Creek–Bilyeu Creek:

Joint grass: *Equisetum*

Ferns: *Dennstaedtia, Dryopteris, Woodwardia*

Conifers: *Metasequoia*

Monocotyledons: *Smilax*

Dicotyledons: *Actinidia, Alangium, Allophylus, Amyris, Aporosa, Cercidiphyllum, Cordia, Dillenites, Ficus, Fissistigma, Hyserpa, Magnolia, Phoebe, Rhus, Ribes, Saurauia, Siparuna, Tripterygium*

Yaquina:

Conifers: *Metasequoia, Pinus, Pseudolarix, Chamaecyparis, Sequoia*

Monocotyledons: *Smilax, Sparangium*

Dicotyledons: *Acer, Aristolochia, Corylus, Cornus, Debeya, Exbucklandia, Fagus, Hydrangea, Laurophyllum, Litsea, Mahonia, Platanus, Prunus, Quercus, Rhamnus*

are composed of the abundant and common *Alangium, Allophylus, Cercidiphyllum, Dillenites, Fissistigma, Magnolia, Metasequoia, Phoebe, Tripterygium,* and the fern *Dennstaedtia.* Among the tropical to subtropical plants in this suite, the earliest known North American occurrence of the genera *Amyris* (rue family) and *Tripterygium* (staff tree family) can be found. The flora features an unusually high number of vines and climbing plants such as *Aporosa, Ficus, Hyserpa, Rhus,* and *Smilax.* Approximately 33 percent of the angiosperms are climbers in contrast to an average of 10 percent in other Oligocene assemblages from Oregon. Equally impressive is the size of individual leaves, with single specimens measuring over a foot in length.

Northeast of Cottage Grove at Lookout Point Reservoir, the Rujada plant locale was first discovered in 1934 by Warren D. Smith of the University of Oregon. The word Rujada is a combination of the names of two local loggers, R. Upton and Jack Anderson, plus USDA. Leaves at this site are found in light yellow tuff layers between basalt flows of the Little Butte Volcanics. A definitive study by Rajendra Lakhanpal, working with Ralph Chaney of the University of California, recognized 40 species. The conifers *Abies, Cunninghamia, Pinus, Pseudotsuga,* and *Tsuga* predominate, while the most common angiosperms are *Alnus carpinoides, Halesia oregona, Quercus consimilis, Rhus varians, Exbucklandia oregonensis,* and *Platanus dissecta.* As the marine embayment withdrew westward, temperate members came to outnumber those of the earlier warmer epoch.

Leaves and petrified stumps in the Little Butte volcanics near Sweet Home and Holley in Linn County are the youngest of the Oligocene floras in Oregon, placed at 24.7 million years in the past. Typifying a coastal, stream-bank bottomland, the Sweet Home is characterized by large leaves with drip points (*Allophylus, Prunus,* and *Platanus*), which are similar to those of the Goshen. The tropical nature is indicated by the genera *Actinidia, Cedrela, Cinnamomum, Meliosma, Reptonia, Schima,* and *Sterculia.* Many of the 54 fossil woods in this remarkably rich flora were buried upright *in situ.* Some of the wood contains quartz pseudomorphs in the cubic crystal shape of the mineral halite. In this case, salt (halite) crystals have been replaced by the more stable mineral quartz. Fossilized wood and stems from near Holley are

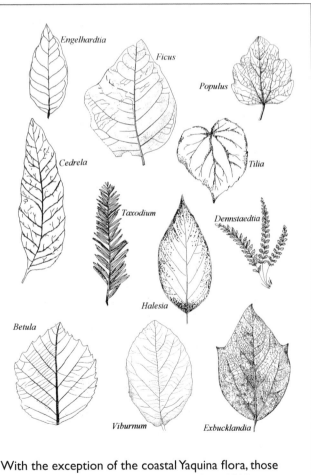

With the exception of the coastal Yaquina flora, those of the Oligocene are found in the northern Willamette Valley. (After Arnold 1937; Eubanks 1960; Gregory 1968; Lakhanapal 1958; McClammer 1978; Meyer 1973.)

prized by collectors as "Holley Blue," a pale blue to lavender agate that takes an especially nice polish. The presence of halite is an indication of a highly saline environment, which may have developed when an arm or inlet of the ocean was cut off or isolated from the open sea. After this restricted water body had substantially evaporated, halite crystallized in the water-soaked wood.

Near Newport a flora of 50 species in mudstones and fine-grained sandstones of the Yaquina Formation were collected by Douglas Emlong while searching for vertebrate fossils. Yaquina strata have been divided into three members. Mollusks are common in the lower and upper marine layers, where the water depths varied, but plant remains are found in the middle, nonmarine unit. James McClammer, who worked with the flora, concluded that most of the plants didn't inhabit the tidal lagoon but probably grew some distance inland. The leaves were carried toward the coastal delta in stream channels. In this subtropical evergreen broad-leafed flora, *Debeya*, an extinct member of the *Aralia* family, is the most numerous plant, followed by *Acer* (maple), *Litsea* (laurel family), *Pinus*, *Prunus* (stone fruit), *Sparangium* (bur-reed), *Sequoia, Exbucklandia* (witch-hazel family)*, and *Cornus* (dogwood). Similar plants grow today in East Asia where summers are wet, and a mean annual temperature ranges around 65 degrees Fahrenheit. The Yaquina is famous for its well-preserved bones of marine mammals such as whales, sea lions, and desmostylids.

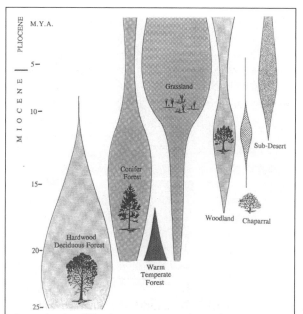

A synthesis of Tertiary vegetation from the Northern Great Basin shows clear drying trends during the late Miocene. (After Axelrod and Bailey 1969.)

The Miocene Landscape

The cool, drier trend initiated during the Oligocene Epoch continued into the Miocene and Pliocene. Short winters, wet summers, and moderate temperatures characterize the early Miocene, as a pervading and ongoing volcanism made its imprint on the Oregon terrain. Heavy falls of ash, lava flows, and lahars, or thick muddy mixtures of water and ash, blocked streams, creating lakes and wetlands, which were optimum for preservation of plant tissue and wood.

Of the many fossil floras in Oregon, probably the best known are those from the Miocene Epoch, spanning the interval between 23.8 to 5.3 million years ago. Compared to modern plant communities, Miocene floras are strikingly varied, representing a wide assortment of prehistoric climates, environments, and physiographic settings. In the Willamette Valley most of the preserved record of Miocene forests is exposed northeast of Salem, whereas to the east of the Cascades more numerous floras are localized in three separate regions: on the Owyhee uplands, in the Blue Mountains, and on the Columbia plateau.

In spite of the remoteness of eastern Oregon in the 1920s and 1930s, floras there held considerable interest for paleobotanists. Travel was primitive, no facilities were available, and visitors customarily stayed at private homes. The pioneer Mascall family often hosted paleontologists on their ranch near Picture Gorge. Thomas Condon visited the Mascall beds at the Van Horn Ranch in 1870, Waldemar Lindgren of the U.S. Geological Survey inspected several localities in 1898, and John Knowlton together with John Merriam stopped at Mascall plant localities near the turn of the century. Ralph Chaney collected in the same region with Merriam in 1920. Naturalist and commercial collector Charles D. Voy from San Francisco sent Mascall plant material to east coast institutions.

Leo Lesquereux of Harvard University and the Smithsonian Institution briefly mentioned plant specimens from the Blue Mountains in 1878. Percy Train, a commercial dealer, contracted to collect 750 items from Rockville and Sucker Creek in 1935 for Chester Arnold, a professor at the University of Michigan. Train later added over 10,000 specimens from Trout Creek. The entrepreneurial Train sold a complete set of fossil plants consisting of 20 species of "leaves and winged seeds—half of which are new to science" from Trout Creek for $20. In 1962 Alan

Two paleobotanists, Roland Brown and Harry MacGinitie, stand out for their studies on regional floral communities. Brown's work in the Northwest, which stretched over 30 years and many summers, involved touring plant locales across Oregon. Knowledgeable in many aspects of natural history, his primary interest was in botany and Tertiary plants. Born in Weatherly, Pennsylvania, Brown received his doctorate in 1926 from Johns Hopkins University, and, after several brief assignments at the Pennsylvania Survey and Forestry Department, he began a job with the U.S. Geological Survey. He held that position until his death in Mauch Chunk, Pennsylvania, in 1961. Roland Brown is on the left and J.C. Reed on the right. (Photo courtesy Archives, U.S. Geological Survey, Denver.)

Harry MacGinitie's paleobotanical career began when he was at Berkeley, where his Ph.D. research was completed in 1935 under Ralph Chaney. Born in Lynch, Nebraska, MacGinitie taught at Humboldt State College, completing floral lists, habitats, and climates for the Trout Creek assemblage. Retiring in 1960, he accepted an appointment at the Berkeley Museum of Paleontology where he continued to do research until his death in 1987. His careful systematic approach and innovative ideas on the vegetational significance and history of western Tertiary floras were largely substantiated by later workers in the field. (Photo courtesy of the Archives, U.S. Geological Survey, Denver.)

Graham treated several aspects of both the Trout Creek and Sucker Creek floras. The Dalles flora was initially brought to the attention of John Newberry from Columbia College in New York when he received nine plant specimens sent by Thomas Condon in 1869, but the plants were virtually ignored for almost 50 years until Ralph Chaney visited the locality in 1916.

Most of the eastern Oregon floral locations came under close scrutiny by Ralph Chaney at one time or another. His classic *Miocene and Pliocene Floras of Western North America* in 1938 and his 1959 *Miocene Floras of the Columbia Plateau*, co-authored with Daniel Axelrod, recorded 40 years of work on Miocene plant populations throughout this area.

The fascinating picture of Miocene time in eastern Oregon features an intricate network of streams and lake basins where thick vegetation grew atop lava plateaus. Recurrent eruptions would smother and destroy the plants, and then preserve them in the sediments. In an ongoing cycle, forests would emerge, only to be demolished by the next volcanic event. Plants from these communities are strikingly similar in age, composition, and ecology, but each is given a separate floral designation and formation as the individual assemblages are confined to separate depositional basins. The strata have yet to be traced continuously from one area to the next. With some variations, all the floras are preserved in upland lake settings, a varied topography of steep hills and valleys, temperate climates, and moderately heavy rainfall. Although there are individual adaptations in age assignments as the basalt flows followed each other and some differences attributed to climate changes, the floras are remarkably comparable. These are the Alvord Creek, the Sucker and Trout creeks, the Stinking Water, the Mascall, the Blue Mountains, the Sparta, and The Dalles.

On the southern edge of the Owyhee plateau, the oldest Miocene plant remains are those from Alvord Creek, dated at 21.3 million years ago. Leaves and pollen within the Alvord Creek Formation were preserved in ash and sediments from a low shield cone. The flora was thoroughly investigated by Harry MacGinitie in 1933 and around 10 years later by Daniel Axelrod. Leaves were shed into quiet lake waters at moderate elevation, bordered by conifer forests, uplands, and distant volcanoes. Rainfall, heaviest in the winter months, was 20 to 30 inches, and seasonal temperatures were more moderate than today, probably averaging 10 degrees warmer annually.

Represented by both leaves and pollen, deciduous plants such as *Acer* (maple), *Amelanchier* (serviceberry), *Juglans* (walnut), and *Rhus* (sumac) populated the Alvord Creek lake shore and slopes, while *Abies* (fir), *Pseudotsuga* (Douglas fir), and *Tsuga* (hemlock) grew at elevations comparable to the California Sierras. Remarkably, only *Quercus* (oak) pollen, but no leaves, has been found, and pollen of *Quercus* and *Abies* is the most frequent.

Of all the Miocene floras from Oregon, the Sucker Creek in Malheur County is perhaps the most famous. Volcanic ash that buried these plants has been named the Sucker Creek Formation, whereas the stream in the same vicinity was changed to its earliest designated popular name, Succor Creek, by the U.S. Board of Geographic Names. A geologic formation retains its official name even though local geographic place-names may change; hence the two different spellings. Chester Arnold's 1937 observations on several aspects of both the Trout Creek

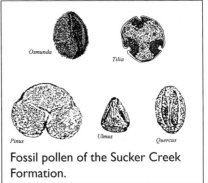

Fossil pollen of the Sucker Creek Formation.

and Sucker Creek floras were followed by that of Alan Graham of Kent State University. Examining the pollen, leaves, fruit, and flowers from 47 separate families, 60 genera, and 69 species, Graham found that lakeside plants were mixed with those growing on slopes and in lowlands during Sucker Creek time. Pollen from high altitude conifers such as *Abies* (fir) and *Picea* (spruce) indicate elevations reaching 2,000 feet above sea level. Minimum winter temperatures did not fall below freezing, as reflected by the presence of *Cedrela pteraformis* (South American cedar). An annual rainfall of over 43 inches supported the large-leafed *Oreopanax precoccinea*, which has a modern equivalent in South American forests. In this flora *Quercus* is the most prevalent leaf fossil.

To the west of Sucker Creek in Harney County, the Trout Creek floral assemblage, although younger, has many parallels. The fossil material consists of leaves, pollen, fruit, stems, roots, and even occasional flowers. When working on his 1933 *Contributions to Paleontology*, Harry MacGinitie produced a complete

A selection of plants from the Owyhee Plateau:

Alvord Creek:

Conifers: *Abies, Juniperus, Keteleeria, Pinus, Tsuga*
Dicotyledons: *Acer, Amelanchier, Amorpha, Arbutus, Carpinus, Ceanothus, Cercocarpus, Mahonia, Photinia, Prunus, Rhus, Rosa, Salix, Sorbus*

Sucker Creek, Trout Creek:

Joint grass: *Equisetum*
Ferns: *Osmunda, Woodwardia*
Maidenhair: *Ginkgo*
Conifers: *Abies, Glyptostrobus, Picea, Pinus, Thuites, Tsuga*
Monocotyledons: *Potamogeton, Typha*
Dicotyledons: *Acer, Ailanthus, Alnus, Amelanchier, Arbutus, Betula, Castanea, Cedrela, Cercocarpus, Cornus, Crataegus, Diospyros, Fagus, Fraxinus, Gymnocladus, Ilex, Juglans, Oreopanax, Magnolia, Mahonia, Nyssa, Oreopanax, Persea, Platanus, Populus, Pterocarya, Pyrus, Quercus, Salix, Sassafras, Tilia, Ulmus, Zelkova*

Stinking Water:

Joint grass: *Equisetum*
Ferns: *Osmunda*
Maidenhair: *Ginkgo*
Conifers: *Abies, Glyptostrobus, Keteleeria, Picea, Pinus, Pseudotsuga, Thuites, Thuja, Tsuga*
Monocotyledons: *Potamogeton, Smilax, Typha*
Dicotyledons: *Acer, Ailanthus, Alnus, Amelanchier, Betula, Cedrela, Crataegus, Diospyros, Gymnocladus, Hydrangea, Mahonia, Nyssa, Oreopanax, Persea, Platanus, Populus, Prunus, Pterocarya, Quercus, Rhamnus, Rosa, Salix, Spirea, Tilia, Ulmus, Vaccinium*

floral list, described the habitat, climate, and associated diatoms and compared the plants to others in the region. His treatment was the first comprehensive account of an upland Miocene plant community on the Owyhee Plateau.

Thirty years after MacGinitie's work, Alan Graham, who examined 13,000 leaves and 40,000 pollen grains, surmised that Trout Creek plants represented a landscape featuring upland lakes adjacent to steep, well-drained slopes at elevations as high as 4,000 feet. Rainfall was an estimated 50 inches per year, minimum winter temperatures were 35 to 40 degrees Fahrenheit, and maximum summer temperatures rarely exceeded 90 degrees. Surrounding the lakes were extensive wetlands as shown by *Equisetum* (horsetail) and *Typha* (cattail) and beyond that a forest of *Acer* (maple), *Amelanchier* (serviceberry), *Quercus* (oak), and *Salix* (willow). Of the 45 species here, only five are coniferous. *Pinus* (pine) and *Thuites* (cypress family) were especially common, producing a mixed forest.

Marking the northern border of the Owhyee uplands in Harney County, lava flows and tuffs in the Juntura basin preserved leaf and fruit impressions as well as diatoms of the middle Miocene Stinking Water and Beulah floras, dated at 12.1 million years of age. Found in the lower Juntura Formation, remnants of these plant communities are similar to those in the more westerly Mascall Formation, although low grade lignitic coals and wetlands, present in the Mascall, are absent here. Ralph Chaney and Daniel Axelrod depict the overall Stinking Water paleolandscape as that of a lowland near sea level and of good drainage, with broad riverine valleys, woodlands and savannas on the upper slopes, and a distant volcanic range. This was predominantly an oak forest, with five species of *Quercus* in addition to *Glyptostrobus oregonensis* (water pine) possibly lining a riverbank. *Alnus* (alder), *Platanus* (sycamore), *Populus* (cottonwood), and *Ulmus* (elm) make up 90 percent of the specimens.

The high silica content of Miocene lakes on the Owyhee plateau, resulting from the chemical makeup of volcanic material, produced conditions favorable for prolific diatom blooms. Microscopic glassy skeletons of these single-celled aquatic plants form entire layers or diatomites in middle Juntura strata. An analysis of diatomite samples at Drinkwater Pass and at Harper pointed to a deep freshwater lake. The presence of *Melosira islandica* and similar species is evidence that the water was oxygen deficient, a condition which occurs often where the circulation is poor, and decaying organic matter uses up the free oxygen. Diatomite is mined commercially at several sites in Oregon

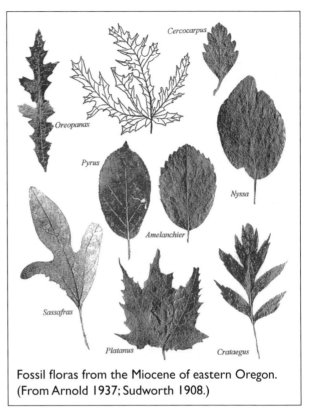

Fossil floras from the Miocene of eastern Oregon. (From Arnold 1937; Sudworth 1908.)

and is used for a variety of industrial purposes such as in filters, for insulation and abrasives, and as kitty litter.

The Blue Mountains experienced a long succession of plant growth, punctuated by intermittent destructive lavas, responsible for the many outstanding fossil communities in the province. The Mascall flora, exposed on the John Day River, the Blue Mountains fossil plant assemblage near Austin in Grant County, and the Sparta in Baker County have many Miocene elements in common. Ash and lava flows of the 15-million-year-old Mascall Formation are known for a wide variety of plants and animals, whereas the younger Rattlesnake Formation, also within the John Day basin, has few plants but numerous mammal bones. Systematically examining the Mascall, Ralph Chaney assessed the forest, sketching the landscape as an area of low relief and uniform temperate climate. Plant remains were deposited on floodplains, in stream channels, or in lakes. The numerous browsing and grazing hoofed mammals entombed in the Mascall imply that these freshwater lakes may have been surrounded by grassy forests and open savannas. Rainfall was 30 inches annually, producing vegetation comparable to that of a modern oak forest in the Ohio Valley.

A selection of Blue Mountain floras:

Mascall, Sparta, Tipton:

Diatoms: *Cocinodiscus, Fragilaria, Melosira, Tetracyclus*

Conifers: *Abies, Cephalotaxus, Glyptostrobus, Keteleeria, Libocedrus, Metasequoia, Picea, Pinus, Pseudotsuga, Sequoia, Taxodium, Thuja, Tsuga*

Monocotyledons: *Nymphaeites, Smilax, Typha*

Dicotyledons: *Acer, Alnus, Amelanchier, Arbutus, Betula, Carya, Celtis, Cercidiphyllum, Diospyros, Fagus, Fraxinus, Hydrangea, Juglans, Laurophyllum, Lindera, Liquidambar, Mahonia, Nyssa, Platanus, Pterocarya, Quercus, Salix, Ulmus, Zelkova*

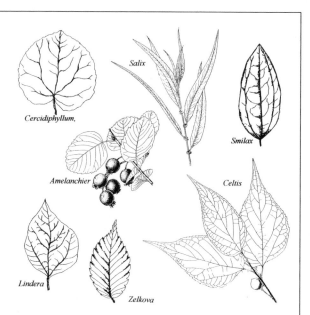

The Miocene *Celtis* (hackberry), *Lindera* (spicebush), *Amelanchier* (serviceberry), *Cercidiphyllum* (katsura), *Salix* (willow), *Smilax* (greenbrier), and *Zelkova* (keaki tree) from the Blue Mountains region show a modern complexion. (After Chaney and Axelrod 1959; Sudworth 1908.)

Melosira granulata from Miocene diatom layers near Vale. (Photo courtesy Oregon Department of Geology and Mineral Industries.)

The presence of *Taxodium dubium* (swamp cypress), *Populus lindgreni* (cottonwood), *Nyssa hesperia* (tupelo), *Typha* (cattail), and *Nymphaeites* (water lily) is an indication of shallow water, where low-grade coal developed. Lower slope plants such as *Carya* (hickory) and *Quercus* (oak) are the most common, whereas those occupying higher regions, such as *Acer* (maple), *Betula* (birch), *Ginkgo, Pinus* (pine), and *Sequoia* (redwood), are sparsely represented and appear to have been restricted in extent.

Although the flora is distinctive, the actual frequency of Mascall leaves is only modest. For example, Oligocene Bridge Creek shales typically yield over 200 leaves per cubic foot of rock examined, whereas the Mascall generally has 10 or fewer. This disparity may reflect the size of the lake basins. Whereas the smaller waterways of the Bridge Creek concentrated leaves, the extensive lakes of the Mascall dispersed them.

Above the Mascall strata at Dayville, the six- to eight-million-year-old Rattlesnake Formation has rare leaf impressions of *Platanus* (sycamore), *Salix* (willow), and *Ulmus* (elm).

Despite the time and care lavished on fossils by private enthusiasts, superb collections are often picked over and lost once some time has passed. This was the case with the Sparta flora northeast of Baker, which was amassed by Leslie R. Hoxie. In 1961 he produced the only account of the floral composition occurring within sediments of the Columbia River basalt. Using evidence provided by the abundant coniferous pollen, he derived a setting of mountain slopes forested by hardwoods with moderate rainfall and temperatures. In addition to pollen, the flora is composed of leaf impressions, fruits, and wood. *Quercus* (oak) is the predominant leaf, but *Acer* (maple), *Carya* (hickory), *Picea* (spruce), *Pinus* (pine), *Platanus* (sycamore), and *Ulmus* (elm) are also numerous. After Hoxie's death, his specimens were divided up and many destroyed.

Blue Mountains floras have long been known from localities near Sumpter, Keating, Austin, and Tipton, small towns in old gold-mining districts in Baker and Grant counties. The first extensive examinations were by Elizabeth Oliver in 1934 and by Ralph Chaney and Daniel Axelrod in 1959. Interlayered within the Columbia River basalts, leaf and fruit impressions of both trees and shrubs, along with pollen, point to a temperate setting around the shores and slopes of an upland lake at 2,000 feet in elevation. Of the seven conifers and 25 deciduous plants, oak (*Quercus*), beech (*Fagus*), black hawthorn (*Crataegus*), and redwood (*Sequoia*) were the most prevalent. Among the diatoms, *Melosira granulata* and *Tetracyclus ellipticus* inhabited freshwater lakes fed by streams with small amounts of dissolved salts. Such Miocene lakebeds in eastern Oregon were so saline that marine diatoms flourished in the water.

Within the Columbia River Plateau, fossil plants near The Dalles in Wasco County are dated between 5.7 to 7.5 million years ago and represent one of the very few late Miocene assemblages in eastern Oregon. Fragmentary vegetation of The Dalles flora is found in flood deposits and volcanic debris making up the formation with the same name. Described by Ralph Chaney in his *Pliocene Floras of California and Oregon,* this community is indicative of a cool, semiarid climate that supported *Acer negundoides,* a new species of box elder, *Amorpha* (indigo plant), *Cercis* (redbud), *Quercus* (oak), and *Ulmus* (elm). The accumulation is composed of only 12 species, and the lack of many broad-leafed deciduous plants and conifers along with the presence of evergreens suggests reduced summer rainfall and a low elevation where the plants grew near a stream bank. The Dalles is notably more modern than other Miocene floras. In a rare occurrence of vertebrates associated with a leaf assemblage, an *Aelurodon* (a primitive wolf) and *Hipparion* (horse) have been found at this locality.

A selected list of Miocene floras from western Oregon:
Collawash, Molalla, Eagle Creek, and Sardine Formation:

Maidenhair: *Ginkgo*
Conifers: *Cephalotaxus, Cunninghamia, Keteleeria, Metasequoia, Pinus, Sequoia, Taxodium*
Monocotyledons: *Smilax*
Dicotyledons: *Acer, Alnus, Amelanchier, Arbutus, Carya, Cercidiphyllum, Cercis, Diospyros, Exbucklandia, Fagus, Fraxinus, Hydrangea, Idesia, Ilex, Juglans, Liquidambar, Magnolia, Nyssa, Platanus, Populus, Prunus, Pterocarya, Quercus, Rhamnus, Rhododendron, Rhus, Rubus, Salix, Sorbus, Sterculia, Ulmus, Zelkova*
Vines: *Berchemia, Cocculus, Vitis*

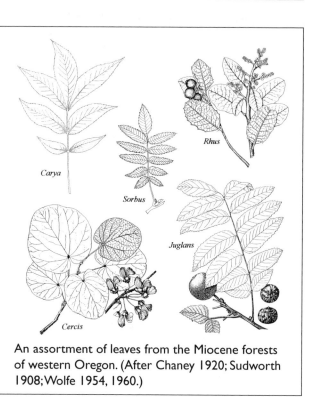

An assortment of leaves from the Miocene forests of western Oregon. (After Chaney 1920; Sudworth 1908; Wolfe 1954, 1960.)

As marine waters withdrew from the Willamette basin during the Miocene, sediments filled the depression. Numerous small plant populations can be found scattered on the western flanks of the Cascades and along the Columbia River. Mudflows and ash layers of the Little Butte Volcanics played a major role in preserving plant communities that include the Molalla and Collawash in Clackamas County and the Eagle Creek near Corbett in Multnomah County.

In spite of similarities in age, the Molalla flora is viewed as subtropical, whereas the Collawash and Eagle Creek are more temperate. Frosts were not severe, and the annual rainfall of 50 to 60 inches was much like that of western Oregon today. The Molalla and the Eagle Creek both flourished close to sea level, but the Collawash developed at a higher elevation. Fagaceae (beech family), Juglandaceae (walnut family), Lauraceae (laurel family), and vines were common in all three, reflecting the continual Miocene cooling trend. Vines (*Berchemia, Rhus, Vitis*) and trees (*Fagus, Quercus*), with few shrubs, were the most abundant genera in the Molalla. The presence of swamp cypress (*Taxodium*), *Quercus* (oak), and *Carya* (hickory) implies adjacent hills around what was probably a wetland produced by copious warm rains.

Among the scattered Miocene plant populations in this part of the state, the Collawash is the most diverse with more than 140 temperate species represented.

Growing close to 2,000 feet in elevation, the luxuriant streamside vegetation was dominated by the broad-leafed trees (*Alnus, Betula, Carya, Liquidambar, Platanus*), the conifers (*Taxodium, Metasequoia*), shrubs (Rosaceae), and vines (*Berchemia, Cocculus, Vitis*) that produced a remarkably rich fossil locality.

The topography of the Eagle Creek varied. *Quercus* (oak), with the highest numerical percentage, grew on the dry ridges at 300 to 500 feet; however, the oak leaves were carried into the moist lower stream valleys where they mixed with maples, elm, and other vegetation. *Ginkgo* is among the most distinctive and abundant plant found in the flora. Some tropical leaves such as those of *Liquidambar* (sweet gum), *Smilax* (lily family), and *Sterculia* (cocoa family) point to a climate substantially warmer than today, and

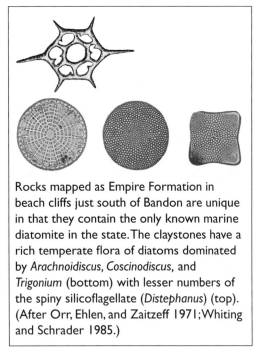

Rocks mapped as Empire Formation in beach cliffs just south of Bandon are unique in that they contain the only known marine diatomite in the state. The claystones have a rich temperate flora of diatoms dominated by *Arachnoidiscus, Coscinodiscus,* and *Trigonium* (bottom) with lesser numbers of the spiny silicoflagellate (*Distephanus*) (top). (After Orr, Ehlen, and Zaitzeff 1971; Whiting and Schrader 1985.)

even though *Cephalotaxus* (yew) is numerous, there is a remarkable absence of evergreen plants.

Some six Tertiary localities along the slope of the Western Cascades from Jackson to Clackamas County yield florules in volcanic rocks of the Sardine Formation. Thirty-three genera include five species of *Quercus* (oak) and four of *Populus* (poplar), the two most abundant, along with the conifers *Abies, Picea, Pinus, Sequoia,* and *Taxodium.*

Rapidly diminishing waters along Oregon's coastline in the mid to late Miocene produced a number of small sediment-filled basins, where two unusual plant specimens have been recorded at Newport and Bandon. Near Newport in Lincoln County small seed cones of a new species of *Pinus berglundii* are preserved in the middle Miocene Astoria Formation. The seeds are similar to modern forms, but distinctive characteristics point to a separate evolutionary lineage. Leaf impressions only occasionally turn up in rocks of the Astoria. Equally exceptional, the only known marine diatomite (algae) in the state was found in beach cliffs and on wave-cut benches south of Bandon in Coos County, Mapped as the Empire Formation, the mudstones have a rich temperate flora dominated by the diatoms *Thalassiosira,* with lesser numbers of *Arachnoidiscus, Coscinodiscus*, and *Trigonium*, deposited in a deep-water environment. In addition to the diatoms, planktonic (floating) and benthonic open ocean foraminifera occur in the mudstones, while silicoflagellate microfloras are present but rare.

Mel Ashwill has contributed to a variety of geologic subjects, but his main interests lie in paleobotany. In 1991 he was honored by the Paleontology Society with the prestigious Strimple Award. Growing up in North Dakota, Ashwill moved with his family to Newberg, Oregon. He received a graduate degree in music at Pacific University and taught for 23 years before retiring. Today he maintains an extensive private fossil collection at his home in Madras where he provides help and knowledge to others from his own resources. (Photo c. 1990; courtesy M. Ashwill.)

The Pliocene Landscape

By late Miocene to early Pliocene time, the modern topography of the Pacific Northwest was well established. The Cascades and Coast Range had been elevated, concurrent with the depression of the Willamette Valley and Puget lowlands. The Pliocene Epoch, between 5.3 and 1.7 million years ago, saw a continuation of cooler, less humid conditions, which were initiated during the Miocene. In addition to the disappearance of warm tropical plants, waning volcanic activity greatly reduced the optimum conditions for fossilization that had been present earlier. Sedimentary basins, so prevalent during the Miocene, diminished along with the dense forests and thick foliage. Pliocene floras, as a consequence, are poorly known, and the climate, topography, and environment are not as well defined as those of earlier epochs of the Oregon Tertiary.

Only the Troutdale flora west of the Cascades and the Deschutes flora to the east of the mountains are assignable to the Pliocene. Even though the Deschutes formation spans the Miocene-Pliocene boundary, the flora is found in the upper mudflow layer of the formation, dated at 5.3 million years old. Separated by the barrier of the volcanic range, these two floras contrast sharply. While the Troutdale reflects heavy rainfall, where the vegetation and climate were dominated by the nearby ocean, the Deschutes grew in a more arid habitat.

In the Deschutes River valley, volcanic sediments, lava flows, and ash derived from regional volcanoes make up the Deschutes Formation. The presence of fossil leaves was first reported by highway engineers during construction of the Vanora highway grade near Madras. In 1938 Ralph Chaney summarized the Deschutes forest as distinguished by *Acer* (maple), *Populus* (aspen), *Prunus* (cherry), and *Salix* (willow) reflecting a semi-arid, cool climate of high relief. Along with the plants, a camel (*Auchenia*) bone was found here earlier by Thomas Condon.

A selection of Deschutes and Troutdale floras:

Joint grass: *Equisetum*

Conifers: *Abies, Chamaecyparis, Sequoia*

Dicotyledons: *Acer, Amorpha, Arbutus, Carya, Cercis, Cornus, Crataegus, Diospyros, Fraxinus, Liquidambar, Mahonia, Persea, Platanus, Populus, Prunus, Pterocarya, Quercus, Rhamnus, Rosa, Salix, Spiraea, Ulmus, Umbellularia, Vitis, Zelkova*

A more recent examination of the Deschutes flora was made by Mel Ashwill during yet another highway widening project in 1990. Ashwill took advantage of the excavation to salvage over 40 huge boulders in which he records up to 8,000 fossil leaves. He added new species, *Populus subwashoensis* (cottonwood) and *Salix florissanti* (willow), to the floral list. After careful examination of the fossils and by comparing them to plants trapped in mudflows associated with the 1980 eruption of Mount St. Helens, Ashwill concluded that the Deschutes material was entombed during a similar dramatic event. The leaves, seeds, branches, and whole trees were carried along in a mudflow and then deposited as a thick carpet of vegetative debris.

In the Portland basin on Buck Creek and above Camp Collins, leaves of the Pliocene Troutdale flora are buried in silty layers between coarse conglomerates of the Troutdale Formation. These deposits were carried by streams flowing from the slopes of nearby volcanoes onto broad floodplains and into deeply dissected valleys. Plentiful rainfall with annual precipitation around 35 inches at a mean temperature of 50 degrees Fahrenheit is indicated by the presence of *Sequoia* (redwood), *Diospyros* (persimmon), *Liquidambar* (sweet gum), and *Pterocarya* (wingnut). The floral composition between Buck Creek and Camp Collins differs considerably, but this may be due to the limited number of specimens so far reported. *Quercus winstanleyi* typified Buck Creek along with the grasses (*Cyperacites*) and willows (*Salix*), while at Camp Collins *Ulmus californica* is by far the most numerous. Species of *Chamaecyparis,*

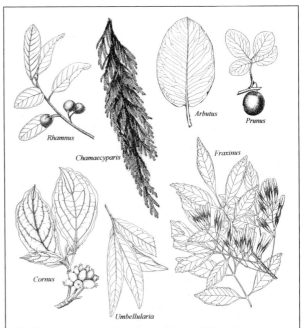

By Pliocene times, the Troutdale and Deschutes floras were essentially modern in complexion. (After Chaney 1938a, 1944b; Chaney and Axelrod 1959; Sudworth 1908.)

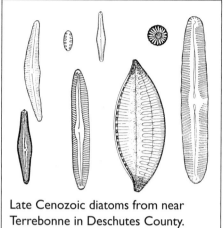

Late Cenozoic diatoms from near Terrebonne in Deschutes County. (After Lohman 1937; Mahood 1981; VanLandingham 1990.)

Cornus, Diospyros, Sequoia, Umbellularia, and *Zelkova* survived the climate change to drier summers, which took place in the late Miocene and brought on the extinction of many species. The presence of *Carya, Persea,* and *Pterocarya* points to the temperate composition of the collection.

The microscopic glassy capsules of aquatic algae, or diatoms, persist after the plant has died, marking the location of old lakebeds. The majority are located in eastern Oregon. Diatoms layers near Terrebonne in Deschutes County and in the Fort Rock basin of Lake County have been assigned variously to the Pliocene and Pleistocene. At Terrebonne, samples from test pits made in 1930 yielded 33 species of diatoms. Since then, the age, origin, and identifications of Terrebonne diatoms have been debated, although Sam VanLandingham concluded in 1990 that there are two different diatom deposits, Terrebonne West, which is Pleistocene, and Terrebonne East, which is Pliocene. Terrebonne is Oregon's most extensively developed diatomite region. Freshwater lake deposits from near the town of Chiloquin, in the Klamath River basin, are similar to those from Terrebonne. Here *Stephanodiscus rhombus* occurs in strata of the Yonna Formation.

The economic possibilities of the diatom-rich layers near Christmas Lake in the Fort Rock basin were explored in 1982 in order to assess the marketing possibilities of absorbent and filtering products from the crushed diatomite. Forty-seven drill core and surface samples revealed diatoms that were typical of those from freshwater Pliocene lakes. Suspended (planktonic) diatoms were more

Pioneering pollen work on the Pleistocene and postglacial periods in the state was carried out by Henry Hansen, professor at Oregon State University, from 1939 onward. Developing an interest in pollen while growing up in a region of extensive bogs around LaCrosse, Wisconsin, he completed a Ph.D. in botany and geology from the University of Washington in 1937. Hansen foresaw the potential for using pollen records to trace vegetational history, and after joining the botany staff at Corvallis his analysis of pollen in bogs throughout the Pacific Northwest formed the foundation for subsequent research. Hansen died in Corvallis in 1989. (Photo courtesy Oregon State University, Archives.)

OREGON CENOZOIC FLORAS

EPOCHS M.Y.A.*

	WESTERN	EASTERN

WESTERN

PLEIST.
1.7

Coquille Fm.

PLIO

Troutdale Flora Troutdale Fm. 1.5-4 mybp.

5.3 5

10

MIOCENE

Molalla Flora Little Butte Volcanics 12.9 mybp

15

Collawash flora Little Butte Volcanics 13-16

Eagle Creek flora Little Butte Volcanics 13-16

Thomas/Bilyeu Creek flora Little Butte Volcanics

Scio flora Little Butte Volcanics

20

Lyons Flora Little Butte Volcanics

23.8

Sweet Home flora Little Butte Volcanics 24.7

25

Rujada flora Little Butte Volcanics 25.1 mybp.

OLIGOCENE

Yaquina flora Yaquina Fm. 28-30.9 mybp.

30

Willamette flora Fisher Fm. 31.1 mybp.

Goshen flora Fisher Fm. 33.4 mybp.

33.7
35

Keasey Fm.

Cowlitz Fm. 38-40 mybp

40

Comstock flora Fisher Fm. 39.9 mybp.

Coos Bay flora Coaledo Fm. 40-45.5 mybp.

Yamhill Fm. 42 mybp

EOCENE

45

50

54

Roseburg/Siletz River 52-56 mybp

55

PALEOCENE

60

EASTERN

Deschutes flora Deschutes Fm. 5.3 mybp.
The Dalles flora The Dalles Group 5.7-7.5 mybp.
Rattlesnake flora Rattlesnake Fm. 6.4 mybp.

Stinking Water flora Juntura Fm. 12.1 mybp.
Sparta flora Columbia River Basalt Group 12.9
Trout Creek flora Trout Creek Fm. 13.1 mybp.
Blue Mts. Flora Columbia River Vol. 13-15
Mascall flora Mascall Fm. 15 mybp.

Sucker Creek flora Sucker Creek Fm. 16.7 mybp.

Alvord Creek flora Alvord Creek Fm. 21.3 mybp.

Bridge Creek flora John Day Fm. 33.6 mybp

Whitecap Knoll John Day Fm. 38.2 mybp
Iron Mountain flora John Day Fm. 38-39 mybp.
Gray Butte flora John Day Fm. 38-39 mybp.
Clarno Mammal Quarry Clarno Fm. 40 mybp.

Clarno Nut Beds Clarno Fm. 43.8 mybp.

Denning Spring flora

*million years ago

Cenozoic plant stratigraphy. (After Orr and Orr 1996; Prothero 2001; Retallack et al. 2004.)

abundant than the bottom-dwelling benthic species, suggesting that the nutrient-deficient lake was relatively deep, but not stagnant. Volcanic ash and pumice, falling directly into the water, provided a ready source for silica, necessary for growth of the glassy diatom skeletons. The most commonly encountered genera were *Cyclotella, Melosira, Stephanodiscus,* and members of the Fragillariaceae family.

Among the several diatoms from the Always Welcome Inn site in Baker City, *Aulacoseira, Fragillaria,* and *Cymbella* lived in shallow waters, while the presence of *Epithemia* indicates stagnant conditions along the shoreline of a lake. Thick diatom layers in the lower section of this locality are evidence that the microscopic plants flourished in the nutrient-rich waters. Similarly, in the nearby Powder River valley, the diatoms *Aulacoseira* and *Tetracyclus* point to a late Miocene marsh and swamp before the waters deepened.

The Pleistocene Landscape

Following the comparatively brief timespan of the Pliocene, the Pleistocene epoch was an even shorter interval of dramatically increased rainfall and colder temperatures as continental ice lobes extended southward from British Columbia, while glacial sheets reached across the tops of the higher mountain peaks.

Work on Pleistocene pollen and plants lagged behind research on floras of the earlier Tertiary. It wasn't until the 1930s that Henry Hansen explored the Pacific Northwest with George B. Rigg, Botany Professor at the University of Washington, initiating a study in the relatively new field of regional bog ecology. Building an extensive reference collection to familiarize himself with pollen from modern forests, Hansen cored 70 wetlands throughout the Northwest. His studies chronicle several distinct forest successions from tundra to modern forests since the last Ice Ages, identifying warm dry intervals. Hansen's pioneering studies formed the basis for later studies tracing vegetation and climate changes during the Pleistocene.

Pleistocene paleobotany attracts few researchers, and the known localities, which have been examined, are scattered through western Oregon from the coast to the Cascades. In Lane County, fragments of wood, needles, and pollen from Lookout Creek, dated at 35,500 years before the present, are

Its presence demonstrating catastrophic tectonic subsidence on the Oregon coast, a rooted conifer stump, dating back thousands of years, is one of several exposed at Moolack Beach by the winter storms of 1998. (Photo courtesy W. Orr.)

from Oregon's youngest fossil plant deposits. Carried downstream from higher elevations and scattered across the valley floor, *Abies* (fir) and *Picea* (spruce) grew in somewhat warmer conditions than did *Pseudotsuga* (Douglas fir) and *Thuja* (arborvitae), which may have populated lower cold valleys.

Paleobotanist Jack Wolfe noted that the Ice Age plant locale at Cape Blanco had only limited remains of *Pseudotsuga* in comparison to frequent appearances of this tree elsewhere in the Pacific Northwest. This observation led him to surmise that Douglas fir did not become dominant here until quite late in the Pleistocene. His conclusions are supported by pollen and plant macrofossil evidence from a bog at Little Lake in Lane County in the central Coast Range, where Cathy Whitlock found that Douglas fir, along with red alder and bracken fern, were established some time after 5,000 years ago.

By compiling Pleistocene pollen data from several bog sites in the Oregon Cascades as well as in adjacent Washington, Whitlock, of Montana State University, presented a picture of the local climate history. Open meadows and a forest of *Picea* (spruce), *Pinus* (white pine), and *Tsuga* (western hemlock) colonized the northern Cascades prior to 12,400 years before the present. As temperatures warmed and rainfall diminished after 10,000 years ago, forests of *Pseudotsuga* (Douglas fir) and *Tsuga* grew at higher elevations, while *Corylus* (hazelnut) and *Quercus* (oak) were found in the lower regions. In the central High Cascades, *Abies* and *Pinus* were predominant. Whitlock concluded that the significant Pleistocene vegetational changes resulted from major alterations in the regional climate.

It wasn't until the 1990s, when standing forests of dead trees on the coast were recognized as evidence of catastrophic changes in sea level, that current attention was focused on Pleistocene and Holocene vegetation. The phenomenon of buried stumps along the coast has been directly related to paleoseismic events, tectonic plate subduction, and earthquakes. A professor at Portland State University, Curt Peterson has shown that as subduction proceeds, the coastal area is slowly elevated when the two plates lock and buckle upward. With a major quake, the upper plate slams down as much as 20 feet or more, bringing the nearshore forests to the surf zone. Peterson examined 275 rooted and buried stumps at 14 localities frequently visible in the surf zone on the central coast. He reported entire fields of buried stumps, roots, and cones of Sitka spruce (*Picea sitchensis*) and western hemlock (*Tsuga heterophylla*) near Newport, as well as moss stems similar to *Scouleria aquatica* at China Creek north of Florence. Extensive Holocene conifer forests, growing on the wave-cut terraces, dunes, or in creek beds were abruptly buried about 1.9 to 4.4 thousand years ago by rising sea levels, coastal migration of sand, and tectonic occurrences. The cycle of terracing, uplift, forestation, burial, flooding, and renewed platform cutting took place repeatedly over the past 1,000 years. The Newport buried forest was exposed in January and May 1995, reburied by October of that year, only to reappear briefly in 1998, and sporadically since then.

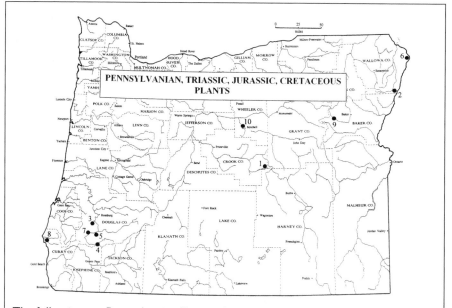

The following are Pennsylvanian, Triassic, Jurassic, and Cretaceous plant localities and a selection of authors who list floras from the locales.

PENNSYLVANIAN

1. Suplee (Mills Ranch), Crook Co. Spotted Ridge Fm. Spotted Ridge flora. Arnold 1945, 1953; Mamay and Read 1956.

TRIASSIC

2. Red Gulch, Wallowa Mountains, Wallowa Co. Hurwal Fm. Flugel, Senowbari-Daryan, and Stanley 1989.

JURASSIC

3. Buck Peak, Douglas Co. Riddle Fm. Diller 1908; Knowlton 1910; Ward 1905.
4. Cow Creek, Douglas Co. Riddle Fm. Diller 1908; Knowlton 1910; Ward 1905.
5. Riddle, Douglas Co. Riddle Fm. Diller 1908; Knowlton 1910; Ward 1905.
6. Snake River Canyon, Wallowa Co. Coon Hollow Fm. Ash 1991a.
7. Thompson Creek, Douglas Co. Riddle Fm. Diller 1908; Knowlton 1910; Ward 1905.

CRETACEOUS

8. Elk River, Curry Co. Rocky Point Fm. Lowther 1967.
9. Lightning Creek, Baker Co. No Fm. Dake 1969; Read and Brown 1937.
10. Mitchell, Wheeler Co. No. Fm. Gregory 1968; Thompson, Yett, and Green 1984.

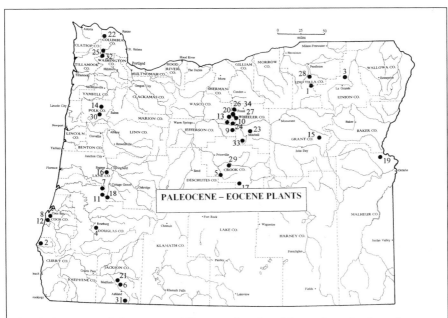

PALEOCENE – EOCENE PLANTS

The following are Paleocene and Eocene plant localities and a selection of authors who list floras from the locales.

PALEOCENE

1. Denning Spring, Umatilla Co. No Fm. Gordon 1985.
2. Floras Creek, Curry Co. Roseburg Fm. Baldwin 1974.
3. Imbler (diatoms), Union Co. No Fm. Van Tassell, McConnell, and Smith 2001.
4. Roseburg, Douglas Co. Roseburg Fm. Brouwers et al. 1995; Bukry and Snavely 1988.

EOCENE

5. Bear Creek valley, Crook Co. Clarno Fm. Clarno Flora. Hergert 1961; Lowry 1940.
6. Bear Creek valley, Jackson Co. Payne Cliffs Fm. Brown 1956; McKnight 1984; Peck et al. 1964.
7. Bear Creek, Douglas Co. Fisher Fm. Comstock Flora. Peck et al. 1964; Retallack et al. 2004; Sanborn 1937.
8. Cape Arago, Coos Co. Coaledo Fm. Coos Bay Flora. Baldwin 1974; Hopkins 1967.
9. Cherry Creek, Jefferson Co. Clarno Fm. Clarno Flora. Chaney 1927, 1956; Hergert 1961.
10. Clarno, Wheeler Co. Clarno Fm. Clarno Flora. Arnold 1952; Bones 1979; Chaney 1927; McKee 1970; Peterson 1964; Retallack, Bestland, and Fremd 1996, 2000; Scott 1954.
11. Comstock, Douglas Co. Fisher Fm. Comstock Flora. Myers, 2003; Retallack et al. 2004; Sanborn 1937; Wolfe 1994.
12. Coos Bay, Coos Co. Coaledo Fm. Coos Bay Flora. Baldwin 1974; Hopkins 1967.
13. Currant Creek, Jefferson Co. Clarno Fm. Clarno Flora. Knowlton 1902.
14. Dallas, Polk Co. Yamhill Fm. Baldwin 1974.
15. Dixie Mtn., Grant Co. Clarno Fm. Clarno Flora. Gregory 1969a.
16. Goshen, Lane Co. Fisher Fm. Goshen Flora. Chaney and Sanborn 1933; Myers 2003; Myers, Kester, and Retallack 2002; Retallack et al. 2004; Wolfe 1994.
17. Hampton Butte, Crook Co. Clarno Fm. Clarno Flora. Hergert 1961; Lowry 1940.
18. Hobart Butte, Douglas County. Fisher Fm. Comstock Flora. Hoover 1963; Peck et al. 1964; Retallack et al. 2004.

19. Huntington, Malheur Co. Clarno Fm. Clarno Flora. Gregory 1969a.
20. Mammal Quarry, Wheeler Co. Clarno Fm. Clarno Flora. McKee 1970; Retallack 1991; Retallack, Bestland, and Fremd 1996, 2000.
21. Medford, Jackson Co. Payne Cliffs Fm. Brown 1956; McKnight 1984; Peck et al. 1964.
22. Mist, Columbia Co. Keasey Fm. Moore 1971.
23. Mitchell, Wheeler Co. Clarno Fm. Clarno Flora. Chaney 1956; Hergert 1961; Manchester, 2000; Retallack, Bestland, and Fremd 2000.
24. Muddy Creek, Jefferson Co., Clarno Fm. Clarno Flora. Retallack, Bestland, and Fremd 2000.
25. Nehalem River valley, Columbia Co. Cowlitz Fm. Warren and Norbisrath 1946.
26. Nut Beds, Wheeler Co. Clarno Fm. Clarno Flora. Bestland et al. 1999; Bones 1979; Manchester 1981, 1994; Retallack 1991; Retallack, Bestland, and Fremd 1996, 2000; Scott 1954.
27. Ochoco Pass, Wheeler Co. Clarno Fm. Clarno Flora. Retallack 1991b.
28. Pilot Rock, Umatilla Co. Clarno Fm. Clarno Flora. Chaney 1956; Hergert 1961; Pigg 1961.
29. Post, Crook Co. Clarno Fm. Clarno Flora. Arnold 1945; Hergert 1961.
30. Rickreall, Polk Co. Yamhill Fm. Allen and Baldwin 1944; Baldwin 1964.
31. Siskiyou Pass, Jackson Co. Payne Cliffs Fm. McKnight 1984; Peck et al. 1964.
32. Timber, Washington Co. Cowlitz Fm. Warren and Norbisrath 1946.
33. West Branch Creek, Wheeler Co. Clarno Fm. Chaney 1956; Hergert 1961; Retallack, Bestland, and Fremd 2000.
34. Whitecap Knoll, Wheeler Co. John Day Fm. Manchester 2000; Retallack, Bestland, and Fremd 2000.

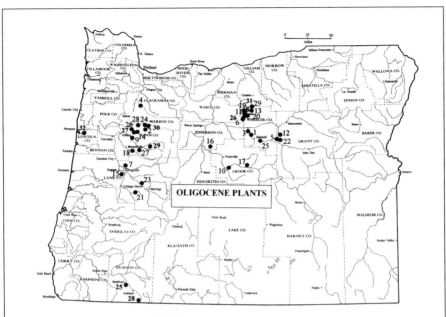

OLIGOCENE PLANTS

OLIGOCENE PLANT LOCALITIES

The following are Oligocene plant localities and a selection of authors who list floras from
the locales.

1. Allen Ranch, NW of Mitchell, Wheeler Co. John Day Fm. Bridge Creek Flora. Chaney
   1927; Meyer and Manchester 1997; Retallack 2000.
2. Bilyeu Creek, Linn Co. Little Butte Volcanics. Bilyeu Creek Flora. Klucking 1964; Myers
   2003; Retallack et al.2004; Wolfe 1994.
3. Bridge Creek valley, Wheeler Co. John Day Fm. Bridge Creek Flora. Chaney 1925; Brown
   1937; Knowlton 1902; Meyer and Manchester 1997; Myers 2003; Retallack, Bestland, and
   Fremd 2000
4. Butte Creek, Marion-Clackamas Co. Scotts Mills Fm. Steere 1959.
5. Butler Basin, Grant Co. Bridge Creek Flora. Chaney 1927; Knowlton 1902; Meyer and
   Manchester 1997.
6. Clarno, Wheeler Co. John Day Fm. Bridge Creek Flora. Chaney 1927; Brown 1959;
   Meyer and Manchester 1997; Retallack, Bestland, and Fremd 2000.
7. Coburg Hills, Lane Co. Fisher Fm. Willamette Flora. Lewis 1950.
8. Cove Creek, NE of Clarno, Wheeler Co. John Day Fm. Bridge Creek Flora. Chaney 1927;
   Meyer and Manchester,1997.
9. Crabtree Creek, Linn Co. Little Butte Volcanics. Scio Flora. Peck 1964; Sanborn 1947.
10. Crooked River valley, Crook Co. John Day Fm. Bridge Creek Flora. Chaney 1925, 1927;
    Meyer and Manchester 1997.
11. Dugout Gulch, NE of Clarno, Wheeler Co. John Day Fm. Bridge Creek Flora. Chaney
    1927; Meyer and Manchester 1997; Myers 2003; Peterson 1964.
12. Foree (4-E), near Dayville, Grant Co. John Day Fm. Bridge Creek Flora. Retallack,
    Bestland, and Fremd 2000.
13. Fossil (high school), Wheeler Co. John Day Fm. Bridge Creek Flora. Brown 1959;
    Knowlton 1902; Manchester and Meyer 1997; Myers 2003; Retallack, Bestland, and
    Fremd 1996, 2000.

14. Franklin Butte, Linn Co. Little Butte Volcanics. Scio Flora. Sanborn 1947.
15. Goshen, Lane Co. Fisher Fm. Willamette Flora. Myers, Kester, and Retallack 2002; Kester 2001; Meyers 2003; Retallack 2004; Vokes, Snavely, and Myers 1951.
16. Gray Butte, Jefferson Co. John Day Fm. Eocene/Oligocene. Ashwill 1983; McFadden 1986; Smith et al. 1988.
17. Gray Ranch, Crook Co. John Day Fm. Chaney 1927; Meyer and Manchester 1997.
18. Holley, Linn Co. Little Butte Volcanics. Sweet Home Flora. Retallack 2004; Richardson 1950.
19. Iron Mountain, Wheeler Co. John Day Fm. Bridge Creek Flora. Chaney 1927; Knowlton 1092; Manchester 2000; Meyer and Manchester 1997; Retallack, Bestland, and Fremd 1996.
20. Knox Ranch, E of Clarno, Wheeler Co. John Day Fm. Bridge Creek Flora. Arnold 1952; Brown 1940; Meyer and Manchester 1997; Peterson 1964; Taylor 1960.
21. Layng Creek, Lane Co. Little Butte Volcanics. Rujada Flora. Lakhanpal 1958; Peck et al. 1964; Retallack et al. 2004; Wolfe 1994.
22. Logan Butte, Crook Co. John Day Fm. Bridge Creek Flora. Retallack, Bestland, and Fremd 2000.
23. Lookout Point Reservoir, Lane Co. Little Butte Volcanics. Rujada Flora. Lakhanpal 1958; Wolfe 1994.
24. Lyons, Linn Co. Little Butte Volcanics. Lyons Flora. Meyer 1973; Wolfe 1994.
25. Medford, Jackson Co. Colestin Fm. Brown 1956; McKnight 1984; Peck, et al. 1964.
26. Painted Hills, Wheeler Co. John Day Fm. Bridge Creek Flora. Bestland and Retallack 1994; Meyer and Manchester 1997; Retallack 1991; Retallack, Bestland, and Fremd 2000.
27. Scio, Linn Co. Little Butte Volcanics. Scio Flora. Sanborn 1947.
28. Siskiyou Pass, Jackson Co. Colestin Fm. McKnight 1984; Peck, et al. 1964.
29. Sweet Home, Linn Co. Little Butte Volcanics. Sweet Home Flora. Enochs, et al. 2002; Retallack, et al. 2004; Richardson 1950.
30. Thomas Creek, Linn Co. Little Butte Volcanics. Thomas Creek Flora. Klucking 1964; Retallack 2004.
31. Twickenham, Wheeler Co. John Day Fm. Bridge Creek Flora. Chaney 1956.
32. Yaquina Bay, Lincoln Co. Yaquina Fm. McClammer 1978.

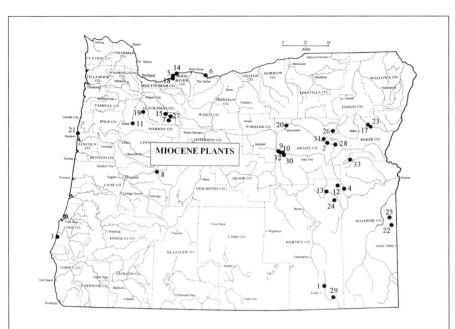

MIOCENE PLANT LOCALITIES

The following are Miocene plant localities and a selection of authors who list floras from the locales.

1. Alvord Creek, Harney Co. Alvord Creek Fm. Alvord Creek Flora. Axelrod 1944; Barnett and Fisk 1984; Chaney 1956.

2. Austin, Grant Co. Columbia River Basalt. Blue Mountain Flora. Chaney and Axelrod 1959; Lohman 1936; Oliver 1934.

3. Bandon (diatoms), Curry Co. Empire Fm. Orr, Ehlen, and Zaitzeff 1971; Whiting and Schrader 1985.

4. Beulah, Malheur Co. Juntura Fm. Beulah-Stinking Water Flora. Chaney and Axelrod 1959.

5. Bonneville, Multnomah Co. Little Butte Volcanics. Eagle Creek Flora. Chaney 1920; Krause 1999.

6. Chenoweth Creek valley, Wasco Co. The Dalles Fm. The Dalles Flora. Chaney 1944a.

7. Collawash River valley, Clackamas Co. Little Butte Volcanics. Collawash Flora. Wolfe 1954, 1960, 1969, 1981.

8. Cougar Reservoir, Lane Co. Sardine Fm. Peck et al. 1964.

9. Dayville, Grant Co. Mascall Fm. Mascall Flora. Chaney 1925b, 1956; Chaney and Axelrod 1959; Wolfe 1981.

10. Dayville, Grant Co. Rattlesnake Fm. Rattlesnake Flora. Chaney 1927, 1956.

11. Drift Creek, Marion Co. Sardine Fm. Peck et al. 1964.

12. Drinkwater Pass (diatoms), Harney Co. No Fm. Hanna 1933; Lohman 1937.

13. Dry Creek, NE of Buchanan, Harney Co. Juntura Fm. Stinking Water Flora. Chaney and Axelrod 1959.

14. Eagle Creek Valley, Hood River Co. Little Butte Volcanics. Eagle Creek Flora. Chaney 1920; Wolfe 1981.

15. Fish Creek, Clackamas Co. Sardine Fm. Peck et al. 1964; Wolfe 1960.

16. Harper (diatoms) Malheur Co. No Fm. Abbott 1970; Kaczmarska 1985; Lohman 1937.

17. Keating, Baker Co. Juntura Fm. Sparta Flora. Chaney and Axelrod 1959; Hoxie 1965.

18. Moffett Creek, Multnomah Co. Little Butte Volcanics. Eagle Creek Flora. Chaney 1920; Krause 1999.

19. Molalla, Clackamas Co. Little Butte Volcanics. Molalla Flora. Durham, Harper, and Wilder 1942; Peck et al. 1964; Wolfe 1960, 1969.

20. Monument, Grant Co. No Fm. Monument assemblage. Wolfe 1969.

21. Moolack Beach, Lincoln Co. Astoria Fm. Miller 1992.

22. Rockville, Malheur Co. Sucker Creek Fm. Sucker Creek Flora. Smith 1932; Wolfe 1969, 1981.

23. Sparta, Baker Co. Columbia River Basalt. Sparta Flora. Hoxie 1965.

24. Stinking Water Creek valley, Harney Co. Juntura Fm. Stinking Water Flora. Chaney and Axelrod 1959; Taggart and Cross 1990.

25. Succor Creek valley, Malheur Co. Sucker Creek Fm. Sucker Creek Flora. Arnold 1937; Graham 1963; MacGinitie 1933; Taggart and Cross 1990; Wolfe 1969.

26. Sumpter, Grant Co. Columbia River Basalt. Blue Mountain Flora. Brown 1937a; Chaney and Axelrod 1959.

27. Three Lynx Power Station, Clackamas River, Clackamas Co. Little Butte Volcanics Peck et al. 1964; Wolfe 1960.

28. Tipton, Grant Co. Columbia River Basalt. Blue Mountain Flora. Brown 1937a; Chaney and Axelrod 1959; Lohman 1936; Oliver 1934.

29. Trout Creek, Harney Co. Trout Creek Fm. Trout Creek Flora. Arnold 1937; Graham 1963; MacGinitie 1933.

30. Van Horn Ranch, Grant Co. Mascall Fm. Mascall Flora. Chaney 1925b; Chaney and Axelrod 1959.

31. Vinegar Creek, Grant Co. Columbia River Basalt. Blue Mountain Flora. Chaney and Axelrod 1959.

32. White Hills, E. of Dayville, Grant Co. Mascall Fm. Mascall Flora. Chaney 1925b, 1956; Chaney and Axelrod 1959; Wolfe 1969, 1981.

33. Unity, Malheur Co. No Fm. Lowry 1940.

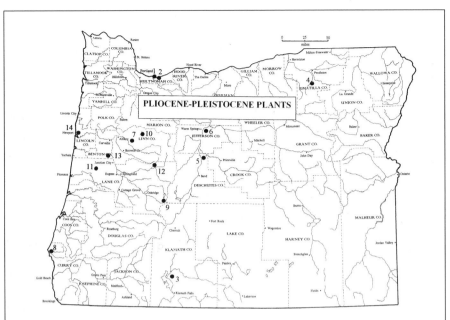

PLIOCENE AND PLEISTOCENE PLANT LOCALITIES

The following are Pliocene and Pleistocene plant localities and a selection of authors who list floras from the locales.

PLIOCENE

1. Buck Creek, Multnomah Co. Troutdale Fm. Troutdale Flora. Chaney 1944b.
2. Camp Collins (on Sandy River), Multnomah Co. Troutdale Fm. Troutdale Flora. Chaney 1944b.
3. Chiloquin (diatoms), Klamath Co. No Fm. Mahood 1981.
4. Pilot Rock, Umatilla Co. No Fm. Denning Spring Flora. Gordon 1985.
5. Terrebonne (diatoms), Deschutes Co. Moore 1937; VanLandingham 1990.
6. Vanora Grade, Jefferson Co. Deschutes Fm. Deschutes Flora. Ashwill 1983, 1996; Chaney 1938, 1956; Chaney and Axelrod 1959.

PLEISTOCENE

7. Beaver Lake, Linn Co. No Fm. Pearl 1999.
8. Cape Blanco, Curry Co. No Fm. Wolfe 1969.
9. Gold Bog Lake, Lane Co. No Fm. Sea and Whitlock 1995.
10. Indian Prairie, Linn Co. No Fm. Sea and Whitlock 1995.
11. Little Lake, Lane Co. No Fm. Worona and Whitlock 1995.
12. Lookout Creek, Lane Co. No. Fm. Gottesfeld, Swanson, and Gottesfeld 1981.
13. Monroe, Benton Co. No Fm. Roberts and Whitehead 1984.
14. Newport, Lincoln Co. Coquille Fm. Baldwin 1950.

INVERTEBRATES

Oregon has long been known for its superb array of fossil invertebrates, a reflection of the state's watery past. The best preserved and most abundant are from the Tertiary Period, between 65 and 2 million years ago, but diverse faunas occur in Paleozoic and Mesozoic rocks of displaced exotic terranes, which are up to 400 million years old.

Defined as animals without backbones, invertebrates comprise a wide range of creatures such as single-celled radiolaria and foraminifera as well as multicelled sponges, corals, arthropods, mollusks, brachiopods, bryozoa, echinoderms, and trace fossils. Vertebrates and plants make up an important part of the modern biota, but their occurrence as fossils is sparse when compared to the numbers of preserved invertebrates. Mollusks alone account for over 100,000 living species. While the Oregon fossil record contains representatives of all of the above invertebrates, the mollusks are by far the most numerous and diverse.

Dating Rocks Using Invertebrates

Because they are so abundant in marine rocks, fossilized mollusks and microfossils are of particular importance for determining the age of strata. However, geologists working in Europe were slow to see the value of fossils as tools to measure geologic time, and it wasn't until 1808 after the stratigraphy of the Paris basin had been deciphered that the utility of mollusks was recognized. European molluscan species were the foundation for the first chronologies, and early efforts to apply these successions elsewhere in the world failed because, like most modern plants and animals, fossils tend to have only limited geographic distribution. Paleontologists soon realized that it would be necessary to establish local successions of fossils for use in a specific region, before they could be correlated between far-flung regions to produce a global chronology.

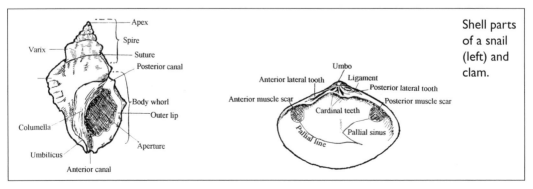

Shell parts of a snail (left) and clam.

William Gabb, seated in the center with the long beard, worked as a paleontologist for the California geology survey. Shown here with his field party in approximately 1888, Gabb was responsible for recognizing, describing, and illustrating great numbers of fossil invertebrate species from the Cretaceous and Tertiary of California, Oregon, and Idaho. Much of the region he and his party explored was almost inaccessible. Considering the incipient state of the science and the limited state of geologic information, pioneering geologists should receive credit for the results they achieved, notwithstanding any inaccuracies perceived by today's workers who have access to a published body of knowledge and the use of technology. (Photo courtesy Archives, Geological Survey, Denver.)

A geologic time standard for the Cenozoic Era, covering the interval from 65 million years ago to the present, was first established in 1833 by the British-born geologist Charles Lyell, who constructed a uniform faunal history of the marine Tertiary in Europe. He based his definition of the "Eocene," "Miocene," and "Pliocene" epochs on the percentage of fossil species in each of these intervals that are still living today. After counting the number of fossil species in a formation, Lyell would calculate what percentage of these had living representatives. A high number of modern equivalents in a fauna would reflect younger strata. Thus, the Eocene would have fewer modern species than the Pliocene. The Paleocene, Oligocene, and Pleistocene were later split off from Lyell's original three epochs.

This percentage strategy was initially applied to the North American West Coast to approximate the Tertiary epochs here. Mollusks were the basis for the first stratigraphic comparisons developed in California. In the early 1900s, Ralph Arnold organized a time scale that relied on the geologic rock formation where the characteristic fossils occurred and not on spans of time or on the fossils present (stages). This led to an unworkable situation where the distinction between time and rock layers became confused. Since rock units or formations frequently extend over more than one time interval or have different ages at various locales, one geologic unit, such as the Eugene Formation, can include the Eocene, but extends into the early Oligocene as well.

Paleontologists continued to work on the problem of fitting local time-rock zones into a global European scheme, and in 1944 a committee of stratigraphers and

Because they were established using rock components and not based on a span of time, early chronologic charts had considerable built-in problems, which paleontologists struggled to resolve. Charles Weaver was one of the paleontologists-stratigraphers who worked to establish standard time intervals. Originally from New York, Weaver took his graduate degree from the University of California before beginning a teaching career at the University of Washington around 1907. Weaver focused on the molluscan stratigraphy of Oregon and Washington, and his first maps and reports described rock formations and accompanying invertebrate faunas from both states. His monumental three-volume work, *Paleontology of the Martine Tertiary Formations of Oregon and Washington*, provided new data and reinterpretations that led to a revision of faunal horizons in the 1940s. (Photo courtesy Oregon Department of Geology and Mineral Industries.)

Although biostratigraphy was initially established using mollusks, eventually microfossils, and mainly foraminifera, came to be recognized for the role they would play in West Coast stratigraphy. Hubert Schenck, who taught the first West Coast course on micropaleontology, saw the usefulness of microfossils as early as the 1920s. After taking geology classes at the University of Oregon, Schenck went on to complete his Ph.D. from Berkeley in 1926. He accepted an appointment at Stanford, where he, along with Siemon Muller, separated time, time-rock, and formations in 1941. Schenck's student, Robert Kleinpell, established the first time scale for the California middle Tertiary based on benthonic foraminifera. In the picture Hubert Schenck, wearing a hat, is standing in the center on the woman's left; he and members of the Bureau of Science, Manila, are conducting surveys in the Philippines in 1921. (Photo courtesy Condon collection.)

Earning a reputation as a rapid but careful worker, Weldon Rau was responsible for much of the biostratigraphic micropaleontology work in the Pacific Northwest from the middle 1960s onward. He earned a Ph.D. from the University of Iowa in 1950 and specialized in benthonic (bottom-dwelling) Tertiary foraminifera. Rau described many isolated faunas along with the local geology while with the Washington Division of Geology and Earth Sciences. (Photo c. 1970; courtesy W. Rau.)

paleontologists, led by Charles Weaver, prepared a landmark master time-stratigraphic framework for all of the Pacific Coast Cenozoic. Pulling together prior work, the committee's graphic presentation of separate stratigraphic sections from California and Mexico, through Oregon, Washington, and into southern British Columbia was a critical first step in the comparison of fossil assemblages (stages) and formations. In this effort, both microfossils and megafossils played a role in precise dating of West Coast formations. However, attempts to correlate mega and microfossils with Lyell's epochs weren't successful, and two different time scales, one for mollusks and one for microfossils, emerged.

From the 1960s to the 1980s, paleontologists came to realize that planktonic microfossils such as foraminifera and radiolaria had advantages for correlation because they occur in marine sediments worldwide, live suspended at shallow water depths, evolve rapidly, and would fit into

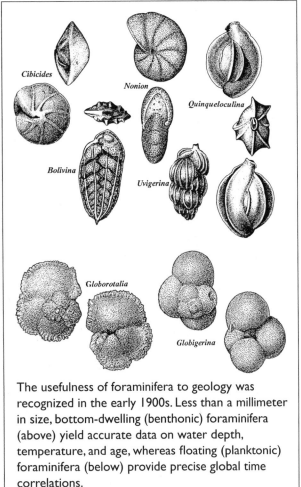

The usefulness of foraminifera to geology was recognized in the early 1900s. Less than a millimeter in size, bottom-dwelling (benthonic) foraminifera (above) yield accurate data on water depth, temperature, and age, whereas floating (planktonic) foraminifera (below) provide precise global time correlations.

Lyell's stages. The time scale was revised in 1981 with the publication *Pacific Northwest Cenozoic Biostratigraphy,* edited by John Armentrout. This landmark work carefully defined and up-dated local chronologies and stratigraphy, placing them into a global framework. The COSUNA chart, "Correlation of Stratigraphic Units of North America," incorporated further data in 1983. These would lay the groundwork for today's stratigraphic standards.

The process of fine-tuning and adjusting rock correlations is ongoing locally as well as internationally, particularly when new methods emerge. Since the early 2000s Donald Prothero at Occidental College has revised many West Coast Tertiary formations using magnetic stratigraphy in conjunction with biostratigraphy and radiometric dates. No doubt even these precise techniques will be improved in the future.

Born in Portland, John Armentrout (below) organized and ran summer programs at the Oregon Museum of Science and Industry while a student at the University of Oregon. For his Master's, Armentrout discovered, mapped, and named outcrops of the Miocene Tarheel Formation near Coos Bay. He earned his Ph.D. from the University of Washington with a dissertation in which he correlated the molluscan paleontology and biostratigraphy of the Lincoln Creek Formation. During a career with Mobil Oil Company, Armentrout's research interests in molluscan paleontology supported his significant contributions to solving Cenozoic stratigraphic problems in the Pacific Northwest. Today Armentrout is retired and lives in Damascus, Oregon, but consults worldwide. (Photo c. 1990; courtesy J. Armentrout.)

Warren Addicott (above) has proven to be one of the most influential contemporary molluscan workers in western paleontology. Receiving a Ph.D. from Berkeley, Addicott began a long career with the U.S. Geological Survey where he specialized in Cenozoic marine mollusks. His many papers fit these invertebrates into the West Coast stratigraphic zones and stages. Currently Addicott lives in Ashland. (Photo c. 1980; courtesy W. Addicott.)

Paleoenvironments and Invertebrates

In addition to their usefulness as geologic time markers, invertebrate fossils play a significant role when reconstructing ancient settings or paleoenvironments. An examination of any modern coast from the beach down across the continental shelf and into deeper water shows profound variations in marine conditions, which include the gradual reduction of wave energy, the light penetration of the water, the oxygen concentration, and the water temperature along with differences in the ocean floor or substrate. Such habitats from prehistoric times can be delineated and extrapolated with fossil invertebrates. Often these deductions can even be made without referring to particular species or genera. For example, unless a clam is a burrowing type, those inhabiting deep quiet water are ordinarily thin-shelled, whereas robust thick shells reflect shallow turbulent conditions. In the highly specialized environments of estuaries, lagoons, and bays, where salinity and temperature may be extreme or may vary rapidly over time and geographic intervals, the diversity or number of invertebrate species tends to be much smaller than in a normal open ocean. Although biomass or diversity may be much higher where the water is rich in nutrients, both are typically lower in the turbulent surf zone, on the outer continental shelf, in deeper water below wave base, and below the light penetration (photic) zone.

A significant part of Oregon's geologic history involves terranes, or slabs of the earth's crust, which have been annexed to the West Coast by plate tectonic processes. Since the reinstatement of the theory of continental drift as plate tectonics in the middle 1960s, fossils have become crucial for deciphering the complex geologic history and biostratigraphy between jumbled terrane rocks, and even the paleolatitude at which terranes originally developed. One of the most exciting areas of paleontology today is the use of fossils to trace the pathways of terranes as they moved across the globe. Identification of invertebrates in the bits and pieces of exotic rocks reveals that many Oregon terranes were originally positioned in the warm tropical Tethyan seaway of the western Pacific Ocean during the late Paleozoic, Triassic, and early Jurassic. By middle Jurassic time, 180 million years ago, the terranes had moved to a mid latitude, and late Jurassic fossils in these rocks reflect a cooler position.

The heat and pressure of metamorphism have altered most terrane rocks, and often only microfossils are available for geologic reconstruction and interpretation. The smallest of fossils, single-celled foraminifera and radiolaria, exist in vast numbers in the water or on the seafloor, and since they are sensitive to temperature and chemistry changes in marine water these microfossils are of particular importance in environmental deductions. Another microfossil, conodonts, the jaws of primitive fish, change color when subjected to elevated temperatures, so they are used to measure the degree and extent of metamorphism in the entombing rock. Increasing temperatures darken conodonts from pale yellow to amber to black, whereas higher temperatures alter them from black to gray to white, or close to crystal clear.

The environments indicated by microfossils are corroborated by megafossil evidence. Since ammonites evolve rapidly and move about globally due

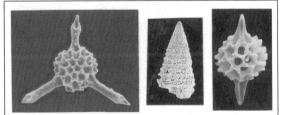

Triassic and Jurassic radiolaria are especially useful for determining the age of the surrounding rocks as well as for identifying the environmental realm. Smaller than 1/32nd inch in diameter, these planktonic microfossils are covered with many spicules that may merge to form a basket-like skeleton of opaline silica.

to their swimming lifestyle, their complex coiled shells are natural tools for deciphering Mesozoic settings. One drawback is that they tend to inhabit the open ocean and are less frequent near reefs or close to shore. When found in terrane rocks from Oregon, strategic Jurassic ammonites such as the Hildoceratidae and Dactylioceratidae show that during this interval Oregon's climate was close to that of present-day Italy and the Mediterranean.

Similarly, in the Tertiary Period the distinctively ornamented pelecypod *Acila* is widely distributed geographically, extending through several geologic epochs and in a variety of marine conditions. These tendencies make it useful for establishing stratigraphic divisions and determining environmental conditions during its past. Today, however, species of this primitive clam are confined to the cool deep waters of the middle and outer shelf.

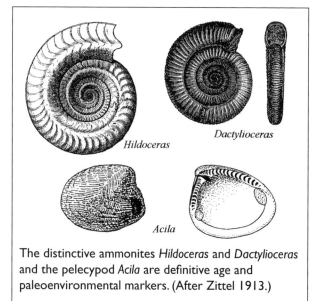

Hildoceras

Dactylioceras

Acila

The distinctive ammonites *Hildoceras* and *Dactylioceras* and the pelecypod *Acila* are definitive age and paleoenvironmental markers. (After Zittel 1913.)

Where other remains are rare, trace fossils have advantages over fossilized body parts for interpreting past geologic events. A trace fossil is an impression, such as a track, trail, or burrow, which records the activity of an organism. Also called ichnofossils, many traces are made by varieties of worms either as permanent burrows or as scavenging tracks when the animal forages across the sea floor. As a rule, a particular animal and its tracks or burrow can't be associated with each other. Indeed, for many soft-bodied invertebrates, ichnofossils may be the only evidence of their presence, providing clues to their activity. Often ichnofossils from one animal number in the thousands, so they are more common than the skeletal part of an organism, which ordinarily makes a single fossil. A burrow might be empty, but the traces may have been filled in with sediment or with material such as wood chips left by the animal's activity. Because the animals move continuously through marine sands and on muddy bottoms, and because their traces cannot be transported after death, they are valuable in environmental reconstruction.

Ichnofossil burrows for both feeding and habitation have thoroughly plowed up these small tracts of shallow marine sediments. (Photos courtesy W. Orr.)

The Paleozoic and Mesozoic Oceans

Beginning in the early Paleozoic 400 million years ago and continuing into the Mesozoic Era, Oregon was covered by wide shallow seas with a shoreline far to the east, across what is now Idaho and Nevada. North America was part of Pangea, a supercontinent formed by the merger of all the world's plates. Eventually the North American plate broke away and separated completely from the other slabs. A most profound event during this period was the arrival of crustal fragments of what had been volcanic island chains, which were annexed to the West Coast of North America. Small volcanic land masses, that rose from the floor of the ancient Pacific Ocean, were being carried slowly eastward as on a conveyer belt to collide with the North American landmass. Today these terrane fragments or volcanic archipelagos form the foundation of the Blue Mountains in northeast and central Oregon and the Klamath Mountains in the southwest. During the earliest intervals these provinces, which projected as low islands above the ocean, hosted dense populations of shallow water clams, snails, corals, ammonites, crabs, and squid. After the smaller landmasses had merged with the West Coast, thousands of feet of sediment and volcanic material blanketed them during the subsequent Cretaceous and Tertiary periods. Bits and pieces of these ancient terranes, which make up Oregon's oldest rocks, can be glimpsed only through erosional breaks or windows in the covering of younger strata.

The Blue Mountains Province

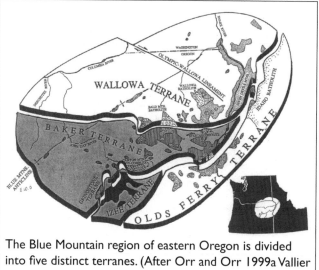

The Blue Mountain region of eastern Oregon is divided into five distinct terranes. (After Orr and Orr 1999a Vallier and Brooks, eds. 1986, 1994.)

The Blue Mountains, across the northeast corner of Oregon, were once part of a large volcanic archipelago called Wrangellia, which originated well out into the Pacific and then moved eastward to accrete against western North America. The oceanic environment of Wrangellia evolved slowly from the Paleozoic through the middle Jurassic as it moved from a tropical to a more temperate setting. Upon arrival at the western continental margin, Wrangellia was sliced up by faults into fragments, which were displaced northward to be scattered from Oregon through Canada to Alaska. This complicated puzzle of Wrangellia's origins was solved once a comparison was made that showed the striking similarity among the rocks and fossils inhabiting the separated pieces. Terrane rocks, forming the building blocks of the Blue Mountains, are the oldest in the state, dating from Devonian time, 400 million years ago, through Jurassic, 145

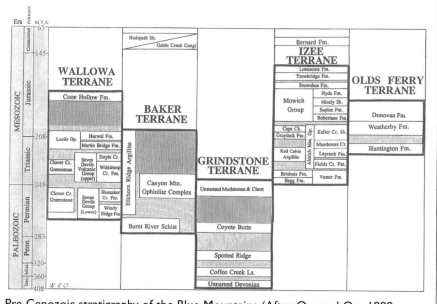

Pre-Cenozoic stratigraphy of the Blue Mountains (After Orr and Orr 1999a; Vallier and Brooks, eds. 1986, 1994.)

As a micropaleontology student of Emile Pessagno at the University of Texas, Charles Blome completed a Ph.D. in 1981 on Triassic radiolaria from eastern Oregon and British Columbia. Since then his research with the U.S. Geological Survey has centered on late Permian and early Triassic radiolaria from mélange sediments of the Grindstone terrane. He notes that faunas from the Oregon Triassic chert are nearly identical to those of the same age found

in Japan. Blome works with Merlynd Nestell of the University of Texas, Arlington, whose graduate degrees were in mathematics but whose expertise extends to microfossils. Charles Blome, left, and Merlynd Nestell, right, are resting on Coyote Butte limestone of the Grindstone terrane. (Photo courtesy M. Nestell.)

Wheat-grain-sized fusulinids *Schwagerina* and *Parafusulina* in the Coyote Butte Formation were recognized as early as 1943 by David Bostwick, the first micropaleontologist to work extensively with Permian fusulinids of eastern Oregon. A professor at Oregon State University, Bostwick examined a float boulder from possible Coyote Butte rocks in Grant County. The 30-pound rock had abundant specimens of a new fusulinid species *Polydiexodina oregonensis*. Even though he searched local outcrops for fusulinids, Bostwick never found this species *in situ*; however, he assigned the boulder to the Coyote Butte two years later when he listed Permian faunas from northwestern United States. Following a career at Oregon State University, he retired in 1978. (Photo courtesy M. Nestell.)

million years in the past. From southwest to northeast, the five distinct terranes composing the Blue Mountains are the Grindstone, Baker, Wallowa, Olds Ferry, and Izee.

In Permian time, the North American continent was still part of the supercontinent Pangea, and those sediments of the Grindstone terrane, which were to become part of Oregon, were deposited well to the west in the tropical Pacific Ocean. Thick marine limestones of the Grindstone, laid down in a shallow seaway, are exposed in present-day Crook, Harney, and Grant counties. A study of radiolaria by Charles Blome and Merlynd Nestell showed the Grindstone rocks to be a mixture or mélange composed of massive slump blocks or olistoliths, which became detached from shallow water limestone banks adjacent to volcanic knolls. The huge slabs then slid downslope to mix with the younger rocks of the nearby Izee terrane. Today the Grindstone terrane is considered by some geologists to be part of the Baker terrane.

The Grindstone, the smallest of the eastern Oregon accreted fragments, has been divided into the oldest unnamed Devonian limestones, followed by the Mississippian Coffee Creek Formation, the Pennsylvanian Spotted Ridge, and the Permian Coyote Butte, capped by a late Permian to early Triassic chert-mudstone layer.

A 200-foot thickness of Devonian limestones, sandstones, and mudstones just southwest of the abandoned village of Suplee in Crook County contains corals (*Battersbyia, Dohmophyllum, Heliolites*), brachiopods (*Atrypa, Emanuella, Grypidula, Schizophoria*), and conodonts (*Icriodus, Ozarkodina, Polygnathus*). The caramel color of the conodonts indicates that they had been altered by temperatures reaching 200 to 400 degrees Fahrenheit during metamorphic processes.

In the same vicinity of Suplee, marine limestones, mudstones, sandstones, and cherts of the marine Coffee Creek are covered by nonmarine layers of the Spotted Ridge Formation. The 350-million-year-old Coffee Creek has a variety of corals (*Dibunophyllum*) and shallow-water brachiopods (*Striatifera, Titanaria*), while the Spotted Ridge Formation is better known for its ferns and fern-like plants, which had inhabited the shoreline and coastal plain.

Sandstones, limestones, and cherts of the Coyote Butte Formation constitute the early Permian interval of the Grindstone terrane. Brachiopods with few fusulinids dominate the upper layers in contrast to the older lower portion that yields abundant corals, fusulinids, and crinoids from a shallow shelf paleoenvironment. This mutually exclusive distribution of brachiopods and fusulinids represents varying water depths. Rex Hanger of the University of Wisconsin described a silicified fauna, in which tiny trilobites and chitons from Twelvemile Creek in Harney County have been replaced by silica. He notes that appearance of the gastropod *Acteonina permiana*, a

A selected listing of Paleozoic fossils from eastern Oregon:

Conodonts: *Icriodus, Neogondolella, Neostreptognathus, Ozarkodina, Polygnathus*

Fusulinids: *Eostaffella, Neoschwagerina, Parafusulina, Polydiexodina, Schwagerina, Yabenia*

Radiolaria: *Albaillella, Folliculuculus, Lentifistula, Pseudoalbaillella*

Amphineura (Chiton): *Arceochiton, Arrochiton, Diadeloplax, Grayphochiton, Homeochiton*

Gastropods: *Acteonina, Tapinotomaria*

Brachiopods: *Antiquatonia, Atrypa, Avonia, Chonetes, Derbyia, Echinoconchus, Emanuella, Grypidula, Kochiproductus, Krotovia, Lingula, Muirwoodia, Neospirifer, Productus, Rynchopra, Schizophoria, Spiriferella, Stenoscisma, Striatifera, Titanaria, Warrenella*

Eurasian genus quite unlike those of the midcontinent of North America, is evidence that the terrane occupied a position in the northern hemisphere during this time.

When paleontologist G. Arthur Cooper examined Coyote Butte brachiopods from Crook County, he observed that the fauna was peculiar because many of the genera were not recorded elsewhere in the early Permian of North America. As early as 1957 he compared the Oregon assemblage to those from Russia, establishing a link between the brachiopods from both regions. However, the significance of this astonishing observation would not be recognized for almost 30 years until the role of plate tectonics was applied to Northwest geology.

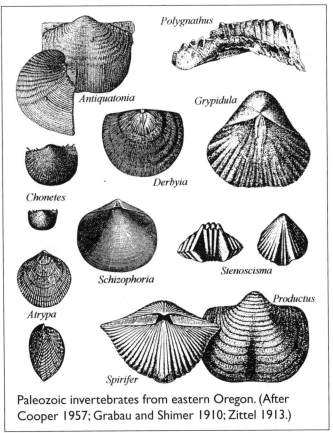

Paleozoic invertebrates from eastern Oregon. (After Cooper 1957; Grabau and Shimer 1910; Zittel 1913.)

The coral *Thysanophyllum* from the Coyote Butte provides additional evidence of the Russian connection. During the Permian a *Thysanophyllum* coral belt stretched along the western margin of North and South America and into the Ural Mountains of Siberia. These land masses were part of the supercontinent Pangea, but the corals were dispersed among terrane fragments when Pangea broke apart. A comparison of species from these ancient reefs established the relationships among the terranes and demonstrated their tropical origin thousands of miles from their present position. Additional Coyote Butte corals such as *Heritschioides, Lithostrotion,* and *Waagenophyllum* reflect shallow tidal conditions in strata near Suplee.

Spread across Crook, Wheeler, Grant, and Baker counties to the Snake River and into Idaho, the Baker terrane is a major component of the Blue Mountain

Selected Paleozoic corals:

Battersbyia, Campophyllum, Dibunophyllum, Dohomophyllum, Heliolites, Heritschioides, Heterophyllia, Hexaphyllia, Lithostrotion, Petalaxis, Thysanophyllum, Waagenophyllum, Zaphrentis

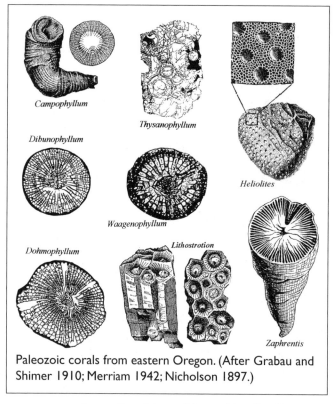

Paleozoic corals from eastern Oregon. (After Grabau and Shimer 1910; Merriam 1942; Nicholson 1897.)

volcanic island chain. Rocks of this terrane were trapped in the subduction zone between massive, converging plates, and, owing to the heat and pressure of metamorphism during collision, few recognizable fossils remain. The chaotic nature of these rocks makes the history of this terrane especially difficult to decipher. The three formations that make up the Baker terrane are the sparsely fossiliferous Permian to Triassic Burnt River Schist, the Permian to Jurassic Elkhorn Ridge Argillite, and the Permian to Triassic Canyon Mountain Complex. While pods within these strata contain radiolaria, conodonts, and fusulinids, the limestones have been altered to marble, making fossils rare.

Remnants of the Wallowa terrane, lying in a wide belt to the northeast of Grindstone exposures in the Blue Mountains, take in the Snake River canyon, most of the Wallowa Mountains, and parts of the Elkhorn and Greenhorn mountains. Composed of numerous formations, these rocks span the late Permian, Triassic,

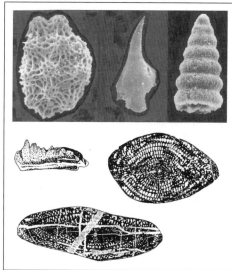

The Triassic conodont *Neogondolella* (center, left) occurs in limestone lenses of the Burnt River schist near Durkee and on the Snake River in Baker County, whereas the Permian radiolaria (*Hegleria* and *Pseudoalbaillella* [top]), and fusulinids (*Parafusulina* and *Neoschwagerina* [center]), along with the Triassic radiolarian (*Canoptum* [top right]), are from the Elkhorn Ridge Argillite. (After Blome 1984; Blome and Nestell 1992; Bostwick and Koch 1962; Morris and Wardlaw 1986; Pessagno and Blome 1982; Wardlaw and Jones 1980.)

and early Jurassic, although the most fossiliferous intervals are in the Triassic. Even though the Wallowa terrane is considered by many to be part of the larger Wrangellia volcanic archipelago, not everyone agrees. At the University of Montana, George Stanley notes that middle Triassic corals from the Wallowa terrane at Pittsburg Landing on the Snake River are quite distinct from those of other accreted Wrangellia fragments. He has proposed that the Wallowa was geographically isolated and not connected to Wrangellia or even part of the warm Tethyan ocean. About 180 million years ago the Wallowa terrane may have been well to the north of Wrangellia, perhaps adjacent to the interior seaway across North America. The position of the Wallowa, closer to the North American continent than others of the Blue Mountains archipelago, makes its tectonic history of particular importance. A faunal exchange between corals and clams of this terrane and North America could have taken place, and evidence presented by the plentiful fossils in its rocks is critical for tracing the geologic history.

This conclusion is supported by the occurrence of spiny productid brachiopods, which dominate the Permian Clover Creek Greenstone within the Wallowa terrane. *Kuvelousia, Waagenoconcha,* and similar brachiopods of boreal (cool) environments led to speculation that the Wallowa terrane was at a northern latitude during the Permian, moving southward to a more tropical region in the Triassic before accreting to North

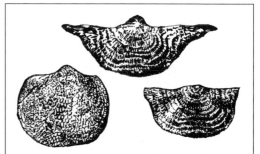

Kuvelousia (right) and *Waagenoconcha* (left) from the Clover Creek Greenstone, and the distinctly ornamented *Megousia* (top) from Hunsaker Creek strata are key brachiopods to interpreting environments of the Wallowa terrane. Most late Permian brachiopods and corals from the Hunsaker Formation are too fragmentary for identification. (After Shroba 1992; Vallier 1967.)

Populations of Triassic mollusks in limestone outcrops along the Snake River at Pittsburg Landing and at the junction of the two forks of Eagle Creek inhabited shallow marine platforms around volcanic islands. A variety of invertebrates such as the flat clams *Daonella* (left) and *Halobia* (right), ammonite molds, and plant fragments from the Wild Sheep Creek Formation at these locations imply one or more basins along an upper shelf or slope. Unfortunately, highway construction has since destroyed the Eagle Creek locality. (After Vallier 1967; White 1994; Zittel 1913.)

America during the late Mesozoic. The fossils further suggest that the Wallowa may not share the same tropical origin as the larger Wrangellia archipelago.

Catastrophic global extinctions at the end of the Permian period, some 245 million years ago, marked the end of the Paleozoic era of ancient life and the beginning of the Mesozoic or middle life, which continued until 65 million years ago. Following the calamitous loss of up to 95 percent of Permian plants and

Paleontologist George Stanley has spent numerous field seasons mapping and collecting in the Blue Mountains. At the geology department at the University of Montana, Stanley's research focuses on Triassic corals in western North America with an emphasis on taxonomy and paleoecology. His extensive travels have contributed to his knowledge and work on global paleogeography. (Photo c. 1980; courtesy G. Stanley.)

A selection of Triassic Martin Bridge and Hurwal invertebrates:

Alga: *Diplopora*
Sponges: *Colospongia, Heptastylis, Stromatomorpha*
Corals: *Astraeomorpha, Coccophyllum, Cyathocoenia, Distichophyllia, Gablonzeria, Kuhnastraea, Maeandrostylis, Reticostastraea, Retiophyllia, Stylophyllopsis*
Ichnofossils (trace fossils): *Chondrites*
Cephalopods: *Aulacoceras*
Pelecypods: *Antiquilima, Astarte, Aviculopecten, Cassianella, Chlamys, Erugonia, Gervillia, Halobia, Liostrea, Lopha, Mysidiella, Mysidioptera, Paleocardita, Parallelodon, Pecten, Plicatula, Septocardita*

animals, the opening of the Mesozoic Era saw an almost entirely new biota. Although the cause of the extinctions isn't known, one possibility may have been profound volcanic activity in Siberia when voluminous flood basalts were extruded across vast regions of northern Asia. Another theory attributes these extinctions to the impact of a major meteorite in the southern hemisphere. After the debacle, plants and animals slowly reestablished themselves in new environments during the Triassic. Within the ocean that covered the Pacific Northwest, the development of stromatolite algae mounds and ultimately of coral reefs signaled the rapid return of a rich diversity of marine life by late Triassic and early Jurassic time.

The most productive fossil interval of the Wallowa terrane is the late Triassic Martin Bridge limestone. Environments ranged from the warm shallow water of an outer shelf to deeper slope settings. Over 1,500 feet of limestone, shales, and conglomerates of this formation were deposited in a narrow basin bordered by lime-rich sandy shoals interspersed with an abundant fauna of well-preserved corals, mollusks, and algae.

Paleontologists argue that the existence of true coral reefs from the Triassic of North America is problematic. Frequently invertebrate shells are so abundant that they construct a complex limestone mass, which projects from the sea floor as a monolith, similar to a reef; but these biostromal mounds, lacking internal structure, do not qualify as true reefs. A reef, or bioherm, has a rigid interlocking internal architecture, much like a skyscraper, with frame-building corals thoroughly cemented together by calcareous encrusting algae. Today living reefs are restricted to warm shallow waters at tropical and subtropical latitudes.

In 1985 George Stanley and Michael Follo located what they considered to be a tropical coral reef, exposed in the Martin Bridge Formation at Summit Point in Baker County. They recognized that the sponges, algae, mollusks, sea urchins, and corals were nearly identical to those from reefs in the central European Alps. This significant discovery was hailed as the first coral-dominated reef of Triassic age in both North America and in the eastern Pacific region. Although late Triassic reefs are common in tropical sediments of the Alps, they were virtually unknown from comparable shallow water deposits of North America. Stanley speculated that the similarity between coral faunas of the Alps and those of Oregon and Idaho may have come about because the numerous volcanic islands that characterized the ancient Pacific were the "stepping stones" by which coral species were dispersed across the ocean. As with similar reef-like structures in western North America, this coral framework originated along the edge of a tropical shallow-water volcanic island. By rapidly budding upward, one solitary coral *Distichophyllia* formed the main network of the reef, which was additionally supported by the branching coral *Retiophyllia*. Other corals, algae, sponges, foraminifera, crinoids, and echinoids made up the remainder of the reef habitat.

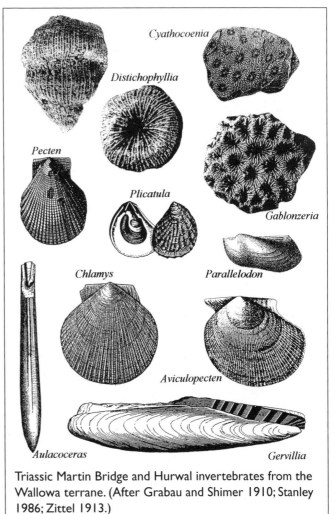

Triassic Martin Bridge and Hurwal invertebrates from the Wallowa terrane. (After Grabau and Shimer 1910; Stanley 1986; Zittel 1913.)

More recently, paleontologist Michael Whalen, of Syracuse University, has taken the position that true late Triassic reefs, similar to those in the European Alps, have yet to be discovered in North America. He concluded that what is found at several locations, such as in the Martin Bridge, are buildups of carbonates and organic debris. A rapidly rising sea level may have hindered extensive reef development, even though the high water was not overwhelming enough to inhibit limestone production entirely.

At the Spring Creek locality in Wallowa County, Martin Bridge sediments were laid down under stormy conditions on a shallow limestone platform surrounding a volcano. This locale is equally rich in corals, sponges, and pelecypods. Stanley and Whalen recorded twenty-one species of Triassic corals and three of sponges, including two new corals, *Maeandrostylis grandiseptus* and *Reticostastraea wallowaensis*, one new genus *Recticostastraea*, and a new sponge *Colospongia whaleni*.

Up to 1,500 feet of late Triassic to early Jurassic carbonate mudstones of the Hurwal Formation are part of the Wallowa terrane. These sediments were derived from local volcanic activity and carried downslope by turbidity currents before settling into the shallow subsiding basin over strata of the Martin Bridge. The Hurwal yields fewer fossils than the Martin Bridge, but the clam *Halobia*, the belemnite *Aulacoceras*, and the trace fossil *Chondrites* are present near Excelsior Gulch and along Eagle Creek in Baker County. A unique blue-green algae within the limestone is the first such discovery in North America.

As the sea advanced landward during the Jurassic, sediments of the Coon Hollow Formation accumulated in a basin atop the Wallowa terrane. The formation experienced several transitions from stream, delta, fan, and nearshore environments to marine sandstones and mudstones. Just north of Pittsburg Landing on the Snake River, the lower Coon Hollow is rich in terrestrial plant remains, but in the upper strata the colonial corals *Coenastraea hyatti* and *Thecomeandra vallieri,* the ammonites *Harpoceras* and *Lupherella,* the pelecypods *Inoceramus, Lima, Mysidiella,* and *Plicatula,* and the radiolaria *Canoptum, Capnodoce, Latium, Triassocampe,* and *Xipha* inhabited deepening ocean waters.

The Izee terrane, at the adjoining corners of Grant, Harney and Crook counties, spans the period from early Triassic through late Jurassic, 245 to 140 million years ago. This small but very important fragment of Oregon was a shallow marine forearc basin, that area lying between the volcanic islands represented by the Wallowa and Olds Ferry terranes and the ocean subduction trench. Erosion of the Olds Ferry provided much of the fossil-rich marine sediment that filled the Suplee-

Broad geologic conclusions can be reached only after years of detailed work by individuals. A systematic study of Paleozoic and Mesozoic rocks near Izee and Suplee was begun by Ralph Lupher in the early 1940s. Lupher received his Master's degree at the University of Oregon, where he met and married Anna Woodward, one of the first women to graduate in geology from that institution. After many summers of fieldwork, he went on to decipher the Jurassic stratigraphy of central Oregon, identifying and naming most of the formations. When a new species of foraminifera, *Pyrgo lupheri,* was named after him, Lupher remarked, "Well, it looks like me, short and fat." (Photo c. 1960; courtesy D. Lupher.)

Ralph Imlay (right), of the U.S. Geological Survey, refined and expanded Lupher's research when he began to work on Mesozoic ammonites of the West Coast in the 1960s. Imlay compiled a detailed biostratigraphy that unraveled a complex series of paleogeographic changes. His conclusions corroborate the early Mesozoic tectonic movement of landmasses that brought together differing terrane fragments of contrasting paleoenvironments. (Photo courtesy Archives, U.S. Geological Survey, Denver.)

Izee depression across central Oregon. Today an erosional opening through the blanket of younger Cenozoic volcanic deposits allows a view of these Mesozoic sediments near Suplee in Crook County and at Izee in Harney County. This key area of some 250 square miles lies between Beaver Creek and the South Fork of the John Day River. A classic study by William Dickinson and Laurence Vigrass in 1965 brings together geologic maps along with summaries of the petrology, stratigraphy, and paleontology of the Suplee-Izee region.

The Izee is significant because it provides a long history of continual deposition that resulted in a one-and-one-half-mile-thick fossiliferous rock sequence divided into a dozen formations. The diversity of ancient marine settings in the Izee is revealed by the rich fauna of brachiopods, ammonites, and pelecypods. Distinctive ammonites and radiolaria show that the terrane originated in the tropics during the late Triassic, but by the middle Jurassic it was sited at a cooler latitude. During the 1960s to 1970s Ralph Imlay pointed out that the early Jurassic ammonites from the Suplee-Izee region are distinctly Tethyan or tropical as is shown by dominant warm-water members of the Hildoceratidae and Dactylioceratidae families.

At Morgan Mountain near Suplee the extinction boundary between the Triassic and Jurassic is marked by the Rail Cabin Argillite and the Graylock Formation above it. Perhaps brought on by a cooling climate, the worldwide late Triassic extinction, over 200 million years ago, involved the demise of 48 percent of the marine organisms, not to mention a similar waning of terrestrial animals. Abundant ammonites of the earlier intervals came close to extinction before recovering again during the Jurassic. Several new species of Jurassic clams (*Agerchlamys, Kalentera*) and ammonites (*Arcestes, Choristoceras, Gabboceras, Sagenites, Vandaites*) appeared in the Graylock following the annihilation.

By Jurassic time, a transgressing advancing sea deepened over the relatively smooth submarine landscape of the Suplee-Izee basin. Within this setting, masses

Thick-shelled oysters, the tidal brachiopod *Lingula* (left), algae, and frame-building corals in the early Triassic Brisbois period attest to calmer waters of a shallow carbonate shelf, whereas the presence of the brachiopod *Discina* (right) in the late Triassic Graylock reflects a quiet brackish embayment. (From Grabau and Shimer 1910.)

A selection of Triassic invertebrates from the Brisbois and Vester formations, the Aldrich Mountains Group, the Graylock Formation, and the Rail Cabin Argillite:

Radiolaria: *Acanthocircus, Canoptum, Capnodoce, Latentifistula, Loffa, Pantanelium, Renzium, Xipha*

Cephalopods: *Alsatites, Arcestes, Arnioceras, Caloceras, Choristoceras, Clionites, Crucilobiceras, Discotropites, Gabboceras, Harpoceras, Juvavites, Paratropites, Placerites, Psiloceras, Sagenites, Sandlingites, Tropites, Vandaites, Vredenburgites, Waehneroceras*

Brachiopods: *Discina, Halorella, Lingula,* Rhynchonellids, Spiriferids, Terebratulids

Pelecypods: *Agerchlamys, Cassianella, Chlamys, Halobia, Kalentera, Monotis, Mysidioptera, Trigonia*

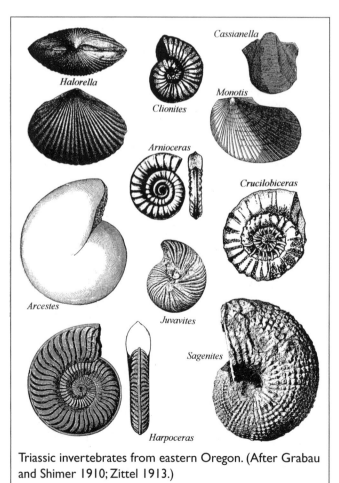

Triassic invertebrates from eastern Oregon. (After Grabau and Shimer 1910; Zittel 1913.)

of pelecypod-rich limestones and tropical shells of the Mowich Group point to a much more southerly latitude for the Izee terrane during this interval. The Mowich Group has been divided into the lowermost Robertson Formation, the Suplee, the Nicely, and the Hyde as the youngest.

Spread across Crook and Grant counties from Beaver Creek to Izee, massive reef-like lens within the Robertson Formation are composed almost entirely of the tightly packed colonies of the oyster-like clam *Lithiotis* (*Plicatostylus*), which is characteristic of the east Pacific and Tethyan tropical middle Jurassic oceans. Often preserved upright in the growth position, *Lithiotis* constructed extensive bioherms across the Suplee shelf during the early Jurassic, but by the late early Jurassic it had become extinct. The bioherms developed best where the reefs consisted of three separate environments (biofacies): an undisturbed life accumulation of *Lithotis* at the core of the reefs, a scattered death assemblage of worn *Lithiotis* shells aligned with the current, and a thick shelly debris of gastropods, brachiopods, and broken pieces of *Lithiotis* on the reef flanks. The internal architecture and succession of changes as the edifice grew and developed elevate these shell accumulations to the status of true reefs or bioherms since they were not merely unstructured masses of shells as biostromes. The brief dominance of *Lithiotis* reefs between the Triassic and late Jurassic may have been due to changes in the chemistry of seawater. Accompanying the *Lithoitis* bioherms, the snail *Nerinea* and the bivalve *Weyla* reflect very shallow open shelf conditions.

The limey, sandy Suplee Formation denotes a nearshore tidal setting, with minimal wave and current activity, similar to that of the Robertson. Numerous specimens of the oyster-like Jurassic clam *Gryphaea,* as well as a large scallop *Weyla*, brachiopods, corals, and the snail *Nerinea* make up the Suplee fauna. *Gryphaea*, characteristic of the Jurassic, is found worldwide in gregarious accumulations on muddy bottoms, whereas *Nerinea* is more typical of limestone shallows above reefs. Ammonites are scarce in the Suplee, but it's possible to collect a few from this stratum as well as from the overlying Nicely Formation, where they are extremely numerous. In the black shales and mudstones of the deeper water Nicely, large carbonate concretions, as much as two feet in diameter, are frequently endowed with shells of oysters, ammonites, and torpedo-like squid (belemnites) at their core. Even more rare are open ocean reptile (ichthyosaur) and fish remains, which turn up in these dark-colored shales. By contrast, the Hyde Formation, the youngest of the Mowich group, has relatively few fossils except for fragmental rhynchonellid brachiopods and the ammonite *Hildoceras*.

A selection of Jurassic invertebrates from the Mowich Group and the Snowshoe, Trowbridge, and Lonesome formations:

Radiolaria: *Parvicingula, Perispyridium, Ristola, Turanta*

Cephalopods: *Arieticeras, Catulloceras, Dactylioceras, Docidoceras, Dorsetensia, Emileia, Fontannesia, Fuciniceras, Gowericeras, Haugia, Hildaites, Hildoceras, Lilloettia, Lytoceras, Normannites, Phylloceras, Psiloceras, Sonnina, Spiroceras, Stephanoceras, Tropites, Witchellia*

Brachiopods: *Rhynchonellids, Terebratulids*

Gastropod: *Nerinea*

Pelecypods: *Astarte, Gervillia, Gryphaea, Lima, Lithiotis, Lucina, Modiolus, Ostrea, Parallelodon, Pholadomya, Pinna, Pleuromya, Posodinia, Trigonia, Weyla*

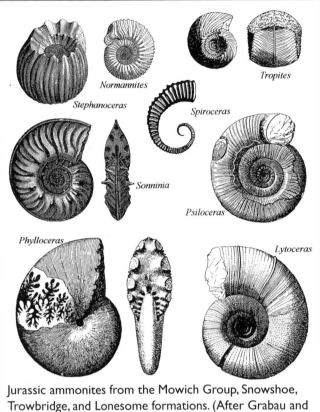

Jurassic ammonites from the Mowich Group, Snowshoe, Trowbridge, and Lonesome formations. (After Grabau and Shimer 1910; Zittel 1913.)

Normannites

Stephanoceras

Spiroceras

Tropites

Sonninia

Psiloceras

Phylloceras

Lytoceras

The remaining sediments of the Izee terrane are a diversity of middle to late Jurassic siltstones, mudstones, sandstones, shales, and volcanic sediments of the Snowshoe, Trowbridge, and Lonesome formations, which have a complex depositional history. Deposited atop the shallow platform in the Suplee district, debris moved downslope into the deeper basin near Izee. The brachiopod *Lingula* and thick-shelled clams *Ostrea* and *Posidonia* substantiate the tidal setting. *Posidonia,* which attached itself to seaweed, is especially common in the Snowshoe, as are a large number of ammonites and radiolaria. Because the ammonites are so numerous, representing a complete range of shallow-water tropical conditions, they are useful in defining biostratigraphic zones in Oregon and elsewhere over wide geographic areas. Tropical Tethyan radiolaria in the Snowshoe are particularly distinctive. The bizarre genera *Perispyridium* and *Turanta* have a remarkably short stratigraphic duration from their first appearance in the middle Jurassic to their disappearance later in the period. In conjunction with the warm-water ammonites *Haugia* and *Catulloceras* they confirm a geographically tropical origin for the Izee terrane. While voluminous collections of ammonites and pelecypods can be made from the Weberg Member of the Snowshoe, the limestones are known for aquatic crocodiles found in the gray to brown strata.

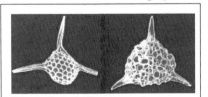

The radiolaria *Turanta* and *Perispyridium*. (After Pessagno and Blome 1982.)

Unlike the Snowshoe, rocks of the Trowbridge and Lonesome formations yield few shells, reflecting deposition in a deeper marine basin that was periodically inundated by catastrophic turbid clouds of mud and sand in submarine landslides.

Exposures of the Olds Ferry terrane curve through eastern Malheur and Baker counties, extend across the Snake River in Oregon, and reach Cuddy Mountain in Idaho. Predominantly volcanic in nature, the Olds Ferry represents an island archipelago close to the North American landmass. Massive volcanic flows are interbedded with sedimentary rocks of the Huntington, Weatherby, and Donovan formations on the flank of the archipelago. Occasional limestone pods within the volcanic debris near Huntington in Baker County yield the Triassic ammonites *Arcestes*, *Clionites*, *Discotropites*, *Sagenites*, and *Tropites*, as well as the flat clam *Halobia*. Ammonites also occur at widely separated locales in Jurassic black shales of the Weatherby. Ralph Imlay traced infrequent

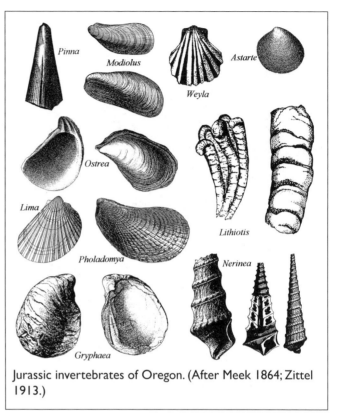

Jurassic invertebrates of Oregon. (After Meek 1864; Zittel 1913.)

exposures of the Donovan Formation to an area less than one-half of a square mile near Beaver Creek and Silvies River in Grant County. Here it is possible to recover a small number of ammonites and pelecypods from the fossiliferous red oxidized strata.

The Klamath Mountains Province

Like the Blue Mountains, southwest Oregon is a complex region containing some of the oldest rocks in the state. An understanding of the local geologic structure began with the mapping of formations and stratigraphy well before the critical role played by plate tectonics was recognized. The geologic history of this province reflects how thought evolved from the 1880s to our present-day knowledge.

The unraveling of the Klamath puzzle started with Joseph Diller, who had begun to examine this area in the late 1800s. Early on he recognized that the Klamath Mountains constituted a separate geologic province, and so he concentrated much of his efforts there. In his map folios of the Coos Bay, Port Orford, Riddle, and Roseburg quadrangles, he delineated and named many formations, covering approximately one thousand square miles; however, his detailed field notes reveal that he examined a far greater area. When faced with deciphering

Joseph Diller spent much of his 40-year professional career in southwest Oregon and adjacent California, covering vast areas that had not been previously explored or mapped. His contributions in paleontology, petrology, volcanology, economic geology, and stratigraphy provided the basis for most of the later studies of the Klamath province. Having joined the U.S. Geological Survey in its fourth year of existence, Diller knew most of the early geologists in America when he retired in 1923. Although his stratigraphy has been repeatedly revised, the volume of his work has been unequaled, and he remains something of a legend to the geologic community. (Photo courtesy Archives, U.S. Geological Survey.)

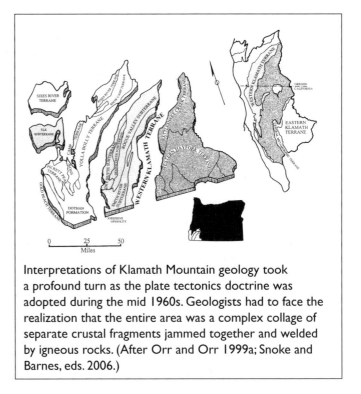

Interpretations of Klamath Mountain geology took a profound turn as the plate tectonics doctrine was adopted during the mid 1960s. Geologists had to face the realization that the entire area was a complex collage of separate crustal fragments jammed together and welded by igneous rocks. (After Orr and Orr 1999a; Snoke and Barnes, eds. 2006.)

Klamath stratigraphy, Diller, of necessity, made some sweeping conclusions. On the basis of the pelecypods *Buchia* (*Aucella*) *crassicollis*, *Buchia piochii*, *Pecten operculiformia,* and *Trigonia aequicostata* Diller assigned the 6,000-foot-thick sandstones, shales, conglomerates, and limestones, which covered broad areas of southwest Oregon, to the Myrtle Formation.

Refining the Klamath stratigraphy in 1959, Ralph Imlay focused on a 2,000-foot exposure near Days Creek in Douglas County where a complete sequence characterized by the slender Jurassic clam *Buchia piochii* is easily distinguished from overlying layers that include the thicker Cretaceous *Buchia crassicollis*. As a result, Imlay deleted the name Myrtle from strata inland and substituted the Jurassic Riddle and Cretaceous Days Creeks formational designations. Rocks mapped as Myrtle on the coast were similarly reevaluated seven years later, when John Koch of the University of Wisconsin replaced the name Myrtle with the Otter Point Formation, the Humbug Mountain Conglomerate, and the Rocky Point Formation, thus eliminating entirely the term "Myrtle."

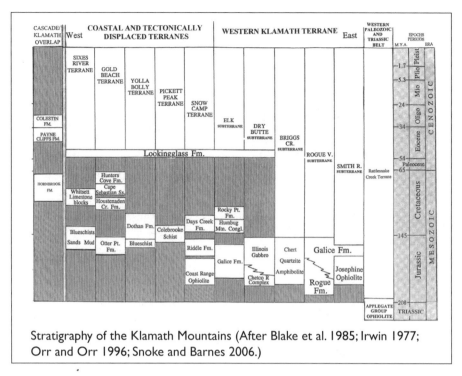

Stratigraphy of the Klamath Mountains (After Blake et al. 1985; Irwin 1977; Orr and Orr 1996; Snoke and Barnes 2006.)

In the 1970s, William Irwin of the U.S. Geological Survey was the first to apply the term "terrane" to help solve the complicated geology in the Klamath Mountains. Terranes are defined as packages of rocks, bounded by faults and confined to a regional area. Each area may have a peculiar fauna and a unique geology differing from that of the neighboring rocks. These separate fragments of rock had been transported eastward atop moving oceanic plates toward the margin of North America to be affixed, or accreted, to that continent by late Mesozoic time. The current view of the structure of the Klamath Province is that a series of sheet-like slabs or terranes are stacked like dominos in an arcuate pattern across southwest Oregon and northern California. Fossils, and especially radiolaria, which are entombed in the rocks, show that the terranes become successively younger toward the west, with the more recent slabs pushed in beneath the older ones. Although most of the major crustal blocks were in place by Cretaceous time, minor terrane movements and fault displacement of rock slabs northward out of California continued well into the Tertiary. Today a geologic picture of the Klamaths is still emerging.

The Klamath Mountains province was initially split into belts, terranes, and subterranes, but with revisions there are currently numerous divisions in both Oregon and California. From east to west, those in Oregon are the Rattlesnake Creek, the Hayfork, the May Creek, and the Condrey Mountain terranes. Adjacent to the Condrey Mountain, the western Klamath terrane is the largest and is composed of five smaller units. In addition, the five coastal terranes from east to west are the Snow Camp, the Pickett Peak, the Yolla Bolly, the Gold Beach, and the Sixes River.

Only a small sliver of the late Triassic–early Jurassic Rattlesnake Creek terrane, formerly part of the Western Paleozoic and Triassic Belt, extends northward from California into Josephine County, Oregon. Cherts and volcanic tuffs, associated with scattered limestone blocks throughout the Rattlesnake Creek terrane, yield radiolaria and conodonts, but most of the microfossil-bearing localities are confined to California. The few fossiliferous rocks are exposed near Waldo and Grants Pass in strata known as the Applegate Group, part of the Rattlesnake Creek. Of the conodonts extracted from 37 samples taken in Oregon and California, just three were from the Triassic of southern Oregon. An analysis of the color changes of the conodonts from yellow to dark amber showed temperatures of these rocks had reached between 500 and 1,000 degrees Fahrenheit. This extreme heat records the accretion process when the slabs were being annealed to North America by volcanic activity.

The western Klamath terrane stretches for more than 200 miles from southwest Oregon into California and is divided into five distinct subterranes, the Smith River, Rogue Valley, Briggs Creek, Dry Butte, and Elk. Of these, only the Smith River, Rogue Valley, and Elk subterranes have yielded stratigraphically significant fossils.

The Smith River and Rogue Valley are composed of volcanic flows and ash of the middle Jurassic Rogue Formation and mineral-impregnated layers of the Josephine ophiolite, covered by sand, shales, and conglomerates of the late Jurassic Galice Formation. While the Smith River built up offshore in a deep marine basin between the volcanic archipelago and the mainland, lavas and ash of the Rogue Valley subterrane accumulated around the periphery of a volcanic island chain. Only a few radiolarian microfossils have been extracted from Rogue Formation rocks, but the Jurassic pelecypod *Buchia concentrica* and the ammonites *Amoeboceras* and *Perisphinctes* (*Dichotomosphinctes*) are present throughout the Galice. Microfossils have come into their own more recently for unraveling terrane relationships, and the tropical nature of the radiolaria from the Rogue indicates that the accreted rocks moved from an equatorial latitude to a cooler boreal position during the middle to late Jurassic.

A selection of Jurassic and Cretaceous invertebrates from southwest Oregon:

Radiolaria: *Acanthocircus, Emiluvia, Hagiastrum, Mirifusus, Perispyridium, Praeparvicingula, Ristola*

Cephalopods: *Aulacotheuthis, Cylindroceras, Olcostephanus, Procanthodiscus, Proniceras, Spiticeras, Thurmanniceras*

Pelecypods: *Buchia*

With his systematic approach to the stratigraphy, Joseph Diller is credited with recording what few Cretaceous fossils are to be found in Rocky Point and Humbug Mountain strata, both now considered as part of the Elk terrane. At the Forks of Elk River in Curry County, layers of concentrated but broken shells and plant debris in the Humbug Mountain and Rocky Point formations are indicative of turbulent, rugged shoreline conditions. Dominated by the Cretaceous clam *Buchia crassicollis*, the fauna also includes the ammonites *Olcostephanus, Phylloceras,* and *Sarasinella* and the fossil squid *Aulacotheuthis* (belemnite).

Five small accreted fragments, adjacent to the Elk terrane, make up the southwest Oregon coast. These are the late Jurassic to Cretaceous Snow Camp, Pickett Peak, Yolla Bolly, Gold Beach, and Sixes River that have been shifted northward by fault movement from California. Fragmental molluscan shells, indicators of beach conditions during the Jurassic, gave way to fine sandstones and mudstones, as the ocean deepened during the Cretaceous. The picture emerging from this time is one of embayments where sands and muds, carried by rivers, accumulated in deltas or on shallow continental shelves.

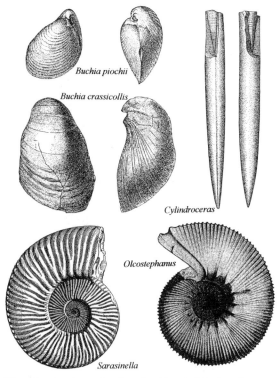

The slender Jurassic clam *Buchia piochii,* the plump Cretaceous *Buchia crassicollis* and additional Jurassic invertebrates. (After Grabau and Shimer 1910; Pessagno et al. 1993; Zittel 1913.)

Tree ferns, ginkgoes, and similar exotic-looking tropical plants, along with *Buchia piochii* from the Jurassic Riddle and *Buchia crassicollis* from the Cretaceous Days Creek, characterize the fauna. Pelecypods, ammonites, and numerous echinoids (sea urchins), provide evidence of an oceanic island chain along a shallow coastline.

Between the Snow Camp and Yolla Bolly terranes, Pickett Peak rocks, altered by the heat and pressure of metamorphism, are devoid of fossils. In southern Curry County, the Yolla Bolly terrane includes Cretaceous sands, muds, and cherts interpreted as a deep-water setting on a marine continental slope adjacent to a volcanic island chain. Few fossils can be found within this terrane, although *Buchia piochii* sometimes occurs imbedded in loose rubble at the mouth of Boulder Creek. The origin of the boulders is uncertain, but they may belong to the Dothan Formation. It is possible to recover worm tubes and carbonized plant remains in rocks mapped as Dothan at Chetco Point near Brookings.

As the ocean encroached over the land, the shallow beach conditions of the Jurassic gave way to deeper waters where sands and muds of the Cretaceous accumulated in embayments or on shallow shelves. Over 10,000 feet of sands, silts, cherts, and volcanic debris of the late Jurassic Otter Point, covered by sands and silts of the Cretaceous Houstenaden Creek, Cape Sebastian, and Hunters Cove formations, make up the Gold Beach terrane, a small elongate slice on the Oregon coast south of Port Orford in Curry County. For the most part, Otter Point strata yield only a meager assemblage of broken plant fragments, the poorly preserved radiolaria *Dictyomitra, Ethmosphaera,* and *Hagiastrum*, a few pelecypods, and ammonites. The oyster-like *Buchia piochii* almost invariably occurs in mudstones, suggesting that the clams may have died as they were swept into deep water by submarine slides or turbidity currents. At Cape Blanco, however, the Otter Point is represented by a spectacular 60-foot-thick mixture of rounded boulders, large chunks of wood, and belemnites (squid), which are thrown together in a muddy sand. The remains of a large fish-like reptile, an *Ichthyosaur*, from this formation at Sisters Rock may have suffered the same fate.

Channel-like sands of the Hunters Cove Formation preserve specimens of the ponderous ammonite *Pachydiscus multisulcatus*, whereas *Inoceramus turgidus* and other clams lived in the agitated marine setting of the Cape Sebastian interval. Layers of this sandstone have abundant organic debris and a diverse selection of trace fossils.

On the southwestern margin of the state, the Sixes River terrane, taking in the watershed of the Sixes River eastward from Cape Blanco, is a mélange or mixture of Jurassic and Cretaceous muds and sands studded with huge gray to white and even pink limestone blocks. The carbonate blocks near Dillard, named the Whitsett limestone, are chaotic, and details of their stratigraphy, age, and origin are not well understood. Foraminifera are numerous in a few of the blocks, but megafossils are absent. In the past, many broken *Buchia* shells were reported in and around the limestones, but today it is difficult to find even a single specimen.

By the end of the Cretaceous Period, the major accretionary events had already annealed the Klamath terranes to southwest Oregon. Based on the similarities between the ammonites *Cleoniceras, Lyelliceras,* and *Oxytropidoceras* and the clams *Inoceramus* and *Megatrigonia* in the Bernard and Hudspeth formations near Mitchell in Wheeler County and those from the Hornbrook Formation at

Grave Creek in Jackson County, a paleographic picture can be drawn. A broad shallow seaway advanced or transgressed, a trend that was taking place globally. As the ocean deepened over most of the Northwest, a diagonally curving shoreline ran from Jackson and Josephine counties to Wheeler County and on to Washington state. Within this marine basin, shelled animal remains are concentrated in the Klamath province, whereas in eastern Oregon they are dispersed throughout Crook, Grant, and Wheeler counties.

From Jackson and Curry counties and into northern California, sediments of the Cretaceous Hornbrook Formation filled a forearc basin west of the volcanic archipelago sited in central Idaho. In this region approximately 100 million years ago, the ocean transgressed landward, covering the varied settings of freshwater streams, calm estuaries, and beaches with ever deeper sediments. Grave Creek in Jackson County marks the southern limit of the seaway, which may have connected to the open ocean in the northwest and to the Ochoco Basin of central Oregon. Ammonites and pelecypods of the Hornbrook are abundant but are often found in fragmented condition where a wave-tossed high energy coastline was present. In fine-grained sandstone layers, horizontal, vertical, or U-shaped burrows of the trace fossils *Skolithos* and *Glossifungites*, along with plant fragments, point to shallow, low-energy mudflats.

The Hornbrook Formation has its share of unusual ammonites. The partially uncoiled *Anisoceras* often reached

Frank Anderson first came into contact with Cretaceous invertebrates while growing up near Medford. Knowledge of Cretaceous rocks of the Pacific Coast advanced significantly when Anderson began to compare West Coast invertebrate faunas and formations. His most comprehensive work, *Upper Cretaceous of the Pacific Coast*, wasn't compiled and published until ten years after his death in 1947. (Photo courtesy Southern Oregon Historical Society.)

A selection of Cretaceous invertebrates from southwestern and eastern Oregon:

Foraminifera: *Bathysiphon, Dicarinella, Globotruncana, Gyroidinoides, Lenticulina, Praeglobotruncana, Rotalipora*

Trace fossils: *Arenicolites, Glossifungites, Ophiomorpha, Planolites, Skolithos*

Scaphopods: *Dentalium*

Cephalopods: *Acanthoceras, Anagaudryceras, Anisoceras, Cleoniceras, Desmoceras, Lyelliceras, Nautilus, Nipponites, Oxytropidoceras, Pachydiscus, Scaphites, Turrilites*

Pelecypods: *Anomia, Anthonya, Aphrodina, Buchia, Cardium, Cercomya, Exogyra, Gervillia, Glycimeris, Inoceramus, Meekia, Megatrigonia, Mytilus, Ostrea, Pholadomya, Pleuromya, Pterotrigonia, Tellina, Trigonia, Yaadia*

Gastropods: *Cypraea, Fusus, Natica, Paosia, Sogdianella, Trochactaeon, Vernedia*

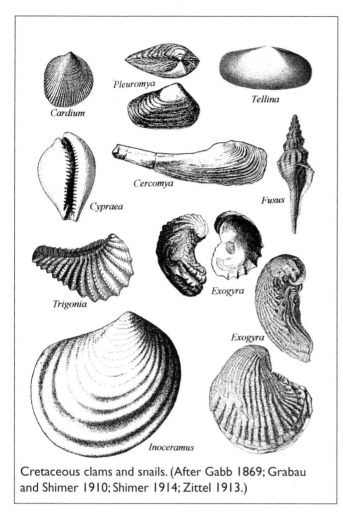

Cretaceous clams and snails. (After Gabb 1869; Grabau and Shimer 1910; Shimer 1914; Zittel 1913.)

more than three feet in length, and its sheer size defies collection as an unbroken piece, whereas the complex coils of *Nipponites* were marginally buoyant, and it existed as a floating form. With a shell up to three feet in diameter, *Pachydiscus* would have been a riveting sight in Cretaceous seas.

Marine Hornbrook deposits in southwest Oregon are comparable to those of the Bernard, Hudspeth, and Gable Creek formations in wide areas of Wheeler, Grant, and Crook counties. The composition of the faunas here is useful in reconstructing the extent of the seaway. A shallow continental shelf is depicted near present day Antone, submarine fans near Mitchell, and a deeper offshore marine basin in the vicinity of the Muddy Creek Ranch (Rajneeshpuram). The uncoiled ammonite *Anisoceras* from the Hudspeth, along with the thick-shelled clams *Trigonia evansana* and the oyster-like *Exogyra* from Bernard strata, suggest muddy wave-washed shallows. The bivalve *Cercomya hesperia* and the gastropods *Vernedia pacifica*, *Sogdianella oregonensis*, and *Trochactaeon allisoni* confirm warm-water Tethyan conditions. A meager fauna of beach worn fossils in the Prairie City area of Grant County represents the eastern-most marine Cretaceous in Oregon.

The Cenozoic Era

The Cretaceous seaway, which covered over two-thirds of Oregon, was the last broad oceanic province in the state before volcanic activity and crustal plate movements combined to generate regional uplift and subsequent withdrawal of the water. Throughout the Cenozoic, the ancient Pacific Ocean was confined to a steadily decreasing shelf west of what are now the Cascades. With the rise of the Cascade volcanic arc, thick layers of Eocene through Miocene sediments poured into the newly created basin. Shorelines fluctuated, advancing and then retreating with some regularity; however, subsequent uplift steadily reduced the seaway so that by middle Miocene time, about 13 million years ago, it was restricted to the western margins of the present Coast Range. At the beginning of the Pliocene Epoch, 5 million years ago, the modern topographic configuration of the state was in place, and all that remained was a final sculpting by Pleistocene glaciers.

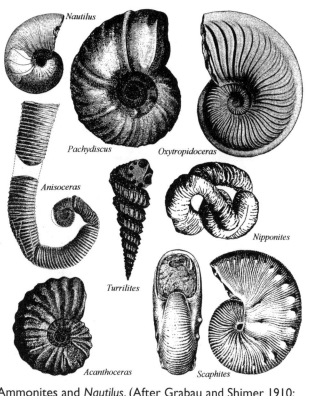

Ammonites and *Nautilus*. (After Grabau and Shimer 1910; Shimer 1914; Zittel 1913.)

The Paleocene Ocean

Spanning 11 million years, the Paleocene is one of the shortest of the Tertiary epochs. Yet this interval in geologic history, beginning around 65 million years ago, is critical because it represents the period of recovery after the wholesale destruction of Mesozoic floras and faunas. Even though western Oregon remained a vast marine environment during this epoch, definitive fossil invertebrates are

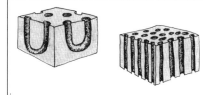

U-shaped burrows of *Arenicolites* and rounded tubes of *Skolithos* trace fossils from southern Oregon.

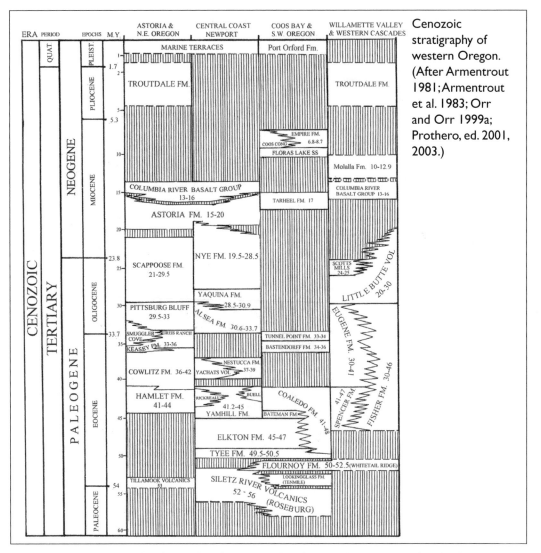

ERA	PERIOD	EPOCHS	M.Y.	ASTORIA & N.E. OREGON	CENTRAL COAST NEWPORT	COOS BAY & S.W. OREGON	WILLAMETTE VALLEY & WESTERN CASCADES

Cenozoic stratigraphy of western Oregon. (After Armentrout 1981; Armentrout et al. 1983; Orr and Orr 1999a; Prothero, ed. 2001, 2003.)

exceedingly rare. Microfossil samples from a site near Bandon in Coos County contain the late Paleocene planktonic foraminifera *Chiloquembelina wilcoxensis, Globonomalina planoconica, Pseudohastigerina wilcoxensis, Subbotina linaperta,* and *Truncorotaloides wilcoxensis.* However, the precise locality has proved to be elusive, and the sample itself can not be found.

In 1995 fossils confirming Paleocene strata were reported from the Tyee basin near Roseburg in Douglas County. A mudstone within the Siletz River volcanics unit yielded diagnostic discoasters (plant microfossils), fragments of the gastropods *Cylichnina* and *Turritella,* along with the bivalves *Brachidontes, Nemocardium,* and *Ostrea* of early Eocene to late Paleocene age.

The Eocene Ocean

In western Oregon, the Eocene Epoch, between 33.7 and 54 million years ago, was marked by great undersea tearing or crustal rifts which exuded massive submarine flows of lava. The combined eruptions built an underwater basalt platform that became the foundation of the Coast Range. Deep marine mud, silt, shelf sand, and coastal deltas accumulated atop the rapidly subsiding submarine slab as the weight of the lavas created an elongate ocean basin populated by invertebrates, fish, and micro-organisms. Ancient beaches delineate the boundaries of the early Eocene seaway near Cape Arago and Coos Bay on the south coast, along the eastern margin of the Willamette Valley from Douglas and Lane counties to Polk and Lincoln counties, and in the far northwest corner of the state in Washington and Columbia counties. By late Eocene time, the ocean had withdrawn considerably toward the west, and at the end of this epoch marine waters were restricted to the Nehalem River basin, across the southeastern Willamette Valley, and at Coos Bay.

In the 1880s the presence of Eocene rocks in Oregon was firmly established by Thomas Condon, geology professor at the University of Oregon, when he recognized the distinctive thick-shelled Eocene clam *Cardita planicosta,* now *Venericardia hornii.* Deciphering the complex stratigraphy of Eocene deposits in the ocean basin across southern Oregon has been ongoing ever since. At the beginning of the twentieth century, Eocene rocks were lumped into the Arago Beds, but this stratum was subsequently subdivided into the Umpqua, Tyee, and Coaledo formations. In 1974 the Umpqua was further refined into the Roseburg, Lookingglass, and Flournoy formations by Ewart Baldwin.

The oldest Eocene exposures of the southern Tyee basin are submarine lavas and sediments of the Roseburg, Lookingglass, and Flournoy formations, found today in Douglas, Coos, and Curry counties. Between the basalts, the sand and silt layers yield a number of microfossils,

Stratigrapher and paleontologist Ewart M. Baldwin spent more than three decades working in southwest Oregon and elsewhere in the state. Born in Pomeroy, Washington, in 1915, Baldwin completed his undergraduate work at nearby Pullman before finishing a Ph.D. at Cornell directed by Charles Merriam. Baldwin subsequently joined the staff of the geology department at the University of Oregon in 1947 where, with the assistance of his many students, he ultimately mapped over 4,000 square miles of the south and central Coast. His *Geology of Oregon* was the first such synthesis since Thomas Condon's book. Baldwin's death in 2009 marks a milestone in the state's geologic history. (Photo 1969; courtesy G. Miles.)

Selected Eocene invertebrates:

South Coast Range: Roseburg, Lookingglass, and Flournoy formations:

Echinoderms: *Eoscutella, Holothurium*

Pelecypods: *Acila, Anomia, Brachidontes, Cardita, Clinocardium, Crassatella, Glycimeris, Microcallista, Nemocardium, Nuculana, Ostrea, Solen, Spisula, Spondylus, Tellina, Venericardia*

Gastropods: *Buccinofusus, Cryptochorda, Cylichnina, Cymatium, Fusinus, Galeodea, Mitra, Olivella, Polinices, Siphonalia, Turritella*

Coos Bay: Coaledo, Elkton, Bateman, Bastendorff, and Tunnel Point formations:

Foraminifera: *Bulimina, Cibicides, Discorbis, Gyroidina, Uvigerina*

Echinoderms: *Eoscutella*

Pelecypods: *Acila, Brachidontes, Gari, Glycimeris, Marcia, Nemocardium, Pachydesma, Pandora, Pecten, Pitar, Saxidomus, Solen, Solena, Spisula, Venericardia*

Gastropods: *Bruclarkia, Calyptraea, Conus, Crepidula, Molopophorus, Polinices, Turritella*

Central Coast Range: Siletz River, Yamhill, and Nestucca formations, and Yachats basalt:

Foraminifera: *Operculina, Pseudophragmina*

Bryozoans: *Crisia, Heteropora, Idmonea, Lichenopora*

Echinoderms: *Briaster, Hetrocentrotus, Lytechinus, Ophiocrossota*

Crustaceans: *Balanus*

Scaphopods: *Dentalium*

Brachiopods: *Gryphus, Terebratulina*

Pelecypods: *Acila, Barbatia, Chlamys, Crassatellites, Glycimeris, Lima, Lucina, Macoma, Mytilus, Nucula, Nuculana, Ostrea, Spondylus, Venericardia, Volsella, Yoldia*

Gastropods: *Acmaea, Calyptraea, Conus, Cymatium, Epitonium, Euspira, Exilia, Homalopoma, Polinices, Turritella*

---

but only occasional invertebrate shells. Planktonic (open ocean) foraminifera from these three formations show a relatively low diversity, dominated by the genera *Morozovella*, *Pseudohastigerina*, *Subbotina*, and *Truncorotaloides*, which indicate tropical rather than temperate ocean waters. In addition to foraminifera, Lookingglass exposures at the mouth of Little River near Glide in Douglas County have outstanding Eocene mollusks. Well-preserved, diverse tropical shells, primarily of pelecypods and gastropods, can be found here. Among them, the large, thick-shelled clam *Venericardia*, along with *Clinocardium* and *Ostrea*, and the snails *Cylichnina, Galeodea,* and *Turritella* inhabited very shallow water on a sandy upper continental shelf. Well to the south near Agness in Curry County, a Flournoy fauna included decapods (crabs) and echinoderms (sea urchins and sand dollars) as well as mollusks. Within the strata, coal seams developed under quiet marine conditions interfingering with nonmarine deltas rich with vegetation. By Flournoy time, the seaway was restricted to a small

The thick-shelled clam *Venericardia hornii* first signaled the presence of Eocene strata in Oregon. (From White 1885.)

region from Newport to Cape Blanco, extending eastward close to the boundary of Coos County.

From the middle to the late Eocene, the dimensions of the south coast basin fluctuated as shorelines moved, and mud, silt, and sand of the Tyee, Elkton, Bateman, Coaledo, and Bastendorff formations flowed into the depression. At its greatest extent, the Tyee-Elkton seaway stretched northward from Cape Blanco, but by the Bateman and Coaledo interval the waters were reduced to an area around Coos Bay.

Tropical to subtropical plants, the clam *Venericardia califia*, and microfossils of the Bateman signaled the final stage of sea withdrawal near Coos Bay before waters again advanced and deepened during the Coaledo and Bastendorff intervals. Within the basin, a sandy Coaledo delta extended north to northwest.

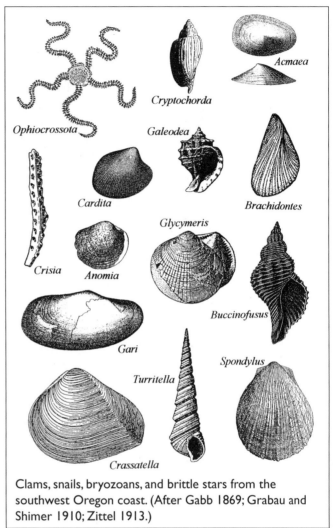

Clams, snails, bryozoans, and brittle stars from the southwest Oregon coast. (After Gabb 1869; Grabau and Shimer 1910; Zittel 1913.)

Partially enclosed by offshore islands, the delta was bordered by a wide coastal plain sprinkled with swamps and marshes. By the beginning of the Oligocene, the ocean had become increasingly shallow with deposition of the Tunnel Point sandstone, when the marine waters were restricted to only one area in the Coos Bay region.

Sediments of these formations reflect a variety of settings in the diminishing waters. Megafossils are present but scarce in the Tyee, Bateman, and Bastendorff, while the Elkton and Coaledo have an abundance of mollusks, crabs, sea urchins, and microfossils (foraminifera, diatoms, ostracods, and radiolaria). Invertebrates of the Elkton are poorly preserved; however, mollusks, along with the more numerous foraminiferal families Lituolidae, Trochamminidae, Buliminidae, and Alabaminidae, suggest a shallow marine to brackish environment for the southern

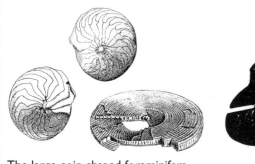

The large coin-shaped foraminifera *Amphistegina* (left) and *Discocyclina* (right) from the Ellendale Limestone Quarry and on Rickreall Creek in Polk County identify warm ocean waters during the Siletz River and Yamhill intervals.

Pleurotomaria, an exotic rare gastropod from the Yamhill Formation, is fascinating in that it seems to occur exclusively in sedimentary layers between submarine basalt flows. These ancient carnivorous snails briefly appeared in the northeastern Pacific Ocean 40 million years ago when water temperatures rose and offshore volcanic seamounts provided rocky habitats. (From Zittel 1913.)

lagoon near Coos Bay. By contrast, foraminifera in the northern basin covering present-day Lincoln County reflect substantial ocean depths of 600 to 2,000 feet, characteristic of outer shelf to slope conditions.

Foraminifera are similarly useful at Yoakam Point, where coal developed during freshwater intervals within the Coaledo. The alternating freshwater and marine episodes can be delineated by the contrasting foraminiferal species, which appeared above (*Uvigerina, Cibicides*) and below (*Discorbis, Gyroidina*) the coal seams. In addition to foraminiferal evidence, the gastropods *Calyptraea, Conus, Crepidula, Polinices* and pelecypods *Brachidontes, Pachydesma*, and *Solen* indicate subtidal and intertidal brackish zones.

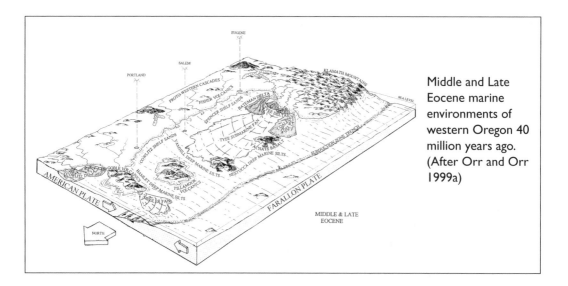

Middle and Late Eocene marine environments of western Oregon 40 million years ago. (After Orr and Orr 1999a)

Even though rare, megafossils in the Bastendorff Formation, such as the paper-thin, broken, and scattered scallops (*Pecten*) and the clam *Acila*, along with the microfossils *Bulimina* and *Uvigerina,* signal deep water and open ocean conditions. Shallowing marine waters, which brought an end to deeper Bastendorff conditions, are marked by the Oligocene Tunnel Point sandstone. Closely packed borings of the clam *Pholas* occur in the upper level, whereas other mollusks are dispersed throughout the formation. *Bruclarkia, Molopophorus, Pachydesma, Spisula,* and *Tellina* from Tunnel Point strata, reflecting an upper marine shelf, denote the end of the Eocene in this region. Today the Tunnel Point is poorly exposed at Coos Head since construction of South Jetty altered and redirected sand deposits at Bastendorff Beach.

Uplift of the Coast Range triggered closure of the marine trough in southwest Oregon, leaving a wide, shallow seaway along what is now the Willamette valley and the northern coast. The central basin was periodically filled with sediments of the Siletz River, Yamhill, and Nestucca formations. The oldest of these is the Siletz River Volcanics, which harbors sparse fragments of megafossils and microfossils within the sedimentary interbeds. The numerous and diverse coin-shaped foraminifera *Pseudophragmina* and *Operculina* and members of the Lagenidae family live today only in tropical waters. Near the western Polk County line, *Lima, Nuculana, Polinices,* and *Turritella* denote inner to middle shelf depths of 100 to 300 feet, but the worn mollusk shells were transported either downslope or parallel to the shore.

Carole Hickman, professor of paleontology at the University of California, Berkeley, summarized data on the Keasey fauna for her Ph.D. research from Stanford. In a 1995 speech about women in paleontology, Carole Hickman relates that her own science is motivated by curiosity, the aesthetics of nature, "the tension of uncertain outcome, and the joy of playing with ideas." (Photo courtesy C. Hickman.)

By contrast, the 6,500 feet of sandstones, siltstones, and conglomerates of the Yamhill Formation in Polk County host diverse populations of mollusks, sharks, crustaceans (barnacles and crabs), bryozoa, corals, algae, and terrestrial plants. The largest faunal assemblage can be found at the Ellendale Limestone Quarry and on Rickreall Creek, where mollusks are the most abundant, and layers of broken shells verify a shallow, turbulent, wave-washed beach. Plants confirm the proximity of the shoreline. Foraminifera from the Yamhill indicate continental slope waters of 1,000 to 6,000 feet in depth and temperatures between 40 and 70 degrees Fahrenheit. Curiously no open ocean (planktonic) foraminifera were found, leading to an assessment that heavy fluvial runoff from the nearby land created a turbid environment deleterious to these microfossils.

Continuing marine deposition across the Coast Range basin into the later Eocene produced as much as 7,500 feet of tuffaceous siltstones of the Nestucca Formation interlayered with sandstones, breccias, and submarine lavas of the Yachats Basalt exposed at Heceta Head. Brackish-water mollusks of the Nestucca dwelt on the shelf just below the intertidal area, but the lack of planktonic foraminifera indicates that the embayment was isolated and of variable depth. Several areas of the inlet may have been separated from the open ocean by barrier islands such as the volcanic highland formed by the Yachats Basalt. Sediments layered between the basalt preserve a splendid shallow-water fauna that inhabited submarine fans on the flanks of an offshore island. Benthic foraminifera and offshore bryozoans (*Crisia, Idmonea, Lichenopora*), brachiopods (*Terebratulina*), pelecypods (*Barbatia, Chlamys, Mytilus*), the tropical gastropod (*Conus*), barnacles (*Balanus*), and sea urchins (*Heterocentrotus, Lytechinus*) are recorded from the basalt.

Across the Nehalem Basin in the northwest corner of the state, the late Eocene Hamlet, Cowlitz, and Keasey formations mirror a wide range of tidal and swampy to deeper marine environments deposited at the margin of the seaway. The 1,000-foot thickness of mudstones, siltstones, sandstones, and conglomerates of the Cowlitz delta was constructed by streams that drained nearby volcanic highlands. Because the formation denotes changing habitats, a great variety of foraminifera, plants, trace fossils, and invertebrates are present in the rocks. The shallow-water mollusks *Acmaea, Glycimeris, Mytilus,* and *Ostrea* in the lower conglomerate layer are succeeded by deeper marine *Acila, Nucula, Propeamussium, Siphonalia,* and *Yoldia,* along with the trace fossil *Phycosiphon incertum* in the upper portion. The brackish marine gastropod *Potamides* is an important component of the Cowlitz and does not occur in the subsequent cooler deeper Keasey Formation. The Cowlitz exhibits the last high diversity Eocene fauna in which nearly all of the gastropods were tropical before faunas were altered by the Oligocene cooling trend.

In contrast to the shelf environments of the Cowlitz, some 2,000 feet of volcanic ash, shales, siltstones, and clays of the Keasey Formation were deposited under continental slope conditions of a deep basin adjacent to volcanic islands. Estimates as to the paleodepth vary, but sedimentation probably took place close to the shore well below wave action. The basin may have ranged as deep as 3,000 feet; however, it is also possible to show that environmental conditions of the Keasey could be met in water less than half that, with the basin depth averaging 600 to 1,500 feet. In certain localities, such as at Rock Creek in Columbia County, the microfauna normally would have inhabited deep water of the lower shelf to upper slope. The light blue to gray Keasey shale in this locality also displays a variety of shells such as *Turricula columbiana* and *Acila nehalemensis*, along with crustaceans and foraminifera.

The Keasey Formation preserves fossils from the deepest marine deposits of the Oregon Tertiary, whose variety and excellent preservation are unmatched. Mollusks, crinoids (sea lily), echinoids (sea urchins), ophioroids (brittle stars), pogonophorans (worms), corals, glass sponges, decapods (crabs), sharks, bony fish, terrestrial plants, algae, and a variety of microfossils are part of this remarkable late Eocene marine environment. At one of the best-known Keasey localities near Mist in Columbia County, crinoids and other echinoderms, such as starfish and sea urchins, are beautifully preserved in an undisturbed life condition. It is unfortunate that much of the site has since been destroyed by commercial collectors.

Elongate tubular trace fossils are present in the Cowlitz Formation but are more widespread in the Keasey. Even though the remains of *Skolithos, Rosselia,* and *Phycosiphon* have been dissolved by acidic groundwater, the burrows and tubes are useful guides to marine environments. *Skolithos* indicates an area of high wave action, whereas *Phycosiphon* points to offshore conditions. Unusual cylinders at Mist were initially interpreted as traces of the deep-water inhabitant *Adekunbiella durhami*, a pogonophoran worm, although it is currently thought that the cylinders may be the work of the mollusk *Dentalium*. Modern pogonophorans were discovered in 1978 living around deep submarine volcanic vents near the Galapagos Islands.

The Keasey was home to several unique deep ocean creatures such as the sea urchin *Salenia,* the sea lily *Isocrinus,* the swimming pteropod snail *Praehyalocylis,* and glass sponges. Even though unusual, inch-long conical shells of the swimming snail *Praehyalocylis cretacea* occur as molds or casts in concretions. Previously unknown in the northwestern United States, these hollow tube-like snails or pteropods appear in the Keasey and from other regional formations in Washington.

At the rich Mist locale, *Salenia schencki* and the coral *Flabellum hertleini* occur in different levels than do the remains of the spectacularly well-preserved

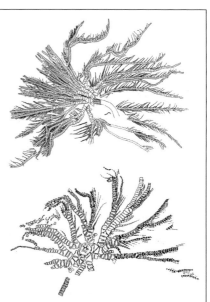

Articulated (whole) crinoid (sea lily) remains are among the most exceptional elements of the Keasey fauna. Stalked Tertiary crinoids are rare elsewhere in the world, and only five species have been recovered from North America. Two of these from Oregon, *Isocrinus oregonensis* and *Isocrinus nehalemensis,* were collected near Mist in Columbia County. (After Burns et al. 2005; Moore and Vokes 1953.)

Cowlitz and Keasey invertebrates. (After Dall 1909; Grabau and Shimer 1910; Zittel 1913.)

Selected Eocene invertebrates from the Cowlitz and Keasey formations:

Ichnofossils (trace fossils): *Adekunbiella, Phycosiphon*
Echinoderms: *Isocrinus, Salenia*
Scaphopods: *Dentalium*
Coelenterates: *Echinophoria, Flabellum*
Pelecypods: *Acila, Acmaea, Barbatia, Glycimeris, Lima, Mytilus, Nemocardium, Nuculana, Ostrea, Pitar, Propeamussium, Solemya, Tellina, Thyasira, Volsella, Yoldia*
Gastropods: *Bathybembix, Bruclarkia, Cancellaria, Conus, Echinophoria, Epitonium, Exilia, Fulgurofusus, Fusinus, Fusitriton, Molopophorus, Natica, Pleurotomaria, Potamides, Praehyalocylis, Ptychosyrinx, Scaphander, Siphonalia, Turricula, Turrinosyrinx, Turritella*

Isocrinus (crinoid) specimens. The unusual concentration of Tertiary crinoids here can be seen in only one or two other places in the world, and it has been speculated that specific ecologic requirements such as carbonate buildup and methane seeps were present for this type of growth and preservation. Stalked crinoids inhabited deeper ocean basins, and, as marine conditions shallowed at the end of the Eocene, they disappeared from the Pacific Northwest.

The uppermost Eocene and early Oligocene interval saw profound changes as the climate altered from tropical to temperate conditions. This environmental transition is reflected in the decline in diversity of both marine invertebrates and land plants. Steady withdrawal of the sea and lower temperatures brought about local extinctions of tropical and subtropical marine populations. The high diversity of species peaked during the latest Eocene but declined steadily as Oligocene climates changed.

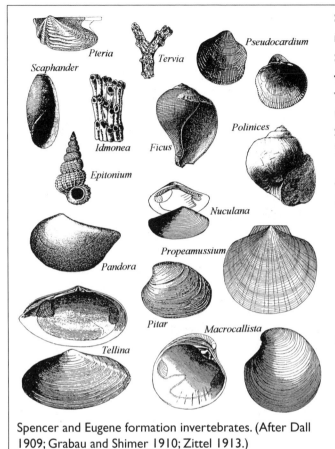

Selected Eocene invertebrates of the Spencer and Eugene formations in the Willamette Valley:

Echinoderms: *Kewia, Salenia*
Scaphopods: *Dentalium*
Bryzoans: *Idmonea, Tembranipora, Tervia*
Cephalopods: *Aturia*
Pelecypods: *Acila, Acmaea, Anadara, Crassatella, Lucinoma, Macoma, Macrocallista, Modiolus, Mya, Mytilus, Nemocardium, Nuculana, Ostrea, Pandora, Parvicardium, Pholadomya, Pitar, Propeamussium, Pseudocardium, Pteria, Semele, Solen, Solena, Spisula, Tellina, Thracia, Yoldia*
Gastropods: *Aceton, Bruclarkia, Calyptraea, Conus, Crepidula, Epitonium, Ficopsis, Ficus, Gemmula, Molopophorus, Natica, Polinices, Scaphander, Sinum, Siphonalia, Turritella*

Spencer and Eugene formation invertebrates. (After Dall 1909; Grabau and Shimer 1910; Zittel 1913.)

Because Keasey sedimentation extends from the Eocene into the early Oligocene, it reflects this important alteration in the ocean populations. Almost three-fourths of the tropical Cowlitz genera died out in the Keasey. This event was followed by a replacement (turnover) fauna of new genera. The gastropods *Cancellaria, Conus, Epitonium,* and *Exilia* and pelecypods *Acila, Pitar, Solemya,* and *Thyasira*, inhabitants of the warm marine ocean, were succeeded by the first appearance of *Bathybembix, Bruclarkia, Fusitriton, Margarites, Ptychosyrinx*, and *Turrinosyrinx*, occupying the cooler early Oligocene Keasey. An even more severe extinction-turnover event took place between the Keasey and the younger Pittsburg Bluff Formation, with a loss of 94 percent of the fauna, but extinctions between the Pittsburg Bluff and the Scappoose were much less severe.

Near Clatskanie and Deer Island in Columbia County, fossiliferous sandstones and conglomerates of the Gries Ranch Formation are from the same time interval and ocean basin as the Keasey but represent a vastly different ecologic setting. Because the Gries Ranch gastropods *Alvania* and *Puncturella* and pelecypods

Arca, Barbatia, Glycimeris, and *Loxocardium* live today in shallow waters, these coarse sands are interpreted as an old shoreline of the Keasey basin.

From the late Eocene and into the early Oligocene, ocean water invaded as far south as Eugene when the Willamette Valley slowly subsided. Sandstones of the Spencer Formation trace the shoreline along the west side of the valley from Eugene to Albany and Dallas in Lane, Benton, and Polk counties. At Spencer Creek in Lane County and Helmick Hill in Polk County, the Eocene pelecypods *Crassatella, Pteria, Spisula, Tellina,* and *Volsella* and gastropods *Ficopsis, Natica,* and *Turritella* reflect an open marine continental shelf away from beaches and embayments. With the transition to cooler marine conditions in the Oligocene, the snail *Ficopsis* disappeared. Recent excavations for a Hyundai factory in west Eugene turned up *Turritella uvasana, Yoldia olympiana, Pitar californiana,* and *Polinices nuciformis* in sediments mapped as the Spencer.

Climate changes are especially notable in the Eugene Formation, which straddles the Eocene-Oligocene boundary. These fossiliferous marine sandstones and siltstones were variously considered Miocene, Oligocene, or Eocene until 2004 when the Eocene-Oligocene boundary was redrawn through the middle of the formation. The lower part of the Eugene Formation displays an abnormally high diversity of nearshore, subtropical faunal members, whereas in the upper strata the faunal variety diminishes and cool-water species appear following an extinction of 94 percent of the genera. *Bruclarkia* and *Molopophorus* that flourished in the lower levels disappeared with the appearance of *Acila, Macoma, Mya, Nuculana,*

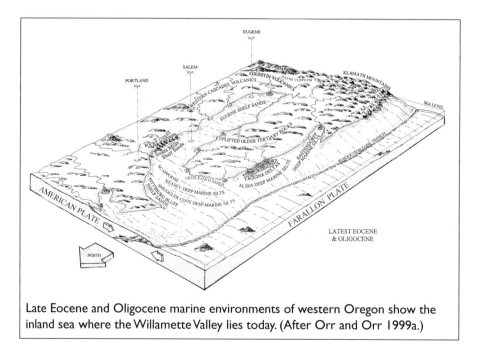

Late Eocene and Oligocene marine environments of western Oregon show the inland sea where the Willamette Valley lies today. (After Orr and Orr 1999a.)

Panopea, and *Spisula* representing cooler conditions, In some regions of the basin, waters may have been relatively calm since many burrowing organisms are preserved in life positions.

With the extinction of most subtropical genera, the Oligocene fauna of the upper Eugene Formation, found at localities in the Coburg Hills and at Brownsville, has fewer species and lacks the warmer and offshore marine elements. Fossil crabs in concretions are particularly distinctive as are bryozoans (*Idmonea, Membranipora, Tervia*), brachiopods (*Eohemithyris, Terebratulina*) and snails (*Acmaea*), and clams (*Chlamys*). *Acila shumardi*, a newly appearing cold-water clam, is prominent.

Today construction within the Eugene urban boundary yields some of the best collections. Even in the 1800s, Thomas Condon gained his first good look at invertebrates in and around the town by monitoring preparations for wells, foundations, and quarries.

The Oligocene Ocean

Well into the Oligocene, spanning 34 to 24 million years ago, the ocean province of western Oregon had become progressively more limited in size and depth with the emergence of the Coast Range and uplift of the Willamette Valley. At the same time, Cascade volcanoes increased in activity. Hemmed in between the two mountain ranges, marine waters lapped up against the east side of the Willamette Valley where the shoreline is delineated by the Scotts Mills Formation in Marion County and the Eugene Formation in Lane County. Deposits on the central and north coast in parts of Lincoln, Tillamook, Columbia, Washington, and Clatsop counties embody the same setting, whereas on the south coast in Coos County the Tunnel Point Formation represents this interval.

Compared to other Tertiary epochs, the Oligocene covers a relatively short time, and marine sediments are more limited when compared to those of the Eocene. At the turn of the twentieth century, the presence of even any Oligocene strata on the West Coast and in Oregon was in question. Whether the Willamette Valley was submerged during the Oligocene and then exposed to erosion during the Miocene had not been established with any certainty. In the 1940s, Bruce Clark of the University of

> **Selected Oligocene invertebrates from the Alsea, Yaquina, and Nye formations of the central Coast Range:**
>
> Foraminifera: *Bolivina, Cassidulina, Cibicides, Gyroidina, Nonion, Uvigerina*
> Scaphopods: *Dentalium*
> Brachiopods: *Terebratulina*
> Pelecypods: *Acila, Aforia, Anadara, Chione, Diplodonta, Ervilia, Katherinella, Lucina, Macoma, Macrocallista, Mytilus, Nemocardium, Nuculana, Panopea, Parvicardium, Pitar, Pseudocardium, Solen, Spisula, Thracia, Venericardia*
> Gastropods: *Architectonica, Bruclarkia, Calyptraea, Crepidula, Cymatium, Exilia, Fulgurofusis, Liracassis, Mitra, Molopophorus, Natica, Polinices, Priscofusus, Siphonalia, Turricula*

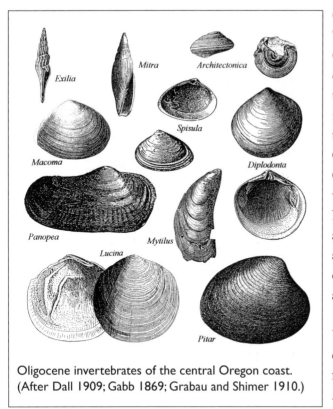

Oligocene invertebrates of the central Oregon coast.
(After Dall 1909; Gabb 1869; Grabau and Shimer 1910.)

California compared hundreds of Tertiary invertebrate species from California to those from Oregon to establish Oligocene boundaries. But the debate surfaced again for a time in the 1960s when it was suggested that there might not even be a marine Oligocene record on the West Coast. This notion was laid to rest when a clear correlation between North American Oligocene rocks and those in Europe was shown by a comparison of planktonic (open ocean) foraminifera from both areas.

Oligocene deposits of the Alsea, Yaquina, and Nye formations are exposed today along the west flank of the central Coast Range. At Newport, where a succession of depositional environments are present, the mollusks *Acila*, *Bruclarkia*, *Macrocallista*, *Parvicardium*, and *Turritella* along with the benthonic (bottom-dwelling) foraminifera *Cassidulina*, *Gyroidina*, and *Uvigerina* of the Alsea indicate cool bathyal to outer shelf water that gradually shallowed with time to beach conditions. Abundant *Mytilus* and *Crepidula* as well as crabs are found in concretions. Similarly, sediments of the Yaquina Formation, illustrating the moderate water depths of a delta, are exposed from Siletz Bay to Seal Rock. An assortment of fish, marine mammals, and plants include the common pelecypods, *Acila*, *Macrocallista*, *Nemocardium*, and *Thracia*. The fragmented condition of the shells suggests transport and wear by tidal currents.

Mudstones of the 4,000-foot-thick Nye taper to less than 500 feet where

Selected invertebrates of the Pittsburg Bluff, Scappoose, and Smuggler Cove formations along the northwest coast:

Ichnofossils: *Chondrites, Cylindrichus, Phycosiphon, Scalarituba, Taenidium, Teichichnus, Zoophycos*
Echinoderm: *Briaster*
Scaphopods: *Dentalium*
Cephalopods: *Aturia*
Pelecypods: *Acila, Anadara, Callista Crenella, Ervilia, Litorhadia Macoma, Macrocallista, Mya, Mytilus, Nemocardium, Nucula, Nuculana, Panopea, Pitar, Solen, Spisula, Tellina, Thracia, Venericardia, Yoldia*
Gastropods: *Acteon, Bruclarkia, Cryptonatica, Molopophorus, Neverita, Opalia, Perse, Polinices, Scaphander, Siphonalia, Taranis, Turritella*

sediments terminated against the rapidly growing Yaquina delta. Extending into the early Miocene, Nye strata have an abundance of deepwater microfossils, fish scales and vertebrae, but only a meager count of mollusks, which are often crushed and distorted. Although microfossil evidence confirms continental slope depths from 1,000 to 2,000 feet, few foraminiferal species overall were found, and less than 10 percent of the total assemblage was planktonic (open ocean). These characteristics of modern-day cool waters led to the conclusion that during the middle Tertiary a southward-flowing cold current and heavy runoff from the land restricted oceanic plankton from nearshore basins.

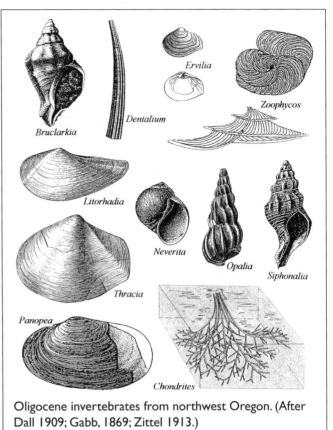

Oligocene invertebrates from northwest Oregon. (After Dall 1909; Gabb, 1869; Zittel 1913.)

In the northern Coast Range, the Pittsburg Bluff, Scappoose, and Smuggler Cove formations are exposed in the roadcuts and creeks of Washington and Columbia counties. Originally placed in the Eocene, the Pittsburg Bluff Formation was assigned to the Oligocene in 1915. Then in 1981 it was positioned at the Eocene-Oligocene boundary; however, currently these sediments have been reassigned to the Oligocene. Sandstone, siltstone, and mudstone of the Pittsburg Bluff and Smuggler Cove strata display a remarkable variety of environments from bathyal slope, to middle/upper shelf water at 300 feet deep, to moderately shallow water about 60 feet in depth, and finally to nearshore intertidal or even terrestrial conditions.

The gastropods *Bruclarkia, Cryptonatica, Molopophorus, Neverita, Perse,* and *Polinices,* and pelecypods *Acila, Callista, Litorhadia,* and *Spisula,* are the most frequent faunal constituents. *Acteon, Pitar, Polinices, Solen,* and *Tellina* typically inhabit sandy mudflats and protected embayments today, whereas the pelecypods *Acila, Ervilia, Nemocardium, Nucula,* and *Yoldia* live at depths over 50 feet or more. Even though the Pittsburg Bluff fauna is well preserved and abundant, it is marked by few species overall. The assortment of mollusks is peculiar in that the shells are broken and show signs of having been transported, yet most display little or no evidence of beach wear. Ellen Moore has speculated that storm waves cast the shells up onto beaches from ocean depths of 75 to 100 feet. In some fossiliferous slabs accumulated in this way, over 50 percent of the matrix is calcium carbonate shell fragments. Sand dollars (*Briaster*) and fish remains—including the ear bones (otoliths) of conger eels, rat tails, and seven genera of shark teeth—are also recorded from this formation.

Of some 48 Pittsburg Bluff species, many occur within hard concretions that formed around the fossil after it was buried. Much of the formation is obscured by thick mature soils and dense vegetation typical of northwest Oregon. Exposures in road cuts, streambeds, and in the sandstone bluffs along the Nehalem River near the town of Pittsburg offer the best collecting spots. Fossils in strata of the overlying Scappoose Formation are similar to those of the Pittsburg Bluff. It has been suggested that the Scappoose might, in fact, represent ancient stream channels or deltas within the Pittsburg Bluff. The Scappoose is exposed along the headwaters of the Clatskanie River where the yellowish sandstones and shales, visible in roadcuts, yield fragmented leaves, wood, and invertebrates.

In the same basin, silty mudstones of the Smuggler Cove Formation bear mollusks, trace fossils, and foraminifera, reflecting rapid deposition in a continental slope environment. Originally designated as the Oswald West Mudstone, the Smuggler Cove includes abundant trace fossils. At Oswald West State Park in Tillamook County, spiraled *Phycosiphon incertum* is scattered through the sandstone beds as are the swirling rooster tail-like burrows of *Zoophycos* and *Teichichnus*, the tubular *Taenidium*, and the multibranched *Chondrites*.

Thomas Condon's assessment of the Willamette Valley, which he called the Willamette Sound, was remarkably perceptive in view of the basin's long marine history. Structurally, the trough resembles both Puget Sound and the Great Valley of California. As the Coast Range rises, the valley sinks, so that well into the future the ocean will reinvade from Portland southward. During the Oligocene a well-marked shoreline of rugged coastal sea cliffs lay along the front range east of Salem near Scotts Mills in Marion and Clackamas counties. Initially called the Butte Creek Beds, these rocks, rich in both plant and invertebrate fossils, were formally designated as the Scotts Mills Formation in the 1980s. Representing the high-water mark of the retreating

Salenia   *Kewia*   *Arbacia*

Despite their fragile nature, sea urchins (*Arbacia*, *Salenia*) and sand dollars (*Kewia*) are well represented in the turbulent, high-energy Scotts Mills sediments. (After Linder, Durham, and Orr 1988; Shroba 1992.)

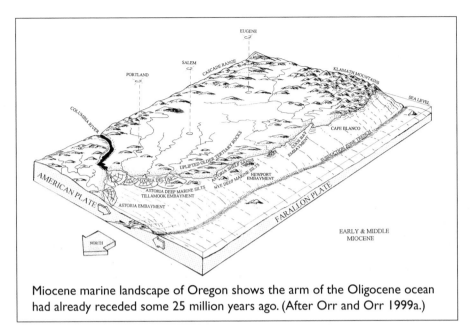

Miocene marine landscape of Oregon shows the arm of the Oligocene ocean had already receded some 25 million years ago. (After Orr and Orr 1999a.)

Oligocene seaway in the Willamette Valley and Western Cascades, the 1,500-foot-thick strata depict a variety of settings from a swampy, poorly drained coastal plain to resistant volcanic headlands making up an irregular coastline of seastacks, embayments, and steep submarine slopes. *Acmaea, Calyptraea, Crepidula, Mytilus,* and *Ostrea* inhabited a shallow rocky ecological niche, whereas *Natica* and *Polinices* point to a muddy bottom. Storm deposits (tempesites) in and around the surf area are reflected by the worn, dispersed shells of mollusks and barnacles. The Scotts Mills and similar limestone lenses can be followed and mapped easily up and down the front range of the Western Cascades by the incidence of oak forests that thrive on the carbonate-rich soil above these rocks.

This marine environment marks the boundary between the subtropical and temperate climate zones on the Pacific Coast. The presence of both cool- and warm-water sea urchins (*Arbacia, Gagaria, Lytechinus, Salenia*) and sand dollars (*Kewia*) in the Scotts Mills Formation places it at mid to low latitudes during the late Oligocene. The tiny dime-sized *Kewia* was remarkable in that it selected and accumulated heavy minerals (magnetite) within its flattened crown (body). In order to live in the high-energy surf zone *Kewia* deliberately ingested sand-sized heavy minerals for ballast. A familiar phenomenon in modern sand dollars, known as a type of "weight belt" strategy, was previously unrecorded in the fossil record. Trace fossils *Cylindrichnus*, preserved in an upright position, represent individual burrows of an invertebrate in the Scotts Mills. These tall, cone-shaped layered sand and clay sheaths, surrounding a central sand-filled tube, have a maximum diameter of three inches.

The Miocene Ocean

At the close of the Oligocene Epoch, 23.8 million years ago, widespread seas across western Oregon had disappeared, and the Miocene ocean occupied a narrow strip extending only slightly east of the present shoreline. Uplift and eastward tilting of the Coast Range and concurrent depression of the Willamette Valley were well under way, defining the limits of the warm-temperate, shallow seaway. Marine Miocene faunas are limited to exposures of the Astoria Formation, representing intermittent embayments along the coast in Lincoln, Tillamook, and Clatsop counties. Similarly, sediments of the Tarheel Formation, the Floras Lake Sandstone, and the Empire Formation filled protected coves in the vicinity of Coos Bay.

By far the most historic and famed marine Miocene strata in the state are sandstones, siltstones, and shales of the Astoria Formation, exposed in narrow strips along the coast. Where it overlies the Yaquina and Nye formations near Newport, the Astoria reaches thicknesses up to 2,000 feet. Carbonized wood, fossil-rich lens of mollusks, small concretions with scallop shells or crabs, and large concretions with marine vertebrates attest to the shallow-water paleosetting of the formation.

Although first named for rocks in and around the town of Astoria, the original fossil collecting sites have been covered by city buildings and wharfs, or by sand dredged from the Columbia River channel, posing ongoing problems for subsequent investigators. Difficulties in finding older locales have been further complicated by the renaming of streets and even by the complete elimination of manmade structures used as reference points. An interesting afternoon can be spent with an article and map by Betty Dodds, in which old fossil sites are tied into historic records. *Fossil Shells from Western Oregon* by Ellen Moore provides geologic excursions, history, and identifications for coastal outcrops.

The Astoria has enjoyed a lengthy past of scientific scrutiny since the early 1800s when shells around the Astoria fur trading post interested explorers and traders. Thomas Condon, who named the "Astoria shales," spent the summers walking the beach looking for shells by breaking open concretions. He sometimes cracked several hundred to reap only two or three worthwhile specimens. Today this formation has been

William Dall's early contribution to Tertiary paleontology of the Pacific Coast was particularly important. His 1909 publication covered Miocene invertebrates, vertebrates, and microfossils from Astoria and Coos Bay. Born in Boston in 1845, Dall began his career working as a naturalist for western explorations organized by the U.S. government. He concentrated on Cenozoic and living mollusks, arranging the collections at the Smithsonian Institution. In all, he named a staggering 5,427 genera, subgenera, and species of fossil and recent mollusks. (Photo courtesy Archives, U.S. Geological Survey, Denver.)

A selection of Miocene Astoria invertebrates from the Coast Range:

Foraminifera: *Bolivina, Buliminella, Epistominella, Nonionella, Plectofrondicularia*

Ichnofossils: *Gyrolithes, Tisoa*

Scaphopods: *Dentalium*

Coelenterates: *Astreopora, Caryophyllia, Lophelia, Stephanocyathus*

Bryozoans: *Membranipora*

Brachiopods: *Frieleia, Terebratulina*

Cephalopods: *Aturia*

Pelecypods: *Acila, Anadara, Chione, Clinocardium, Delectopecten, Dosinia, Gari, Katherinella, Lucinoma, Litorhadia, Macoma, Modiolus, Nucula, Panopea, Patinopecten, Propeamussium, Solen, Spisula, Tellina, Thracia, Thyasira*

Gastropods: *Bruclarkia, Cancellaria, Crepidula, Cryptonatica, Cylichnina, Ficus, Molopophorus, Neverita, Priscofusus, Psephaea, Liracassis, Sinum, Turritella*

---

examined repeatedly, and the locales have been visited by many West Coast molluscan paleontologists.

Astoria mollusks include some 97 species and 73 genera. The composition is exceptional in that just a few species occur in phenomenal numbers. Shell layers of *Anadara devincta, Katherinella angustifrons,* and *Patinopecten propatulus* dominate the fauna and can be traced for miles in beach cliff exposures. Most of the clams have their valves still together and closed, showing little wear or abrasion and suggesting they were not transported far before burial. This is typical of organisms living on a soft muddy seafloor at a moderately shallow 500-foot depth. Bryozoans, corals, fish, crabs, plants, mammals, and microfossils contribute to the diverse assemblage. Commonly occurring in the Astoria, individual tubes of the trace fossil *Tisoa* are lined with secretions of sulfide minerals. Scratch marks in the

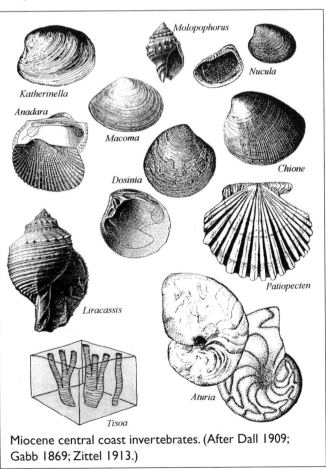

Miocene central coast invertebrates. (After Dall 1909; Gabb 1869; Zittel 1913.)

Ellen Moore was one of the few women majoring in geology at Oregon institutions during World War II, completing her degree from Ewart Baldwin. Writing to a number of institutions for job opportunities, Moore received responses such as that from Yale University—"Our department is completely manned." The curator of another museum responded, "Will be glad to take you out for dinner but can't hire you." Although she initially accepted a position in Portland with the Army Corps of Engineers, most of her career was at the U.S. Geological Survey. Moore currently lives and works in Corvallis at Oregon State University. (Photo courtesy E. Moore.)

burrows are thought to have been made by a shrimp-like arthropod, even though none of the tubes contain arthropod remains.

At the popular Beverly Beach State Park site in Lincoln County, some 35 Astoria gastropods and pelecypods include members of the Naticidae and Muricidae families, boring gastropods that preyed on mollusks by drilling a hole in the shell to rasp out the soft internal tissue. Of the two molluscan communities at Beverly Beach, one is dominated by the scaphopod *Dentalium* and the other by the clams *Acila* and *Macoma. Cryptonatica*, a predatory snail of the Naticidae family, had drilled most of the shells, preferring the clams over snails, and *Macoma albaria* above all others. Perhaps this was due to *Macoma*'s shallow habitat or slow response to predators. The shell size may have also been a factor, and curiously the extremely small or large mollusks were not attacked by the shell-boring snails. Snails targeted the highest point on their prey, the hump or umbo, for drilling.

Fossil burrows seen at nearby Moolack Beach in Lincoln County were initially thought to have been made by a small shrimp, *Gyrolithes*. Seven-inch-long tight spirals are preserved as internal fillings in sandy muds of the Astoria Formation. Even though it is not known with certainty what animal produced the burrows, an examination of present-day tidal flats shows that a scavenging soldier crab, in digging downward, would rotate its body, producing screw-shaped tubes. A second interpretation suggests that the burrows were made by a shallow marine polychaete worm, an elongate, segmented member of the phylum annelida. The most familiar annelids are earthworms and leeches.

Middle Miocene strata in the vicinity of Coos Bay were unknown until 1949 when mollusk-impregnated siltstones were brought to the surface and piled up near North Spit during dredging operations in the channel. The specimens that had been unearthed were well-preserved and free of any rock matrix, but, since they were scattered over the surface of the spoils piles, contamination was possible. In comparing the mollusks to others in the region, Ellen Moore, a University of Oregon student at the time, concluded the assemblage was heretofore unreported

Selected Miocene invertebrates from the Empire and Tarheel formations and Floras Lake Sandstones on the south coast:

Crustaceans: *Balanus*

Scaphopods: *Dentalium*

Bryozoans: *Eurystomella, Membranipora*

Pelecypods: *Anadara, Clinocardium, Cryptomya, Delectopecten, Diplodonta, Dosinia, Glycimeris, Katherinella, Lucinoma, Macoma, Modiolus, Mya, Mytilus, Ostrea, Panope, Patinopecten, Pecten, Saccella, Siliqua, Solen, Spisula, Tellina, Thracia, Thyasira, Yoldia*

Gastropods: *Amauropsis, Antiplanes, Bruclarkia, Cancellaria, Chlorostoma, Crepidula, Cryptonatica, Cylichnina, Epitonium, Fusinus, Liracassis, Mitrella, Neverita, Nucella, Olivella, Polinices, Priscofusus, Scaphander, Sinum, Turvia*

and that the Miocene age was new to Coos Bay. Ultimately she determined that the specimens were similar to those from the Astoria Formation.

By 1966 the shells from the dredge piles had been carried away by waves, but John Armentrout, at the University of Oregon, discovered identical outcrops between Pigeon Point and the old Sitka Dock. Fossils at this locale were sparse but unworn and beautifully preserved, and he formally named the exposure the Tarheel Formation. Approximately 70 percent of the fauna consisted of the clam *Dosinia whitneyi*, also found at Astoria in Clatsop County and at Scotts Mills in Marion County. A total of 18 genera were recovered. Of these *Bruclarkia oregonensis* and *Katherinella angustifrons* are

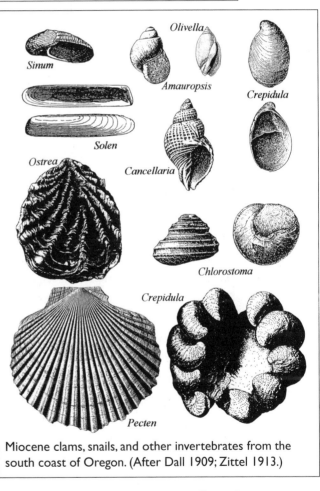

Miocene clams, snails, and other invertebrates from the south coast of Oregon. (After Dall 1909; Zittel 1913.)

extinct, and *Dosinia* is extinct locally. *Patinopecten oregonensis cancellosus* is the most common pelecypod found within the concretions. The ancient setting of the Tarheel was that of warm to temperate marine waters at shallow to moderate upper shelf depths reaching to 180 feet. The lack of brackish tidal shells indicates that the setting was not part of an estuary or directly adjacent to land.

Southward from Coos Bay, exposures of the Miocene Empire Formation appear as far as Cape Blanco in Curry County. The Empire was assigned to the Pliocene until the Tertiary epochs were revised in 1983, and the base of the Pliocene boundary was moved upward from 10 to 5 million years before the present. Near Charleston the formation forms prominent cliffs and consists of around 3,000 feet of coarse sandstones studded with concretions over a foot in diameter. Within the strata, the Coos Conglomerate, long considered as a separate unit, is presently regarded as part of the Empire. Easily among the most fossil-rich strata in the state, it has been popular among collectors since the early 1900s. At Fossil Point, the Coos Conglomerate lens is fully exposed only at low tide, and the site is privately owned. An area of less than one acre, it comprises beach material eroded from shallower layers of the Empire and redeposited downslope as fill in a narrow submarine trough. Most of the shells show the wear and abrasion of transport, and scattered bone fragments of marine vertebrates such as whales, sea lions, walruses, and seals are common within the layers.

Empire invertebrates are generally intact, although most of the specimens, visible on the surface, are badly weathered. Mollusks, inhabiting what was a warm to temperate embayment, are abundant in comparison to the numbers of microfossils and vertebrates. Of the 72 genera and 129 species of mollusks, 50 genera and 38 species still live along the West Coast. John Armentrout, who studied the fauna in 1967, concluded that the late Miocene climate of Coos Bay varied little from that of today. *Delectopecten, Dosinia, Katherinella, Panopea, Nuculana, Sinum,* and *Tellina* are common from the Miocene and the Recent. Preserved intact, colonies of the slipper shell, *Crepidula princeps* (snail), indicate an embayment of calm shallow water.

At Floras Lake and Blacklock Point near Cape Blanco in Curry County, the Empire Formation has been subdivided into the lower Sandstone of Floras Lake, while the name Empire was retained for the upper 300 feet. Contrasting faunas characterize the upper and lower Floras Lake Sandstone layers. The basal section has only a few species, but a dramatic increase in faunal diversity is notable toward the top. The pelecypod *Mytilus,* the barnacle *Balanus,* and the gastropod *Nucella* in the lowest intervals verify a wave-washed tidal setting close to a rocky shoreline. But the appearance of the snail *Bruclarkia* and the clams *Macoma* and *Spisula* toward the middle of the sequence, followed by the scaphopod *Dentalium,* the pelecypods *Katherinella* and *Patinopecten,* and gastropod *Liracassis* near the top reflect water deepening to 60 feet. At the highest level the sandstone contains scattered *Dentalium* and unbroken shells of the clam *Spisula.*

The Pleistocene Ocean

Late Miocene and Pliocene marine sediments have not been recorded from the central and northern Coast Range, and Pleistocene exposures are limited to the extreme western margins of the state. Sea level history along the West Coast is complicated by at least three independent geologic processes. During Pleistocene glacial advances, sea levels dropped off profoundly by up to several hundred feet when water was locked up elsewhere as ice. Simultaneously, the coastline, topped by stairstep terraces, was elevated in stages because of tectonic plate subduction. In yet another geologic phenomenon, periodic catastrophic earthquakes dropped whole sections of the coast by several feet in 300- to 500-year cycles.

Joining the faculty of the University of Southern California, micropaleontologist Orville Bandy was a diligent researcher, who focused on foraminifera. Studying with Earl Packard and receiving a degree from Oregon State University, he was one of the first to investigate faunas from Cape Blanco. After military service during World War II, he worked under J.J. Galloway at Indiana University for a Ph.D.. His reputation was that of a tireless worker, who wrote copiously on foraminifera. Separating the Port Orford and Elk River formations in 1950, he observed that the shallow-water foraminifera *Elphidium hannai* was the most common species. (Photo c. 1965; courtesy University of Southern California, Archives.)

The Pleistocene experienced ice sheets of continental glaciers that projected from Canada southward into Washington along with almost continuous ice caps across the higher mountain ranges. When glacial ice melted and sea level rose, large tracts of the coastal plain became submerged, providing new habitats for invertebrates. Shell deposits at Cape Blanco, which takes its name from their profuse whiteness, form the crest of a sequence that begins with Port Orford sediments, covered in turn by the Elk River Formation, and topped by terraces. Coarse sandstones and conglomerates of the Port Orford Formation, that grade upward to silt at the top, contain scattered trace fossils, molds of shells, and foraminifera.

In contrast to the Port Orford Formation, the Elk River has an exceptional invertebrate fauna. These shell-rich conglomerates and sandstones have been variously assigned to the Pliocene and Pleistocene; but since the megafossils consist almost entirely of species still living today, the formation is presently assigned to the Pleistocene. Over one hundred organisms include mollusks, arthropods, foraminifera, trace fossils, sponges, fish, and mammal remains. The clam *Clinocardium meekianum baldwini* is the most abundant fossil encountered. Casts of *Clinocardium* and *Macoma*, burrows of annelids (worms), and sponges such as *Cliona* are encountered throughout the lower levels, whereas shell beds of the small bivalve *Psephidia* are profuse near the top. Accumulations of *Psephidia*

A selected Pleistocene fauna from the Port Orford and Elk River formations and from terraces of southwest Oregon:

Foraminifera: *Elphidium*

Ichnofossils: *Macaronichnus*

Crustaceans: *Balanus*

Porifera: (sponges): *Cliona*

Bryozoans: *Balanophyllia, Celleporina, Costazia, Heteropora*

Pelecypods: *Cardita, Chlamys, Clinocardium, Hiatella, Hinnites, Macoma, Modiolus, Mya, Nuculana, Psephidia, Saxidomus, Schizothaerus, Siliqua, Thracia, Tresus, Zirfaea*

Gastropods: *Acmaea, Amphissa, Buccinum, Calliostoma, Calyptraea, Crepidula, Epitonium, Fusitriton, Lepeta, Margarites, Mitrella, Nassarius, Ocenebra, Odostomia, Olivella, Polinices, Thais*

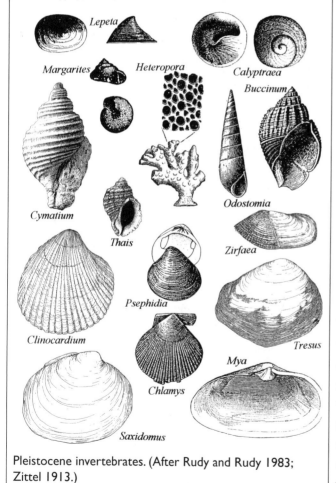

Pleistocene invertebrates. (After Rudy and Rudy 1983; Zittel 1913.)

are concave-up, and experiments show that these layers of shells were formed by small sand ripples migrating across the sea floor overturning them.

Identifying eleven shallow-water species of foraminifera in the Elk River, Orville Bandy interpreted a marine setting dominated by cool-water elements. While plant fossils at the base of the formation accumulated in what was probably a channel of the ancestral Sixes and Elk River drainages, invertebrates in the younger upper layers occupied the shallow water of a muddy estuary. The basin gradually shallowed to less than 100 feet in depth as the land was elevated and dissected by coastal streams.

Elevated terrace deposits cap the Port Orford and Elk River formations. Collision and subduction of the Juan de Fuca plate beneath the western edge of North America dramatically raised and tilted the Oregon coastal area to form step-like terraces. These can be seen as discontinuous belts ranging from 20 to 50 feet in thickness and reaching elevations inland up to 1,600 feet. Because the rate of elevation varies from place to place, the age and relationship of individual terrace fragments to each other are difficult to determine. More recently these surfaces have been accurately dated by using the uranium-thorium decay ratios in the skeletons of the corals *Balanophyllia* and *Celleporina,* the bryozoan *Heteropora,* and the sea urchin *Strongylocentrotus*. Amino acid decay within the shell calcite of the large clams *Mya truncata* and *Saxidomus giganteus* also provides precise dates, while oxygen isotopes yield paleotemperature information. Although these methods of chemical paleontology are innovative and remarkably quantitative, they will not completely replace standard biostratigraphic techniques for dating because they involve expensive and time-consuming laboratory procedures.

Pleistocene invertebrates from terrace deposits are meager in Oregon except for faunas at Cape Blanco in Curry County and at Bandon in Coos County. This is in sharp contrast to the prolific occurrence of fossil invertebrates from terraces along the coasts of British Columbia, Washington, California, and Mexico. Reflecting a sandy, shallow, offshore ocean setting, mollusks and barnacles, as well as echinoids, bryozoans, and foraminifera, are preserved in sea cliff terraces at Cape Blanco. Water temperatures were cooler, as 20 of the fossil species are known only from regions considerably north of present-day Cape Blanco. Reflecting a very slow rate of deposition, abundant paired valves of the large clams *Saxidomus giganteus, Schizothaerus capax,* and *Macoma irus* in upright life position imply that the shells were deposited below the level of vigorous wave action and not transported any significant distance after death.

Terraces near Bandon at Grave Point and Coquille Point are notable for their high fossil concentrations, representing up to 58 species. Plates and valves of the barnacle *Balanus cariosus* were recorded by Victor Zullo as the most conspicuous by bulk, but they were exceeded in number by the pelecypod *Hiatella arctica,* the gastropods *Fusitriton oregonensis*, *Margarites pupillus,* and *Oenopota tabulata,* and by the bryozoans *Costazia* and *Heteropora*. Of the 37 invertebrates from Coquille Point, seven have not been recovered at Grave Point. Today *Mya japonica*, found at Coquille Point, is extinct in the eastern Pacific, and its presence here is the most southerly known to date. Although the Bandon shells are probably the same age as those from Cape Blanco, only 30 percent of the species are found in both sites because of the contrasting habitats. Shallow water along a rocky coast at Bandon protected the mollusks from waves, while at Cape Blanco the shells were tossed up and sorted in the turbulent water by an advancing sea.

Freshwater and Terrestrial Invertebrates

Even though terrestrial and freshwater invertebrates are known from several intervals of Oregon's fossil record, they are not preserved in the same diversity and numbers as ocean-dwelling clams and snails because of their fragile shells. Freshwater clam shells are ordinarily thicker than those from the oceanic realm

G. Dallas Hanna, who completed a degree in paleontology and zoology from George Washington University in 1918, spent his career on the West Coast. As a curator for 51 years at the California Academy of Sciences, he worked on most of Oregon's freshwater invertebrates. Hanna not only published on freshwater invertebrates; his interests extended to living and fossil marine mollusks, diatoms, silicoflatellates, and foraminifera. (Photo 1923; courtesy California Academy of Sciences.)

A relic of the Pleistocene Epoch, the freshwater clam *Pisidium ultramonatauum*, living today in northeastern California and southcentral Oregon, had a wider habitat across a region extending to southeast Idaho and adjoining Utah. The distribution pattern of *Pisidium*, of the snail *Carinifex*, and of the freshwater sucker *Chasmistes* shows that a previously connected chain of lakes and drainage basins reached for hundreds of miles from Nevada and the Pit River in California to Klamath and Fossil lakes and the Malheur basin in Oregon, ultimately stretching to the Snake River and beyond.

and are often composed entirely of the mineral aragonite in contrast to the calcite of marine shells. With burial and time, the aragonite shell, or "mother of pearl," is unstable and tends to recrystallize to calcite. In this degeneration process, the volume change and resulting fractures in the shell structure make it vulnerable to solution by groundwater. Even if they remain intact, the effects of recrystallization often cause them to flake and crumble as they are removed from the entombing rock matrix.

The oldest land-dwelling mollusks in Oregon are from Oligocene and Miocene deposits in the John Day and Malheur basins, while the majority of freshwater invertebrates are from Pliocene and Pleistocene lake deposits in the Great Basin across the south-central part of the state.

Fossil terrestrial snails entombed in John Day strata belong to the same genera as those living today. In spite of the presence of numerous shells, many are too broken for exact identification. Nearly 120 specimens were collected by Thomas Condon, and almost all are of the snail *Polygyra*. However, nine genera and eleven species are represented: *Ammonitella, Epiphragmophora (Helix), Gastrodonta, Helicina, Oreohelix, Polygyra, Polygyrella, Pyramidula,* and *Rhiostoma*. Of these, *Rhiostoma* currently lives in the tropics of Asia. The remains of the freshwater clam *Unio condoni* and a pond snail *Lymnaea* also occur with the fauna.

Near Oregon's southeastern border, middle Miocene freshwater mollusks from the Juntura basin in western Malheur and Harney counties are usually incidental to collectors mainly interested in vertebrates. Three species of clams and six of snails suggest a perennial lake that reached depths up to 30 feet around 15 million years ago. At that time, lakes were created by lava flows that dammed streams. The most common mollusks found are *Carinifex, Fluminicola, Radix, Sphaerium,* and *Viviparus*. A similar gastropod fauna from tuffs, diatomites, and streambeds in the Powder River valley confirms the presence of freshwater ponds during the late Miocene to early Pliocene.

By the Pleistocene Epoch, increased precipitation and lowered temperatures created vast pluvial lakes across the Great Basin. Invertebrates from these ancient bodies of water are found at Fossil and Summer lakes in the Warner valley, and near Harney Lake in Harney County. Mollusk

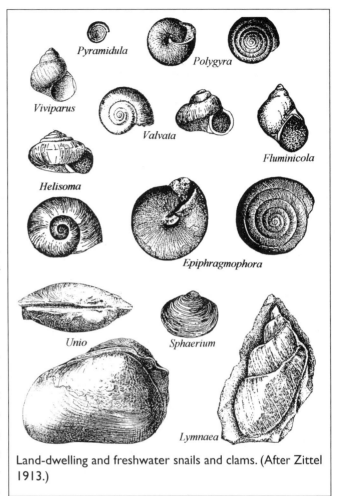

Land-dwelling and freshwater snails and clams. (After Zittel 1913.)

shells litter the surface at Fossil Lake in such abundance that they lighten the sand. The white shell of *Carinifex newberryi* is particularly distinctive. *Gyraulis, Helisoma, Limnophysa, Pisidium, Sphaerium,* and *Valvata,* which are found here, are all still living. At nearby Summer Lake, the Chewaucan marshes, and Lake Abert, thin beds of shells and shell fragments make up an astonishing 90 percent of the strata. Wells drilled to depths over 700 feet penetrated layers of compacted mollusks, indicating a much longer history of occupation by Pleistocene lakes than previous believed.

Freshwater clams and snails from Warner and Harney lakes as well as in places along the Snake River valley have been thoroughly compressed, crushed, and cemented together making extraction and identification difficult. Only the shell of the robust snail *Viviparus* is preserved to any degree, while other the gastropods *Carinifex, Parapholyx, Valvata, Vorticifex* and the clam *Pisidium* occur as casts.

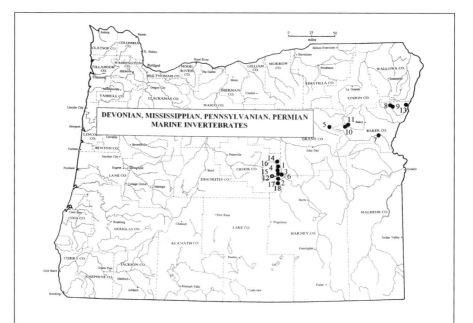

DEVONIAN TO PERMIAN INVERTEBRATE LOCALITIES

The following are Devonian, Mississippian, Pennsylvanian, and Permian marine invertebrate localities and a selection of authors who list faunas from the locales.

DEVONIAN

1. Suplee (Weberg Ranch), Crook Co. No Fm. Blome and Nestell 1991, 1992; Blome and Reed 1992; Danner 1968; Johnson and Klapper 1978; Kleweno and Jeffords 1962; Morris and Wardlaw 1986; Savage and Amundson 1979; Sorauf 1972.
2. Twelvemile Creek (Berger Ranch), Harney Co. No Fm. Blome and Nestell 1992; Johnson and Klapper 1978; Morris and Wardlaw 1986.

MISSISSIPPIAN

3. Coffee Creek, Crook Co. Coffee Creek Fm. Blome and Nestell 1992; Dutro 1985, 1989; Merriam 1942; Merriam and Berthiaume 1943; Sada and Danner 1973.

PENNSYLVANIAN

4. Suplee (Mills Ranch), Crook Co. Spotted Ridge Fm. Mamay and Read 1956; Merriam 1942; Merriam and Berthiaume 1943.
5. Greenhorn Mountains (Vinegar Hill), Grant Co. Morris and Wardlaw 1986.

PERMIAN

6. Big Flat, Grant Co. Coyote Butte Fm. Bostwick and Nestell 1965.
7. Burnt River, Baker Co. Burnt River Schist. Morris and Wardlaw 1986.
8. Clover Creek, Baker Co. Clover Creek Greenstone Permian/Triassic. Gilluly 1937; Smith and Allen 1941; Shroba 1992.
9. Eagle Creek, Baker Co. Clover Creek Greenstone. Gilluly 1937; Shroba 1992.
10. Elkhorn Ridge, Baker Co. Elkhorn Ridge Argillite. Blome et al. 1986; Pessagno and Blome 1986; Bostwick and Nestell 1967; Morris and Wardlaw 1986; Yochelson 1961.
11. Granite Boulder Creek, Grant Co. Elkhorn Ridge Argillite. Blome et al. 1986; Merriam 1942; Morris and Wardlaw 1986.

12. Grindstone Creek, Crook Co. Coyote Butte Fm. Blome et al. 1986; Cooper 1957; Merriam 1942; Murchey and Jones 1994; Wardlaw, Nestell, and Dutro 1982.
13. Homestead, Baker Co. Hunsaker Creek Fm. Shroba 1992; Smith and Allen 1941; Vallier 1967; Vallier and Brooks 1970.
14. Suplee (Bernard Ranch), Crook Co. Coyote Butte Fm. Blome and Nestell 1992; Bostwick and Nestell 1967; Cooper 1957; Merriam 1942; Skinner and Wilde 1966; Stevens and Rycerski 1983.
15. Triangulation Hill, Crook Co. Coyote Butte Fm. Blome and Nestell 1992; Wardlaw, Nestell, and Dutro 1982.
16. Trout Creek, Crook Co. Coyote Butte Fm. Blome et al. 1986; Morris and Wardlaw 1986; Sada and Danner 1973.
17. Tuckers Butte, Crook Co. Coyote Butte Fm. Blome and Nestell 1992; Cooper 1957; Merriam 1942; Merriam and Berthiaume 1943; Wardlaw, Nestell, and Dutro 1982.
18. Twelvemile Creek, Harney Co. Coyote Butte Fm. Blome and Nestell 1991; Blome and Reed 1992; Cooper 1957; Hanger, Hoare, and Strong 2000; Hanger and Strong 1998; Merriam 1941; Merriam and Berthiaume 1943; Murchey and Jones 1994; Wardlaw and Jones 1980; Wardlaw, Nestell, and Dutro 1982.

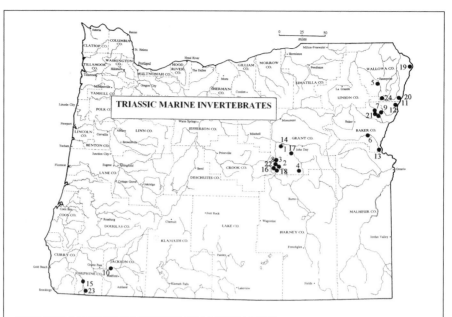

TRIASSIC MARINE INVERTEBRATE LOCALITIES

The following are Triassic marine invertebrate localities and a selection of authors who list faunas from the locales.

1. Aldrich Mountains, Grant Co. Aldrich Mtn. Group (Triassic/Jurassic). Blome 1984; Blome and Nestell 1992; Blome et al. 1986; Dickinson and Vigrass 1965.

2. Beaver Creek, Grant Co. Begg Fm. Blome and Nestell 1991, 1992; Blome et al. 1986; Dickinson and Vigrass 1965; Washburne 1903.

3. Begg Creek, Grant Co. Begg Fm. Dickinson and Vigrass 1965.

4. Big Flat, Grant Co. Begg/Brisbois Fms. Blome and Nestell 1992; Blome et al. 1986; Dickinson and Vigrass 1965; Schenck 1931.

5. Black Marble Quarry, Wallowa Co. Martin Bridge Fm. Smith and Allen 1941.

6. Durkee, Baker Co. Burnt River Schist. Morris and Wardlaw 1986.

7. Eagle Creek, Baker Co. Martin Bridge Fm. McRoberts 1993; Newton 1987; Ross 1938; Smith 1912; Smith and Allen 1941.

8. Elkhorn Ridge, Baker Co. Elkhorn Ridge Argillite. Blome and Nestell 1992; Blome et al. 1986; Bostwick and Koch 1962; Bostwick and Nestell 1967; Dickinson and Vigrass 1965; Pessagno and Blome 1986.

9. Excelsior Gulch, Baker Co. Hurwal Fm. Follo 1986 1992 1994; LaMaskin et al. 2008; Smith and Allen 1941.

10. Gold Hill, Klamath Co. Applegate Group. Irwin, Wardlaw, and Kaplan 1983; Wardlaw and Jones 1980.

11. Hells Canyon, Wallowa Co. Martin Bridge Fm. Newton 1983, 1986; Newton et al. 1987; Stanley 1979; Stanley and Whalen 1989; Whalen 1988; Vallier 1967.

12. Homestead, Baker Co. Grassy Ridge Fm. Vallier 1967; Vallier and Brooks, eds. 1986.

13. Huntington, Baker Co. Huntington Fm. Blome et al. 1986; LaMaskin 2008.

14. John Day River (South Fork), Grant Co. Brisbois Fm. Blome 1984; Blome and Nestell 1992; Dickinson and Thayer 1978; Dickinson and Vigrass 1965.

15. Kerby, Josephine Co. Applegate Group. Irwin, Wardlaw, and Kaplan 1983.

16. Little Dog Creek, Grant Co. Elkhorn Ridge Argillite. Blome and Nestell 1992; Blome et al. 1986; Bostwick and Nestell 1965.
17. Miller Mountain, Grant Co. Fields Creek Fm. Blome and Nestell 1992; Blome et al. 1986.
18. Morgan Mountain, Grant Co. Rail Cabin Argillite. Blome et al. 1986; Dickinson and Vigrass 1965; Pessagno and Blome 1986; Taylor and Guex 2002.
19. Pittsburg Landing, Wallowa Co. Doyle Creek/Wild Sheep Creek Fms. Goldstrand 1994; LaMaskin et al. 2008; White 1994; White and Vallier 1994; Vallier 1967; Vallier and Brooks, eds. 1986.
20. Spring Creek, Baker Co. Martin Bridge Fm. Newton 1983; Newton et al. 1987; Stanley and Beauvais 1990; Stanley and Whalen 1989; Senowbari-Daryan and Stanley 1988.
21. Summit Point, Baker Co. Martin Bridge Fm. Newton et al. 1987; Stanley and Senowbari-Daryan 1986.
22. Virtue Hills, Baker Co. Elkhorn Ridge Argillite. Blome and Nestell 1992; Blome et al. 1986; Bostwick and Koch 1962; Bostwick and Nestell 1965.
23. Waldo (Takilma), Josephine Co. Applegate Group. Irwin, Wardlaw, and Kaplan 1983.
24. Wallowa Mountains, Wallowa Co. Martin Bridge Fm. Follo 1994; Newton et al. 1987; Smith and Allen 1941; Stanley 1986.

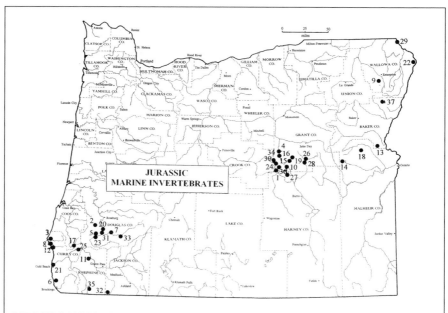

JURASSIC MARINE INVERTEBRATE LOCALITIES

The following are Jurassic marine invertebrate localities and a selection of authors who list faunas from the locales.

1. Beaver Creek, Grant Co. Mowich Group/Snowshoe/Donovan Fms. Dickinson and Vigrass 1965; Imlay 1967, 1968, 1973; Lupher 1941; Lupher and Packard 1930.

2. Buck Peak, Douglas Co. Riddle Fm. Diller 1907, 1908; Imlay et al. 1959; Koch 1966.

3. Cape Blanco, Curry Co. Otter Point Fm. Dott 1966, 1971; Koch 1966.

4. Cow Creek, Grant Co. Mowich Group. Dickinson and Vigrass 1965.

5. Cow Creek, Douglas Co. Riddle Fm. Imlay et al. 1959.

6. Crook Point, Curry Co. Otter Point Fm. Koch 1966.

7. Days Creek, Douglas Co. Riddle Fm. Diller and Kay 1924; Imlay and Jones 1970; Imlay et al. 1959.

8. Elk River, Curry Co. Galice Fm. Diller 1903; Imlay and Jones 1970; Koch 1966.

9. Enterprise, Wallowa Co. Hurwal Fm. Follo 1994; Imlay 1968; Smith and Allen 1941.

10. Flat Creek, Grant Co. Mowich Group/Snowshoe/Trowbridge Fms. Dickinson and Vigrass 1965; Imlay 1968, 1973; Lupher 1941; Pessagno, Whalen, and Yeh 1986; Smith 1980.

11. Galice, Josephine Co. Galice Fm. Irwin, Jones, and Kaplan 1978.

12. Humbug Mountain, Curry Co. Otter Point Fm. Koch 1966.

13. Huntington, Baker Co. Weatherby Fm. Imlay 1973, 1980, 1986; Wagner et al. 1963.

14. Ironside, Malheur Co. No Fm. Imlay 1973; Wagner et al. 1963.

15. Izee, Grant Co. Mowich Group/Snowshoe/Trowbridge Fms. Blome and Nestell 1991, 1992; Dickinson and Vigrass 1965; Fraser, Bottjer, and Fischer 2004; Imlay 1968, 1973, 1980, 1981; Pessagno and Blome 1982; Pessagno, Whalen, and Yeh 1986; Smith 1980.

16. John Day River (South Fork), Grant Co. Mowich Group/Snowshoe/Trowbridge/Lonesome Fms. Dickinson and Vigrass 1965; Imlay 1968, 1981; Lupher 1941, Pessagno and Blome 1982; Pessagno and Whalen 1982.

17. Johnson Creek, Coos Co. Galice Fm. Dott 1966; Koch 1966.

18. Juniper Mountain, Malheur Co. Weatherby Fm. Fouch 1968; Imlay 1973, 1980, 1986; Wagner et al. 1963.
19. Morgan Mountain, Grant Co. Graylock Fm. Blome and Nestell 1992; Dickinson and Vigrass 1965; Imlay 1968, 1986; Pessagno and Whalen 1982; Taylor and Guex 2002.
20. Myrtle Creek, Douglas Co. Riddle Fm. Diller 1907; Imlay et al. 1959; Koch 1966; Popenoe, Imlay, and Murphy 1960.
21. Otter Point, Curry Co. Otter Point Fm. Koch 1966.
22. Pittsburg Landing, Wallowa Co. Coon Hollow Fm. Goldstrand 1994; Imlay 1981, 1986; Stanley and Beauvais 1990; White 1994; White and Vallier 1994; White et al. 1992.
23. Riddle, Douglas Co. Riddle Fm. Diller 1907, 1908; Diller and Kay 1924; Imlay and Jones 1970; Imlay et al. 1959.
24. Robertson Ranch, Grant Co. Robertson Fm. Dickinson and Vigrass 1965; Fraser, Bottjer, and Fischer 2004; Imlay 1968; Lupher 1941; Nauss and Smith 1988.
25. Rogue River, Curry Co. Galice/Rogue Fms. Dott 1966; Pessagno and Blome 1990.
26. Seneca, Grant Co. Mowich Gr./Snowshoe Fm. Blome and Nestell 1992; Dickinson and Vigrass 1965; Imlay 1968, 1973, 1986; Lupher 1941; Pessagno and Blome 1980, 1982; Pessagno, Whalen, and Yeh 1986.
27. Silvies River, Grant Co. Mowich Group. Dickinson and Vigrass 1965; Imlay 1973; Lupher 1941; Pessagno and Whalen 1982; Pessagno and Blome 1986; Smith et al. 1988.
28. Silvies River, Harney Co. Donovan Fm. Dickinson and Vigrass 1965; Imlay 1973; Lupher 1941.
29. Snake River Canyon, Wallowa Co. Coon Hollow Mudstone. Goldstrand 1994; Morrison 1964; Stanley and Beauvais 1990; White and Vallier 1994 .
30. Suplee, Crook and Grant Cos., Mowich Group/Snowshoe/Trowbridge Fms. Dickinson and Vigrass 1965; Fraser, Bottjer, and Fisher 2004; Imlay 1973, 1980, 1981, 1986; Lupher 1941; Pessagno and Blome 1982, 1986; Pessagno, Whalen, and Yeh 1986; Smith 1980; Taylor 1982.
31. Thompson Creek, Douglas Co. Riddle Fm. Diller 1907, 1908.
32. Turner-Albright Mine, Josephine Co. Josephine ophiolite. Pessagno et al. 1993.
33. Umpqua River, Douglas Co. Riddle Fm. Imlay et al. 1959.
34. Wade Butte, Crook Co. Snowshoe Fm. Imlay 1973; Lupher 1941.
35. Waldo (Takilma), Josephine Co. Applegate Group. Irwin, Jones, and Kaplan 1978; Irwin, Wardlaw, and Kaplan 1983.
36. Warm Springs, Grant Co. Mowich Group. Dickinson and Vigrass 1965; Imlay 1986; Lupher 1941; Smith 1980.
37. Wallowa Mountains, Wallowa Co. Hurwal Fm. Follo 1994; Imlay 1968, 1973; Smith and Allen 1941.

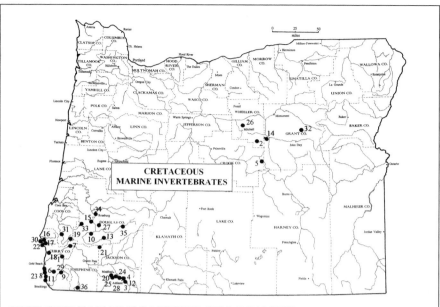

CRETACEOUS MARINE INVERTEBRATE LOCALITIES

The following are Cretaceous marine invertebrate localities and a selection of authors who list faunas from the locales.

1. Agness, Curry Co. Days Creek Fm. Popenoe, Imlay, and Murphy 1960.

2. Antone, Wheeler Co. ?Hudspeth Fm. Jones 1960; Kleinhaus, Balcells-Baldwin, and Jones 1984; Popenoe, Imlay, and Murphy 1960.

3. Ashland, Jackson Co. Hornbrook Fm. Anderson 1905, 1958; Nilsen 1984; Packard 1921; Packard and Jones 1965; Peck, Imlay, and Popenoe 1956; Popenoe, Imlay, and Murphy 1960; Sliter, Jones, and Throckmorton 1984; Ward and Westerman 1977.

4. Bear Creek, Jackson Co. Hornbrook Fm. Gaona 1984; Nilsen 1984; Packard 1921.

5. Beaver Creek, Crook Co. Bernard Fm. Dickinson and Vigrass 1965; Packard 1921; Popenoe, Imlay, and Murphy 1960; Schenck 1936; Squires and Saul 2002.

6. Blacklock Point, Curry Co. Cape Sebastian/Hunters Cove Fm. Dott 1971.

7. Boulder Creek, Curry Co. Dothan Fm. Ramp 1969.

8. Cape Sebastian, Curry Co. Cape Sebastian/Hunters Cove Fms. Bourgeois and Leithold 1984; Dott 1971; Howard and Dott 1961.

9. Chetco River, Curry Co. Dothan Fm. Baldwin 1973; Diller 1907; Ramp 1969.

10. Cow Creek, Douglas Co. Days Creek Fm. Diller and Kay 1924; Imlay et al. 1959; Popenoe, Imlay, and Murphy 1960.

11. Crook Point, Curry Co. Cape Sebastian/Hunters Cove Fms. Dott 1971; Howard and Dott 1961.

12. Dark Hollow, Jackson Co. Hornbrook Fm. Anderson 1958; Nilsen 1984; Popenoe, Imlay, and Murphy 1960; Sliter, Jones, and Throckmorton 1984; Squires and Saul 2002.

13. Days Creek, Douglas Co. Days Creek Fm. Diller and Kay 1924; Imlay et al. 1959; Koch 1966.

14. Dayville, Grant Co. Hudspeth Fm. Jones 1960; Popenoe, Imlay, and Murphy 1960; Squires and Saul 2002.

15. Dillard, Douglas Co. Whitsett Limestone. Anderson 1958; Diller 1898; Johnson 1965; Imlay et al. 1959; Popenoe, Imlay, and Murphy 1960.

16. Elk River, Curry Co. Days Creek Fm. Diller 1903; Imlay and Jones 1970; Koch 1966; Popenoe. Imlay, and Murphy 1960.

17. Elk River, Curry Co. Humbug Mtn./Cape Sebastian Sandstone. Diller 1903; Dott 1971.

18. Euchre Creek, Curry Co. Rock Point Fm. Diller 1903; Imlay and Jones 1970; Koch 1966.

19. Foggy Creek, Coos Co. Days Creek Fm. Baldwin 1973; Imlay et al. 1959 Popenoe, Imlay, and Murphy 1960.

20. 49 Placer Mine, Jackson Co. Hornbrook Fm. Jones 1960; Peck, Imlay, and Popenoe 1956; Popenoe, Imlay, and Murphy 1960; Lupher and Packard 1930; Sliter, Jones, and Throckmorton 1984.

21. Grave Creek, Jackson Co. Hornbrook Fm. Jones 1960; Popenoe, Imlay, and Murphy 1960; Sliter, Jones, and Throckmorton 1984.

22. Humbug Mountain., Curry Co. Humbug Mountain Conglomerate/Rocky Point Fm. Koch 1966.

23. Hunters Cove, Curry Co. Hunters Cove Fm. Dott 1971; Howard and Dott 1961; Koch 1966.

24. Jacksonville, Jackson Co. Hornbrook Fm. Anderson 1958; Nilsen 1984; Packard 1921; Sliter, Jones, and Throckmorton 1984; Squires and Saul 2002.

25. Medford, Jackson Co. Hornbrook Fm. Gaona 1984; Nilsen 1984; Sliter, Jones, and Throckmorton 1984; Ward and Westerman 1977.

26. Mitchell, Wheeler Co. Hudspeth Fm. Jones 1960; Kleinhaus, Balcells-Baldwin, and Jones 1984; Packard 1921; Packard and Jones 1962; Popenoe, Imlay, and Murphy 1960; Schenck 1936; Wilkinson and Oles 1968.

27. Myrtle Creek, Douglas Co. Rocky Point/Days Creek Fm./Humbug Mountain Conglomerate. Anderson 1958; Diller 1907; Imlay et al. 1959; Koch 1966; Popenoe, Imlay, and Murphy 1960.

28. Phoenix, Jackson Co. Hornbrook Fm. Anderson 1958; Lupher and Packard 1930; Peck, Imlay, and Popenoe 1956; Popenoe, Imlay, and Murphy 1960; Squires and Saul 2002.

29. Pistol River, Curry Co. Days Creek Fm. Baldwin 1974; Popenoe, Imlay, and Murphy 1960.

30. Port Orford, Curry Co. Rocky Pt. Fm./Humbug Mountain Conglomerate. Diller 1903; Imlay et al. 1959; Koch 1966.

31. Powers, Coos Co. Days Cr. Fm. Popenoe, Imlay, and Murphy 1960.

32. Prairie City, Grant Co. Bernard Fm. Koch 1966; Mobley 1956.

33. Riddle, Douglas Co. Days Creek Fm. Anderson 1958; Diller 1908; Diller and Kay 1924; Popenoe, Imlay, and Murphy 1960.

34. Roseburg, Douglas Co. Whitsett Limestone. Anderson 1958; Diller 1907; Imlay et al. 1959; Popenoe, Imlay, and Murphy 1960.

35. Umpqua River (south), Douglas Co. Days Creek Fm. Anderson 1958; Imlay et al. 1959; Popenoe, Imlay, and Murphy 1960.

36. Waldo, Josephine Co. Days Creek Fm. Imlay et al. 1959; Popenoe, Imlay, and Murphy. 1960.

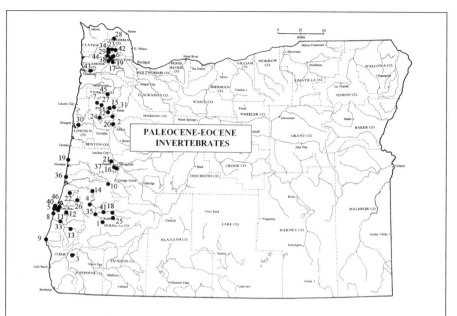

PALEOCENE AND EOCENE MARINE INVERTEBRATE LOCALITIES
The following are Paleocene and Eocene marine invertebrate localities and a selection of authors who list faunas from the locales.

PALEOCENE
1. Roseburg, Douglas Co. Roseburg Fm. Baldwin 1974; Brouwers et al. 1995.

EOCENE
2. Albany, Benton Co. Spencer Fm. Weaver 1942.
3. Agness, Curry Co. Flournoy Fm. Baldwin 1974; Berglund and Feldmann 1989; Kooser and Orr 1973; Orr and Kooser 1971; Turner 1938.
4. Basket Point, Douglas Co. Elkton Fm. Baldwin 1973, 1974; Thoms 1965; Turner 1938; Weaver 1942.
5. Bastendorff Beach, Coos Co. Bastendorff Fm. Baldwin 1973, 1974; Dall 1909; Detling 1946; Dott 1966; Rooth 1974.
6. Buxton, Washington Co. Keasey Fm. Moore 1963, 1976; Warren and Norbisrath 1946.
7. Buell, Polk Co. Yamhill Fm. Baldwin 1964; Boggs, Orr, and Baldwin 1973; Steere, ed. 1977.
8. Cape Arago, Coos Co. Coaledo Fm. Baldwin 1973, 1974; Blake and Allison 1970; Steere 1955, 1977; Turner 1938.
9. Cape Blanco, Curry Co. Roseburg Fm. Bandy 1944.
10. Comstock, Douglas Co. Fisher Fm. Baldwin 1973, 1974; Hoover 1963; Merriam 1942; Turner 1938.
11. Coos Bay, Coos Co. Coaledo Fm. Baldwin 1973, 1974; Burns and Mooi 2003; Detling 1958; Dott 1966; Moore 2000; Rooth 1974; Steere 1955; Turner 1938; Vokes, Norbisrath, and Snavely 1949; Weaver 1942.
12. Coos Bay, Coos Co. Tyee/Elkton Fms. Baldwin 1974; Bird 1967.
13. Coquille River, Coos Co. Roseburg/Lookingglass/Flournoy Fms. Baldwin 1973, 1974; Hoover 1963; Merriam 1942; Miles 1977; Thoms 1965; Turner 1938; Vokes, Norbisrath, and Snavely 1949; Weaver 1942.
14. Elkton, Douglas Co. Tyee/Elkton/Bateman Fms. Baldwin 1973, 1974; Bird 1967; Thoms 1965.

15. Ellendale Quarry, Polk Co. Siletz River Volcanics. Baldwin 1964; Hickman 1976; Steere, ed. 1977; Thoms 1965.
16. Eugene, Lane Co. Eugene Fm. Burns and Mooi 2003; Hickman 1969, 2003; Retallack et al. 2004; Schenck 1923; Shroba 1992; Squires 2003; Washburne 1914.
17. Gales Creek, Washington Co. Keasey Fm. Schenck 1936; Steere 1957.
18. Glide, Douglas Co. Roseburg/Lookingglass Fms. Baldwin 1973, 1974; Brouwers et al. 1995; Hoover 1963; Merriam 1942; Miles 1977; Turner 1938; Weaver 1942.
19. Heceta Head, Lane Co. Yachats Basalt. Shroba 1992; Shroba and Orr 1995.
20. Helmick Hill, Polk Co. Spencer Fm. Baldwin 1964; Cushman, Stewart, and Stewart 1947; Schenck 1936; Steere, ed. 1977.
21. Hyundai Factory (Eugene), Lane Co. Eugene Fm. Retallack et al. 2004.
22. Johnson Creek, Coos Co. Tyee Fm. Allen and Baldwin 1944; Weaver 1942.
23. Keasey, Columbia Co. Keasey Fm. Burns, Campbell, and Mooi 2005; Hedeen 1999; Hickman 1975, 1976, 1980; Moore 1976; Schenck 1936; Steere, ed. 1977; Warren and Norbisrath 1946.
24. Little Luckiamute River, Polk Co. Yamhill Fm. Baldwin 1964; Warren, Norbisrath, and Gravetti 1945.
25. Little River, Douglas Co. Lookingglass Fm. Merriam 1942; Packard 1923; Turner 1938; Weaver 1942.
26. Matson Creek, Coos Co. Tyee Fm. Baldwin 1974.
27. Mill Creek, Polk Co. Tyee/Yamhill Fm. Baldwin 1964; Baldwin et al. 1955; Schenck 1936.
28. Mist, Columbia Co. Cowlitz/Keasey Fms. Adegoke 1967; Burns and Mooi 2003; Burns, Campbell, and Mooi 2005; Hedeen 1999; Hickman 1980; Moore 1976; Moore and Vokes 1953; Schenck 1936; Squires 2003; Warren and Norbisrath 1946; Zullo 1964.
29. Nehalem River valley, Columbia Co. Cowlitz/Keasey Fms. Hedeen 1999; Mumford 1989; Squires 2003; Steere, ed. 1977; Warren and Norbisrath 1946; Warren, Norbisrath, and Grivetti 1945; Weaver 1942.
30. Newport, Lincoln Co. Tyee Fm. Baldwin 1974; McKeel and Lipps 1972; Moore and Moore 2002; Snavely, MacLeod, and Rau 1969.
31. Oregon Portland Cement Co., Polk Co. Yamhill Fm. Baldwin 1964; Boggs, Orr, and Baldwin 1973; Hickman 1976; Schenck 1936; Steere, ed. 1977.
32. Rickreall, Polk Co. Siletz River Volcanics/Yamhill Fm. Allen and Baldwin 1944; Baldwin 1964; Linder 1986; Schenck 1936.
33. Riverton, Coos Co. Coaledo Fm. Baldwin 1973, 1974; Weaver 1942.
34. Rock Creek, Columbia Co. Cowlitz/Keasey Fms. Hedeen 1999; McDougall 1975; Mumford 1989; Niem et al. 1994; Schenck 1936; Steere, ed. 1977; Warren and Norbisrath 1946.
35. Roseburg, Douglas Co. Lookingglass Fm. Baldwin 1974; Brouwers et al. 1995.
36. Sacchi Beach, Douglas Co. Tyee/Coaledo/Elkton Fms. Baldwin 1974; Bird 1967; Stewart 1956.
37. Spencer Creek, Lane Co. Spencer Fm. Vokes, Snavely, and Meyers 1951.
38. Sunset Tunnel, Washington Co. Keasey Fm. Hedeen 1999; Steere, ed. 1977.
39. Timber, Washington Co. Cowlitz/Keasey Fm. Hickman 1975, 1976, 1980; Moore and Vokes 1953; Schenck 1936; Steere, ed. 1977; Warren and Norbisrath 1946; Warren, Norbisrath, and Grivetti 1945; Weaver 1942.
40. Tunnel Point, Coos Co. Tunnel Point Sandstone. Baldwin 1973; Moore 1976.
41. Umpqua River, Douglas Co. Roseburg/Tyee/Bateman Fms. Baldwin 1973, 1974; Thoms 1965; Turner 1938.

42. Vernonia, Columbia Co. Keasey Fm. Hedeen 1999; Hickman 1975, 1976; Moore 1976; Moore and Vokes 1953; Niem et al. 1994; Schenck 1936; Warren and Norbisrath 1946; Warren, Norbisrath, and Grivetti 1945.

43. Wilson River, Tillamook Co. Nestucca Fm. Warren, Norbisrath, and Grivetti 1945.

44. Wolf Creek, Washington Co. Cowlitz/Keasey Fm. Hickman 1974; Moore and Vokes 1953; Warren and Norbisrath 1946.

45. Yamhill River, Yamhill Co. Nestucca Fm. Baldwin et al. 1955.

46. Yoakam Point, Coos Co. Coaledo Fm. Detling 1946, 1958; Steere, ed. 1977.

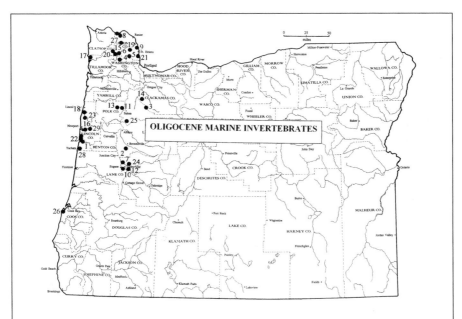

OLIGOCENE MARINE INVERTEBRATE LOCALITIES

The following are Oligocene marine invertebrate localities and a selection of authors who list faunas from the locales.

1. Alsea, Lincoln Co. Alsea/Yaquina Fms. McKeel and Lipps 1975; Moore 1976; Snavely and Vokes 1949; Snavely, MacLeod, and Rau 1969; Snavely et al. 1975.

2. Brownsville, Linn Co. Eugene Fm. Hauck 1962; Hickman 1969; Retallack et al. 2004; Shroba 1992.

3. Butte Creek, Marion/Clackamas Co. Scotts Mills Fm. Burns and Mooi 2003; Durham, Harper, and Wilder 1942; Harper 1946; Linder 1986; Linder, Durham, and Orr 1988; Miller and Orr 1986, 1988; Orr and Miller 1984; Peck et al. 1964; Retallack et al. 2004; Washburne 1914.

4. Buxton, Washington Co. Pittsburg Bluff/Scappoose Fm. Moore 1963, 1976; Warren and Norbisrath 1946.

5. Clatskanie River, Columbia Co. Scappoose/Gries Ranch Fms. Niem et al. 1994; Schenck 1936; Warren and Norbisrath 1946; Weaver 1942.

6. Coal Creek, Columbia Co. Pittsburg Bluff Fm. Moore 1976.

7. Coburg, Lane Co. Eugene Fm. Hickman 1969; Retallack et al. 2004; Shroba 1992.

8. Conyers Creek, Columbia Co. Gries Ranch Fm. Moore 1976; Vokes, Norbisrath, and Snavely 1949; Warren and Norbisrath 1946; Warren, Norbisrath, and Grivetti 1945; Wilkinson 1959.

9. Deer Island, Columbia Co. Gries Ranch Fm. Durham 1942, 1944; Moore 1976; Vokes, Norbisrath, and Snavely 1949.

10. Eugene, Lane Co. Eugene Fm. Hickman 1969, 2003; Moore 1976; Retallack et al. 2004; Schenck 1936; Shroba and Orr 1995; Squires 2003; Steere, ed. 1977.

11. Eola Hills, Polk Co. Eugene Fm. Baldwin 1964; Hickman 1969; McKillip 1992; Schenck 1936; Shroba 1992; Steere 1959; Washburne 1914.

12. Glenwood (Eugene), Lane Co. Eugene Fm. Retallack et al. 2004; Shroba 1992.

13. Holmes Gap, Polk Co. Eugene Fm. Baldwin 1964; Hickman 1969; Schenck 1936; Steere, ed. 1977.

14. Marquam, Clackamas Co. Scotts Mills Fm. Durham, Harper, and Wilder 1942; Harper 1946; Linder 1986; Linder and Orr 1983; Linder, Durham, and Orr 1988; Orr and Miller 1982; Shroba and Orr 1995.

15. Nehalem River valley, Columbia Co. Pittsburg Bluff Fm. Diller 1896; Moore 1976; Moore and Vokes 1953; Steere, ed. 1977; Warren and Norbisrath 1946.

16. Newport, Lincoln Co. Alsea/Yaquina Fms./Nye Mudstone. McKeel and Lipps 1975; Moore 1963, 2000; Snavely et al. 1975; Vokes, Norbisrath, and Snavely 1949; Weaver 1942.

17. Oswald West State Park, Tillamook Co. Smuggler Cove Fm. Diller 1896; Niem et al. 1994.

18. Otter Rock, Lincoln Co. Yaquina Fm. Moore 1963; Steere 1954; Vokes, Norbisrath, and Snavely 1949 .

19. Pittsburg, Columbia Co. Pittsburg Bluff Fm. Burns and Mooi 2003; Diller 1896; Moore 1976; Warren and Norbisrath 1946; Schenck 1936; Washburne 1914; Weaver 1942.

20. Rock Creek, Columbia Co. Pittsburg Bluff Fm. Moore 1976; Schenck 1936; Steere 1957; Steere, ed. 1977; Warren and Norbisrath 1946; Weaver 1942.

21. Scappoose, Columbia Co. Scappoose Fm. Niem et al. 1994; Trimble 1963; Warren and Norbisrath 1946 .

22. Seal Rock, Lincoln Co. Yaquina Fm. Vokes, Norbisrath, and Snavely 1949.

23. Siletz Bay, Lincoln Co. Yaquina Fm. Addicott 1966; McKeel and Lipps 1975.

24. Springfield, Lane Co. Eugene Fm. Hickman 1969; Retallack et al. 2004; Steere 1958.

25. Talbot, Marion Co. Eugene Fm. Hickman 1969.

26. Tunnel Point, Coos Co. Tunnel Point Sandstone. Baldwin 1973; Hickman 1969; Moore 1963; Moore and Vokes 1953; Schenck 1936; Warren, Norbisrath, and Grivetti 1945; Warren and Norbisrath 1946; Weaver 1942 .

27. Vernonia, Columbia Co. Pittsburg Bluff Fm. Moore 1976; Moore and Vokes 1953; Schenck 1936; Warren and Norbisrath 1946.

28. Yachats, Lincoln Co. Alsea Fm/Yachats Basalt. Shroba 1992; Shroba and Orr, 1995; Snavely et al. 1975.

29. Yaquina Bay, Lincoln Co. Alsea/Yaquina Fms./Nye Mudstone. Addicott 1966; McKeel and Lipps 1975; Snavely et al. 1975; Steere 1954; Vokes, Norbisrath, and Snavely 1949; Weaver 1942.

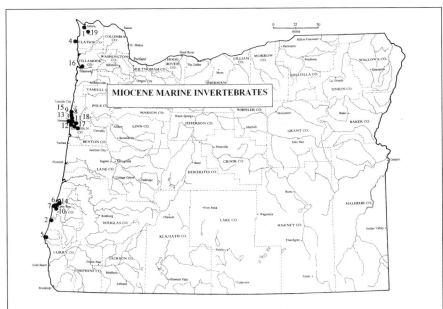

MIOCENE MARINE INVERTEBRATE LOCALITIES

The following are Miocene marine invertebrate localities and a selection of authors who list faunas from the locales.

1. Astoria, Clatsop Co. Astoria Fm. Addicott 1964; Colbath 1985; Dall 1909; Dodds 1963; Moore 1963.

2. Bandon, Coos Co. Empire Fm. Ehlen 1967; Terry 1968.

3. Beverly Beach, Lincoln Co. Astoria Fm. Colbath 1985; Moore 1963; Moore and Moore 2002.

4. Cannon Beach, Clatsop Co. Astoria Fm. Moore 1963; Moore and Vokes 1953; Washburne 1914.

5. Cape Blanco, Curry Co. Floras Lake Sandstone/Empire Fm. Addicott 1980, 1983; Bourgeois and Leithold 1984.

6. Coos Bay, Coos Co. Tarheel/Empire Fms. Addicott 1980, 1983, Armentrout 1967; Dall 1909; Diller 1901; Howe 1922; Moore 1963, 2000; Weaver 1942.

7. Coos Head, Coos Co. Empire Fm. Armentrout 1967.

8. Depoe Bay, Lincoln Co. Astoria Fm. Moore 1963; Steere 1954 .

9. Fogarty Creek, Lincoln Co. Astoria Fm. Moore 1963; Steere 1954; Steere 1954, 1977.

10. Fossil Point, Coos Co. Tarheel/Empire Fms. Armentrout 1967; Baldwin 1973; Dall 1909; Diller 1901; Moore 1963; Steere, ed. 1977; Terry 1968; Weaver 1942, 1945.

11. Moolack Beach, Lincoln Co. Astoria Fm. Rooth 1987.

12. Newport, Lincoln Co. Astoria Fm. Moore 1963, 1971; Moore and Moore 2002; Packard and Kellogg 1934; Schenck 1936; Snavely, Rau, and Wagner 1964; Snavely, MacLeod, and Rau 1969; Steere 1954; Weaver 1942.

13. Otter Rock, Lincoln Co. Astoria Fm. Addicott 1966; Diller 1896; Moore 1963; Steere 1954; Vokes, Norbisrath, and Snavely 1949.

14. Pigeon Point, Coos Bay, Coos Co. Tarheel/Empire Fms. Armentrout 1967; Moore 1963; Weaver 1945.

15. Spencer Creek, Lincoln Co. Astoria Fm. Moore 1963; Packard and Kellogg 1934; Steere 1954.

16. Tillamook, Tillamook Co. Astoria Fm. Dall 1909; Moore 1963; Moore and Moore 2002; Washburne 1914.
17. Wade Creek, Lincoln Co. Astoria Fm. Moore 1963; Packard and Kellogg 1934.
18. Yaquina Bay, Lincoln Co. Astoria Fm. Diller, 1896; McKeel and Lipps 1975; Moore 1963; Snavely, Rau, and Wagner 1964; Snavely et al. 1964; Steere, ed. 1977; Stewart 1956; Vokes, Norbisrath, and Snavely 1949; Weaver 1942.
19. Youngs River, Clatsop Co. Astoria Fm. Frey and Cowles 1972.

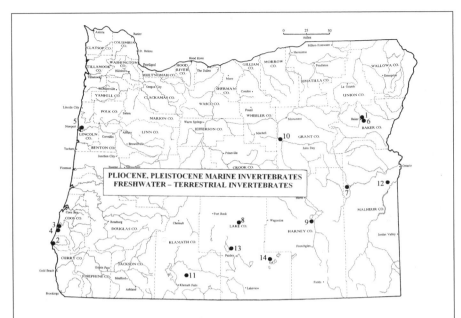

PLIOCENE, PLEISTOCENE MARINE INVERTEBRATES
FRESHWATER – TERRESTRIAL INVERTEBRATES

PLIOCENE AND PLEISTOCENE MARINE AND FRESHWATER INVERTEBRATE LOCALITIES

The following are Pliocene and Pleistocene marine invertebrate localities and a selection of authors who list faunas from the locales.

PLIOCENE

1. Always Welcome Inn (Baker City), Baker Co. No Fm. Van Tassell et al. 2007.

PLEISTOCENE

2. Cape Blanco, Curry Co. Elk River/Port Orford Fms. Addicott 1964; Baldwin 1945; Bandy 1944, 1950; Clifton and Boggs 1970; Diller 1902, 1903; Moore 2000; Muhs et al. 1990; Roth 1979.
3. Coquille Point, Coos Co. No Fm. Addicott 1964; Muhs et al. 1990; Zullo 1969.
4. Grave Point, Coos Co. No Fm. Addicott 1964; Zullo 1969.
5. Newport, Lincoln Co. No Fm. Coquille Fm. Baldwin 1950 .

FRESHWATER

6. Always Welcome Inn (Baker City), Baker Co. No Fm. Pliocene. Van Tassell et al. 2007.
7. Black Butte, Malheur Co. Juntura Fm. Miocene. Taylor 1960.
8. Fossil Lake, Lake Co. No Fm. Pleistocene. Allison 1966; Taylor 1960.
9. Harney Lake, Harney Co. No Fm. Pleistocene. Hanna 1922, 1963.
10. John Day River Valley, Grant Co. John Day/Mascall Fms. Oligocene/Miocene. Hanna 1920, 1922; Stearns 1902, 1906.
11. Wilsons Quarry Pit, Klamath Co. Yonna Fm. Pliocene. Newcomb 1958.
12. Mitchell Butte, Malheur Co. No Fm. Miocene. Taylor 1963.
13. Summer Lake, Lake Co. No. Fm. Pleistocene. Hanna 1922, 1963.
14. Warner Lake, Lake Co. No. Fm. Pleistocene. Hanna 1922, 1963.

ARTHROPODS

TRILOBITES, CRABS, AND INSECTS

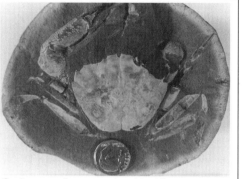

Crabs are often preserved inside concretions where the legs, carapace, and even antennae are usually intact. (Photo courtesy W. Orr.)

The vast array of living arthropods includes creatures such as crabs, shrimp, lobsters, barnacles, scorpions, spiders, and insects, a group that has been eminently successful in dominating the planet. In the words of paleontologist Don Prothero, "[B]y almost any criterion, the flies, cockroaches, and other arthropods rule the earth." However, Oregon's fossil record doesn't reflect this great diversity. Trilobites, a fossil arthropod common during the Paleozoic Era, are among the most sought after by collectors, but they have been found in only one locality in the state. Insects are the single most diverse group of arthropods, but they are similarly limited as fossils in Oregon, while spiders, mites, and so on are unknown. Crustacea, shrimp-like arthropods, are the most abundant and occur across western Oregon in the same marine strata as the mollusks. Lesser numbers of crayfish inhabited freshwater streams and ponds of eastern Oregon during the Tertiary.

The oldest and most rare arthropods in Oregon are fingernail-sized trilobites from the Permian Coyote Butte Formation, part of the Grindstone terrane. First discovered in 1967, trilobites proved too fragmentary to be named specifically. However, in 2000 close to 40 trilobite specimens from near Twelvemile Reservoir in Crook County were identified as *Cummingella oregonensis*. The head (cephalon), tail (phygidium), thoracic segments, and spines are preserved as silicified pieces. That is, they have been replaced by silica. Extracting the fossils from the entombing limestone requires the careful immersion of the rock in a weak acid solution to dissolve the matrix. Found along with microfossils, corals, crinoids, and brachiopods, they confirm a shallow oceanic shelf environment for Coyote Butte strata.

Remains of crustaceans are numerous in the state, in contrast to the record for other arthropods. This extremely diverse collection of shrimp, crabs, crayfish, barnacles, and sow bugs lives in an assortment of environments from fresh and marine water to land. By far, the most common of these in the Oregon fossil record are the Decapoda, or lobsters, crabs, and shrimp. In the coastal waters today these ten-legged crustaceans successfully occupy a striking variety of ecologic niches. Some are ocean bottom dwellers, where they scavenge for food, but many swim intermittently to feed on plants and small invertebrates. Decapods display a remarkable range of variations, and many have the tail or legs modified for swimming. Most species are easily sexually differentiated by the body size or shape of the segmented tail that typically is thin and pointed on males and broad and rounded on females.

Crabs and shrimp go through multiple molting or "ecdysis" stages as they grow and add segments. Each new growth stage is accompanied by peeling away the old shell and the hardening in place of a new one. While the castoff body armor, or carapaces, are often destroyed by wave action or by other scavengers, they may be preserved as fossils. The remains of crabs frequently occur in dense masses of sedimentary rock called "concretions," which actually form around the fossil after death and burial. These are rounded pebble-like accumulations of mineral material where a particle such as wood or shell acts as a center of cementation. The process of forming concretions is like making candles by repeatedly immersing a wick (the fossil) in hot wax (the cement). Cementing material, such as calcium carbonate in solution moving through porous rock, will accumulate around a crystallization point such as a fossil. Ordinarily concretions are harder than the entombing material, and may collect as cobbles at the base of a slope as the surrounding rock weathers and crumbles. To expose the fossils, concretions are split along the widest diameter with sharp blows of a hammer. Crustaceans preserved in concretions require little preparation, and often a vibra-drill tool can be used to improve the exposure. In the case of crabs, all of the limbs and even hair and antennae may be preserved.

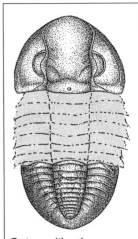

Quite unlike those from the mid-continent of North America, the diminutive trilobite *Cummingella* is of Eurasian origins, further demonstrating the exotic nature of the Grindstone terrane. These arthropods became extinct by the end of the Permian period, a quarter of a billion years ago. (After Hahn et al. 2000.)

Parts and pieces of appendages, carapaces, and spines of crabs, extracted from fine-grained sandstones and siltstones of the Cretaceous Days Creek Formation, are among Oregon's oldest crustaceans. These specimens from near Agness were collected by Joseph Diller of the U.S. Geological Survey during the late 1800s. Diller placed them in the Smithsonian Museum, and over 50 years later they were examined by Rodney Feldmann, who realized they represented a new species. Naming them *Hoploparia riddlensis,* Feldmann concluded that this was the earliest occurrence of the family Nephropidae in North America.

From the same location, several species of crabs were named in the 1970s by Professor William Orr of the University of Oregon. *Lophopanopeus baldwini*

The most ambitious work on crustaceans of the United States was by Mary Rathbun, who was one of the exceptional women in American science. From a distinguished New England family, Rathbun was employed at the U.S. National Museum to organize and catalog their huge collection of marine animals in 1884. Hence, she became involved in decapods from around the world, and at age 66 she completed her monograph, *The Fossil Stalk-Eyed Crustacea of the Pacific Slope of North America*. Rathbun continued to work until her death in 1943. (Photo courtesy Smithsonian Institution, Archives.)

Rodney Feldmann of Kent State University is another among the small number of paleontologists specializing in fossil crustaceans. Feldmann's research on the paleoecology and biogeography of crabs treats the global distribution of populations. A native of North Dakota, Feldmann received his Ph.D. in paleontology in 1967 from the University at Grand Forks. In addition to decapod biogeography and paleoecology, he has also published extensively on Antarctic invertebrates. (Photo 2007; courtesy R. Feldmann.)

Similar in size to a modern crayfish, *Hoploparia riddlensis* occurs in Cretaceous rocks of southwest Oregon.

and *Zanthopsis rathbunae* were discovered in the early Eocene Lookingglass Formation, while *Plagiolophus weaveri* and *Raninoides washburnei*, along with *Cancer*, were recorded in the Flournoy. The crustaceans were embedded in concretions an inch or two in diameter, and the most common species, *Plagiolophus weaveri,* was usually intact. Yet another new ranid crab, *Rogueus orri ,* occurring at the same location, was named by Rodney Feldmann and Ross Berglund, a Boeing engineer from Washington state. Berglund was presented the Strimple Award by the Paleontology Society for his outstanding work as a private paleontologist. Superficially ranid crabs resemble crayfish, having an elongate carapace but with spines on the front edge. *Rogueus orri* is characterized by peculiar paired forked spines. Fossil ranid crabs of this type are rare in deposits worldwide, because they often live offshore in deep water where chances of preservation are diminished. A fauna of clams, snails, and small echinoderms, also from the Agness locale, mark the boundary of the early Eocene Flournoy seaway that had diminished to a small area in southwest Oregon.

Throughout the early Tertiary Period, a subsiding ocean basin across what is now the Oregon Coast Range was filled with layers of sediments as the shape and depth of the trough and shoreline changed before the province was raised by tectonic uplift. This paleosetting was home to a great variety of marine animals, and arthropod remains are present in most of the sediments from Eocene through the Pleistocene.

Eocene Decapod remains are present in coastal deposits of the Elkton, Nestucca, Yamhill, and Keasey formations. With the exception of the deep-water Keasey, all represent uppermost shelf conditions. Claw fragments of the ghost shrimp, *Callianassa*, are present at Elkton, while near Dallas new species of *Zanthopsis perforatus*, *Raninoides dallasi*, and *Ranina giganteus* are to be found in the Yamhill Formation. From the northwest corner of the state, Rodney Feldmann listed the crabs *Brachioplax, Eumorphocorystes, Portunites, Raninoides*, and *Zanthopsis*, and an isopod (sow bug) from bathyal levels of the Keasey ocean. He also noted 44 specimens of *Lyreidus alseanus*, a ranid crab from the Nestucca Formation near Waldport and on Teal Creek in Tillamook County. This crustacean first evolved in shallow water before moving into deeper marine environments.

Oligocene crustaceans can be obtained along the coast as well as inland in the Willamette Valley. The Yaquina Formation near Newport bears *Zanthopsis vulgaris* as the most commonly occurring crab, with lesser specimens of *Eucrate, Mursia*, and *Portunites*. These arthropods inhabited offshore shelf and even slope environments. A varied fauna from the Eugene Formation in the Willamette Valley was originally described by Mary Rathbun as *Calappa, Persephona, Palehomola, Portunites, Raninoides,* and *Zanthopsis*. More recently, Rodney Feldmann identified the ghost shrimp (*Callianassa oregonensis*), the mud crab (*Raninoides washburnei*), and the swimmer crab (*Megokkos magnaspina*) as distributed in the shallow-water tropical Eocene strata of the Eugene Formation. In contrast, *Calappa laneensis, Persephona bigranulata,* and *Zanthopsis vulgaris* are from the more temperate upper Eugene strata located near the Springfield railroad junction. Greg Retallack places these Oligocene crabs in an offshore setting near a rocky coast with pocket beaches. Along with mollusks, barnacles, echinoderms, and fish, these crabs were inhabitants of the Tertiary ocean that occupied the Willamette Valley as far south as Eugene.

Miocene arthropods are found on both sides of the Cascades. Along the coast, marine crustaceans were occupants of the warm shallow seaway, whereas in eastern Oregon freshwater arthropods occupied the lakes and ponds dotting the Columbia River Basalt plateau. In one of the earliest references to Miocene crustaceans, a single ghost shrimp, *Callianassa oregonensis*, and barnacle fragments from the Astoria Formation were reported by James Dana, who accompanied the Wilkes Exploring Expedition from 1838 to 1842. Ellen Moore, in her classic paper on Astoria invertebrates, records the mud shrimp *Upogebia*. This worldwide stalk-eyed crustacean, which inhabits rock burrows, has been found from the Jurassic to the present, but Moore's report from the Astoria is the first from Miocene rocks. Thirteen new decapods, including *Cancer, Macrocheira*, and *Tymolus*, reported by Rodney Feldmann and his students, show that crustaceans maintained a high diversity through the warm middle Miocene, after which cooler water circulation brought on a decline later during that epoch in the northeastern Pacific Ocean.

During the Miocene epoch, streams, lakes, and ponds of eastern Oregon served as sites of outstanding preservation for animals and plants, and broken crayfish remains are not uncommon at the numerous freshwater lakes throughout the area. *Astacus chenoderma*, a crayfish, cited by Mary Rathbun from near Vale and Malheur City, occurs in probable Juntura Formation sediments. Arnold Shotwell mentions crayfish from the Juntura basin sediments but does not provide identifications.

Pleistocene barnacles and crabs at Cape Blanco were the focus of detailed studies by Victor Zullo, a geology professor at the University of North Carolina. In one unique occurrence, he noted that the tiny pea crab, *Pinnixa faba*, lived in a symbiotic association within the valves of the clam, *Tresus capax*. Symbiotic relationships of this type are extremely rare finds in the fossil record because of postmortem dispersal by waves. The Cape Blanco clams, preserved upright in "life position" with valves intact, were apparently

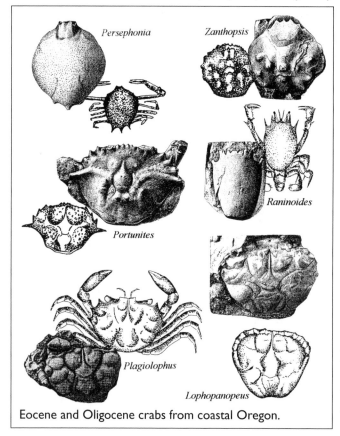

Persephonia

Zanthopsis

Raninoides

Portunites

Plagiolophus

Lophopanopeus

Eocene and Oligocene crabs from coastal Oregon.

overcome and killed quickly, followed by a rapid burial in what was a calm embayment.

Slightly south and west of Bandon in fossiliferous pockets at Coquille Point and Grave Point, pelecypods, gastropods, crustaceans (*Cancer*, *Pugettia*), barnacles (*Balanus*), echinoderms, and bryozoa inhabited what were the shallow protected reaches of a cool water rocky coast during the Pleistocene. Based on the boreal (cool) components of the fauna, Zullo judged that today the animals would be residents of south Alaska or British Columbia.

Barnacles are sessile (attached) crustaceans, with a circular wall or armor of plates, commonly associated with mollusks in Tertiary marine sediments. The distinctive grooved triangular plates are easily recognized and make up significant portions of some sediments. In the Marquam Member of the Scotts Mills Formation they make up over 75 percent of the limestone lenses that characterize this late Oligocene unit in Clackamas County. Northwest of Salem, small barnacles *Balanus* are clustered together on fossilized wood in deposits mapped as the Eugene Formation. Barnacles live in shallow water, where their remains are scattered and show beach wear; however, they also

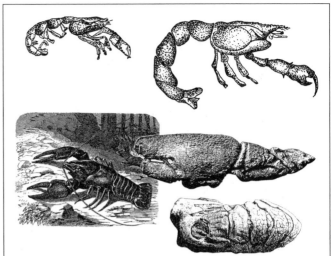

During the Miocene, the ghost shrimp *Callianassa* (left), and mud shrimp *Upogebia* (right) lived along the Oregon coast, whereas crayfish such as *Astacus* (below) inhabited freshwater streams and lakes in eastern Oregon. (After Steele and Jenks 1887; Rathbun 1926.)

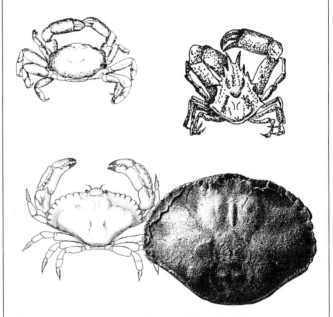

The tiny pea-crab *Pinnixa* (top, left) lived within the valves of a large clam. *Pinnixa*, along with the dungeness crab *Cancer* (bottom) and the kelp crab *Pugettia* (top, right) were common on the southern Oregon coast during the Ice Ages. (After Rudy and Rudy 1983.)

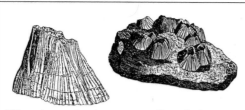

Worn barnacle plates may reflect their shallow marine habitat in the high energy surf zone. (From Zittel 1913.)

inhabit areas of great depths, and their presence is not proof of a particular ecological niche.

Ostracods, or "bean bugs," are among the smallest of fossil arthropods, measuring less than one-twenty-fourth inch in length. Similar to a flea in appearance and size, ostracods come equipped with a defensive clam-like shell of paired valves. They live in both marine and freshwater sediments, routinely turning up in the extraction process for foraminifera. In the North American Gulf Coast they have been used extensively in the petroleum industry for age and paleoenvironment analyses. Eye spots or eye tubercles on the ostracods are indicators of water depths within the light (photic) zone of the continental shelf. At less than 600 feet, eye structures are elaborate, while in deeper levels these features are lacking.

Another group of arthropods, the isopods, which include sow bugs and wood lice, have invaded a wide variety of environments on land as well as in both fresh and marine waters. In spite of being pervasive, their fragile carapace makes them exceedingly rare as fossils. A single sow bug was reported in 1964 from the Eocene Keasey Formation by H. K. Brooks, a specialist at the University of Florida, who wrote that "the specimens are the remains of isopods. No other fossils have been described from the United States though they are very common today. . . . Specimens have been reported from Mesozoic and Cenozoic of Europe."

The few recorded examples of fossilized insects in Oregon were in Clarno and John Day deposits where the remains are frequently found among those of plants. The discovery of these arthropods is exceptional, and their past record, like that of snakes, toads, and frogs, doesn't begin to match their present-day widespread abundance and diversity. Insects feeding on wood and leaves tend to leave visible evidence behind. Tubular burrows or traces are signs that wood has been bored by insects, and when the tunnels have been subsequently filled in with silica during

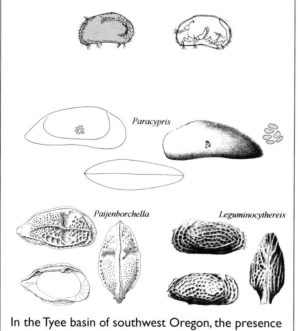

Paracypris

Paijenborchella   *Leguminocythereis*

In the Tyee basin of southwest Oregon, the presence of key ostracods, mollusks, and coccolith floras delineates Paleocene and Eocene stratigraphy.

the petrification process, details of the insect passageway are perfectly preserved. Many fossil soils (paleosols) show where insects, such as beetles, have dug to the silty clay to leave vertical unbranched tubular burrows or traces.

In the 1920s paleobotanist Ralph Chaney collected gall midges on leaves of maple and alder, the wings of dragonflies (Odontata), the jaws of tropical sugar beetles (Coleoptera), and common midges (*Cecidomyia*) at Post in Crook County in the Eocene Clarno Formation. Some 70 years later Steve Manchester noted small, oblong capsules from the Clarno Nut

Photomicrograph of silicified poplar wood showing a tunnel filled with chips made by a wood borer in an Eocene Clarno specimen. (After Gregory 1969b; 2.5x enlargement.)

Beds, which turned out to be the eggs of stick insects (*Eophasma*). The one-eighth-of-an-inch-long fossils, representing three species, were completely replaced by silica. In other instances, rounded sand-sized pellets have been interpreted as the fecal remains from termites, whereas elliptical clusters, which are abundant in the Clarno, are of uncertain origin. They may also have been made by termites or ants. Tiny chips from wood borers, that filled tunnels in Eocene *Populus* (poplar) and *Salix* (willow) specimens from the Clarno, can be found when preserved in silica. The strong mandibles of the borers carved out crescent-shaped pieces of wood that were then deposited in the burrow before being sealed in during the petrifaction process.

Signs of insect activity on fossil leaves in John Day strata have also been reported at Dugout Creek (Iron Mountain) in Wheeler County. *Trichoptera*, caddis-fly larval cases, consist of tiny cylinders which are frequently constructed from leaves and pieces of *Metasequoia* (dawn redwood) needles or grains of sand. Random patterns, made by larvae, are often visible on fossil leaves of the middle Big Basin Member of the John Day Formation. Indications of such leaf mining extend back to the Paleozoic Era in other regions of the United States.

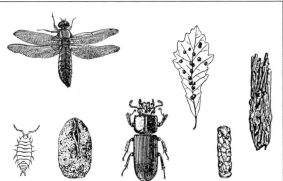

The variety of insects in Oregon's fossil record is represented by a dragonfly, leaf galls, caddis fly larva enclosed in *Metasequoia* needles, caddis fly larva with sand grains, a sugar beetle, the egg of *Eophasma*, a stick insect, and an isopod (clockwise from left). (After Cockerell 1927; Lutz 1935; Peterson 1964; Sellick 1994.)

The leaves of *Quercus simulate* (oak) from Miocene shales of the Blue Mountain flora near Tipton, as well as in Nevada, have been discovered with galls still attached. The newly described gall *Antronoides cyanomontanus* is elongate, spindle-shaped, and about 7.0 millimeters long. It's produced when oak gall wasps (Cipinidae) penetrate the leaves to lay eggs. Adjacent to the elongate galls, circular areas on the leaves may represent places where the leaf reacted to the wasp action by blocking the gall from forming or where fungi and bacteria were present.

The most unique insects are often those captured and entombed by tree sap in miniature tar-pit-like environments. All manner of airborne particles, most commonly pollen and small seeds, are also found in tree sap along with insects. In time, a process of slow drying chemically converts the sap to amber. Although there have been many possible candidates for specimens in amber, to date no examples of this mode of preservation are known from Oregon.

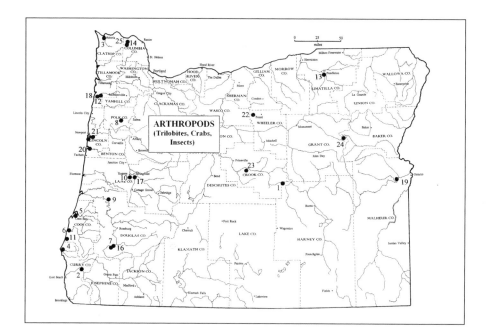

ARTHROPODS
(Trilobites, Crabs, Insects)

ARTHROPOD LOCALITIES
The following is a list of Oregon fossil arthropod (trilobites, crabs, insects) localities and a selection of authors who list faunas from the locales.

TRILOBITES
1. Twelvemile Creek (Berger Ranch), Harney Co. Coyote Butte Fm. Permian. Hahn et al. 2000.

CRABS
2. Agness, Curry Co. Days Creek/Lookingglass/Flournoy Fms. Cretaceous/Eocene. Berglund and Feldmann 1989; Feldmann 1974; Kooser and Orr 1973; Orr and Kooser 1971.
3. Astoria, Clatsop Co. Astoria Fm. Miocene. Dall 1909; Moore 1963; Rathbun 1926.
4. Cape Blanco, Curry Co. Elk River Beds. Pleistocene. Nations 1970; Zullo and Chivers 1969.
5. Coos Bay, Coos Co. Empire Fm. Miocene. Nations 1970; Weaver 1942.
6. Coquille Point, Coos Co. No. Fm. Pleistocene. Zullo and Chivers 1969.
7. Cow Creek, Douglas Co. Days Creek Fm. Cretaceous. Feldmann 1974.
8. Dallas, Polk Co. Yamhill Fm. Eocene. Baldwin 1964; Stokesbary 1933.
9. Elkton, Douglas Co. Elkton Fm. Eocene. Rathbun 1926; Weaver 1942.
10. Eugene, Lane Co. Eugene Fm. Eocene. Hickman 1969; Rathbun 1926.
11. Grave Point, Coos Co. No Fm. Pleistocene. Zullo and Chivers 1969.
12. Hebo, Tillamook Co. Nestucca Fm. Eocene. Feldmann 1989.
13. McKay Reservoir, Umatilla Co. McKay Fm. Miocene. Rathbun 1926..
14. Mist, Columbia Co. Keasey Fm. Eocene. Rathbun 1926; Steere 1957; Zullo 1964.
15. Newport, Lincoln Co. Yaquina/Astoria Fms. Oligocene/Miocene. Nyborg 2002; Rathbun 1926.
16. Riddle, Douglas Co. Days Creek Fm. Cretaceous. Feldmann 1974.
17. Springfield Junction, Lane Co. Eugene Fm. Hickman 1969; Retallack et al. 2004; Weaver 1942.
18. Teal Creek, Tillamook Co. Nestucca Fm. Eocene. Feldmann 1989.
19. Vale, Malheur Co. No Fm. ?Miocene. Rathbun 1926.
20. Waldport, Lincoln Co. Nestucca Fm. Eocene. Feldmann 1989.
21. Yaquina, Lincoln Co. Yaquina Fm. Oligocene. Rathbun 1926; Weaver 1942.

INSECTS
22. Iron Mountain, Wheeler Co. John Day Fm. Oligocene. Peterson 1964; Retallack, Bestland, and Fremd 2000; Sellick 1994.
23. Post, Crook Co. John Day Fm. Oligocene. Cockerell 1927.
24. Tipton, Grant Co. Columbia River Basalt. Miocene. Erwin and Schick 2007 .

ISOPOD
25. Mist, Columbia Co. Keasey Fm. Eocene. Peterson 1964.

AMPHIBIANS AND REPTILES

Sea creatures frolicking in an imaginary ocean. Although ichthyosaurs (left) have been known from Oregon for over a century, plesiosaurs (right) were only discovered here in 2004. (From Fugier 1891.)

Amphibians and reptiles are treated together because they constitute the intermediary vertebrate stage between fish and mammals. The shift from aquatic to terrestrial habitats began in the Devonian some 400 million years ago with the appearance of ancestral amphibians, an evolution completed in the Paleozoic 250 million years in the past with the emergence of primitive reptilians. Birds and mammals evolved from these reptiles during the early Mesozoic. By the Cenozoic, changes in aquatic amphibian and reptile diversity in the western interior of North America began with a sharp decline across the Eocene-Oligocene boundary. The decrease in aquatic species was probably related to a developing arid climate, where flowing streams and lakes diminished.

Although the evolutionary stage between fish and amphibians is critical, fossil evidence of this event is almost nonexistent. In contrast, the transition from reptiles to mammals is well documented. Preserved remains of both amphibians and reptiles in Oregon are decidedly spotty since the preservation of the smaller vertebrates poses special problems. Predators and scavengers may simply eat the entire carcass of the animal, destroying any chance that it will become part of the fossil record. In addition, the fragile nature of the small bone material for both amphibians and reptiles works against a complete record of either group. The skeletons easily disarticulate, and whole specimens are rare.

Amphibians

Fossil evidence of amphibians, such as frogs, toads, salamanders, and newts, is scarce in Oregon's record. Of these, the salamanders and newts most closely resemble the earliest primitive amphibians. Frogs and toads are comparatively recent additions and did not appear until quite late in amphibian history. Even though they are key indicators of wetlands, frogs are mentioned only in passing by researchers when investigating vertebrates. The delicate bones of the frog *Rana* are not uncommon in Tertiary sediments, but most are too fragmentary for identification. Their disarticulated remains are frequently recovered from Miocene lakebeds of the Rattlesnake, McKay, and Juntura formations of eastern Oregon, and they are plentiful at the Pleistocene Woodburn bog site in Marion County.

By contrast with frogs, accounts of salamanders are more complete. In spite of their fragile nature, which makes them among the rarest of fossils, there are several salamander remains from both western and eastern Oregon. *Palaeotaricha oligocenica*, a new Oligocene species, was recovered from dark-gray, leaf-bearing shales of the Fisher Formation near the town of Goshen in Lane County. Although lacking a small part of the

The rare Eocene salamander *Palaeotaricha*. It took months of painstaking preparation with a needle and drill to expose the fragile intact skeleton. (After Van Frank 1955; photo courtesy Condon collection.)

tail and several bones, the black carbonized five-inch-long articulated skeleton is exquisitely preserved. Since that discovery, the Goshen locale has yielded a number of additional specimens. Known for its outstanding plant preservation, the early Oligocene interval of the Fisher represents forest floor, stream, and lake settings that would have been attractive to amphibian life. Exposures of the Fisher along the highway were cut back in 2007, revealing the uniform thin beds of a lake. In carbon-rich blocks, a salamander was discovered in 2008 that clearly displays the outline of fleshy tissue on the limbs, tail, and body.

Because John Day Formation habitats were comparable to those at Goshen, amphibians are similarly found in conjunction with fossil leaves. Near Fossil in Wheeler County, a new species of salamander was collected in 1977. Around two inches in length, the amphibian *Taricha lindoei* consists of a nearly complete skeletal impression.

Reptiles

Reptiles are cold-blooded vertebrate animals, which have more in common with birds than with frogs and salamanders. Unlike amphibians, most reptiles do well on land and breathe with lungs. Many, such as snakes and lizards, have long retractable tongues that provide an important organ for smell. Reptiles appear in the fossil record during the Carboniferous Period, more than 300 million years ago. They quickly diversified and became dominant during the Mesozoic Era, when dinosaurs, the largest of all reptiles, reached remarkable sizes and numbers, second only to the blue whale in geologic history. Dinosaurs were extinguished about 65 million years ago, while reptiles adapted to an extraordinary degree, fully exploiting terrestrial environments as well as successfully invading and populating the oceans. Pterosaurian reptiles even solved the complicated equation of heavier-than-air flight by evolving splendid air foils (wings) for gliding.

The majority of Oregon's fossil reptiles are represented by aquatic animals such as turtles, crocodiles, plesiosaurs, and ichthyosaurs that lived, died, and were entombed in a variety of settings from lakebeds and shallow marine sands to offshore deep water. The few terrestrial reptilian fossils are limited to single occurrences of a dinosaur, a lizard, and a pterosaur, whereas marine, freshwater, and terrestrial turtles are far more numerous.

Despite our current knowledge of modern marine turtles, information about their past, which dates back to the early Cretaceous Period, is not well documented, and their remains from the Mesozoic are even more rare than those of ichthyosaurs. Skeletal material tends to be intact, although individual bony plates of the shell are often scattered with other vertebrate bones. Marine turtles in Oregon are recorded from Eocene and Miocene sediments, where they mark the ocean shoreline along the east side of the Willamette Valley.

The most recently unearthed, but oldest, marine turtles from the state are from the Tyee and Coaledo formations. An 18-inch-long carapace (shell), complete with skull, limb bones, and breast plates (plastron) of a forerunner of green turtles, was preserved in 45-million-year-old Coaledo sediments at Coos Bay. Removed from

Earl Packard, shown inspecting the skull of a leatherback turtle from the Miocene Astoria Formation, was a professor first at the University of Oregon and then at Oregon State University. He promoted paleontology research in the state for more than 40 years. His diverse interests ranged from Paleozoic invertebrates to Tertiary vertebrates. For most of his teaching career Packard remained in the Oregon system, employed as a geology professor at Eugene before becoming Dean of Sciences after the State Board of Higher Education transferred those departments to Corvallis in 1932. Retiring in 1946, he moved to Palo Alto, where he died in 1983. (Photo courtesy Archives, Oregon State University.)

a cliff at Cape Arago in 1985, it was identified by William Orr at the University of Oregon as belonging to the order Chelonia, or green turtles. A decade later, a similar green turtle turned up in the Tyee Formation near Scottsburg in the southern Coast Range. The Tyee find is significant because, as a submarine deepwater fan, this formation is notably poor in fossils of any kind. By contrast, the Coaledo delta, which was constructed within a partially enclosed basin, is rich in invertebrates.

Northward along the coast, turtle fragments are not uncommon in the Miocene Astoria and Nye formations. One of the few specimens of a well-preserved skull and shell (carapace) fragments of a marine turtle was recovered in sandstones of the Astoria in 1940 by Earl Packard. The skull of the *Psephophorus oregonensis* measured over a foot in length, suggesting the animal may have reached well over six feet. Thick bony plates of the carapace fit tightly together like a mosaic, while the skull is substantially more narrow and elongate than that of the living counterparts. The family of this Miocene turtle, Dermochelydae, includes the modern leatherbacks that attain a length of eight feet and a weight over half a ton.

An ancient sea turtle from the 45-million-year-old Coaledo sediments on the south coast is beautifully preserved. (Photo courtesy W. Orr.)

Freshwater and terrestrial turtles are more common in eastern Oregon, where ongoing volcanic activity and numerous stream and lake basins of the Eocene and Miocene enhanced preservation. Entire carapaces of freshwater pond and box turtles have been found in lake sediments of the Miocene Rattlesnake and Mascall beds. Although fragmentary, breast plates and carapaces of four specimens were identified as *Clemmys*. Only one of the four was completely aquatic, while the others were partially terrestrial. The incidence of *Clemmys owyheensis* from near Rome in Malheur County is indicative of a wetlands and bird habitat once typical of this desert region.

Land turtles or tortoises typically occur in Tertiary sites where mammal bones and leaves are present. Fragments of land turtle skeletons have a distinctive shape but are difficult to identify with confidence unless most of the shell is present. The single plate of a thick-shelled tortoise *Hadrianus* is among those from the Eocene Clarno Nut Beds collected by Professor Edward Cope in the late 1880s. In some cases, complete turtle carapaces have been encountered in the Oligocene John Day intervals, although generally only small pieces of scutes, or plates, are found. Several entire shells of the aquatic *Stylemys capax*, up to two feet in length, were

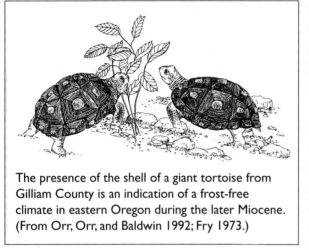

The presence of the shell of a giant tortoise from Gilliam County is an indication of a frost-free climate in eastern Oregon during the later Miocene. (From Orr, Orr, and Baldwin 1992; Fry 1973.)

collected at the famous Turtle Cove in Grant County by Cope. These locales now lie within the John Day National Monument, which is closed to public collecting.

At three feet in length, two giant land tortoises from near Arlington in Gilliam County rival those living in the Galapagos Islands today. In 1971 the massive, nearly intact, and mineralized carapace, limbs, and lower breastplate of a *Geochelone orthopygia* were found at two separate locations, one in a pumice quarry and the other in a scablands channel. Of probable middle Miocene age, the turtles were uncovered together with the bones of an unidentified horse, camel, mastodon, and rodents. These huge turtles may have become extinct as the eastern Oregon climate cooled rapidly during the late Tertiary.

Lizards and snakes are the most widely distributed of the modern reptiles, and their ancestry can be traced as far back as the Permian and Triassic. However, owing to their fragile skeleton and conditions unfavorable to preservation, they have a scanty and incomplete record. Vestiges of just one snake and one lizard have been reported in Oregon. Six vertebrae of the snake *Ogmophis oregonensis* from along the John Day River were identified by Professor Edward Cope in the 1870s. *Ogmophis*, whose modern-day equivalent is the boa, is roughly the same size as those living in the area today.

Specialists in amphisbaenian reptiles are almost as rare as the reptiles themselves, and David Berman of the Carnegie Museum described a new genus

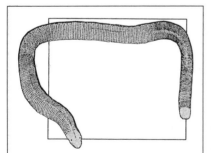

This limbless lizard, resembling a worm, lives today only in tropical areas of Florida, but about 25 million years ago it was present in eastern Oregon.

of limbless burrowing lizard from the Turtle Cove tuffs at Camp Creek in Crook County. The eight specimens of *Dyticonastis rensbergeri* represent the first record of Tertiary amphisbaenians west of the Rocky Mountains. Members of the Amphisbaenidae family have a wedge-shaped skull and tightly bound bone elements that form the burrowing organ. Today similar lizards live only in the tropics and characteristically lack limbs, giving them a worm-like appearance. Several of the tiny vertebrae, ribs, and inch-long skulls, one with the lower jaw intact, were found in the same strata as large numbers of burrowing gophers.

In evolutionary terms, crocodiles and alligators are among the most conservative of vertebrate creatures. Nevertheless, they survived well beyond the Age of Reptiles (Mesozoic) but show little modification from their primitive ancestors. Capable of living in freshwater or marine environments, crocodiles can readily be traced back to the middle Triassic, 200 million years ago, so that their rich geologic record provides valuable clues about the relationship between reptiles and mammals. Ancient marine crocodiles of the Teleosauridae family turned up frequently in Jurassic sediments of Europe and Africa but were unrecorded in North America until a skull, jaw, several vertebrae, and limb bones, embedded in blocks of sandy limestone from the Weberg Member of the Snowshoe Formation, were collected by Professor Earl Packard in the 1930s. Found near Suplee in Crook County, the shallow-water crocodile was six feet long, heavily armored, with an elongate narrow snout studded by slender curved teeth. Two other crocodiles have come from the same sediments. David Taylor identified a small specimen, similar to Packard's teleosaurid and assigned it to the Metriorhynchidae family. Common in Jurassic seas, this crocodile had limbs modified as paddles and a fin-like tail.

In 2007 the complete articulated skeleton of a similar Jurassic crocodile from the same region was still in remarkable condition when worked out of the limestone by members of the Portland-based North American Research Group. These crocodiles are especially significant in that they were recovered from rocks of the Izee terrane, land fragments that developed elsewhere in the Pacific Ocean but were transported to Oregon by crustal plate mechanisms.

When it was recovered in 1962 by paleobotanist Jane Gray, an alligator from the Eocene Clarno Formation in eastern Oregon was the first such fossil reptile in the Pacific Northwest. Almost 90 percent of the jaw and teeth of the beast

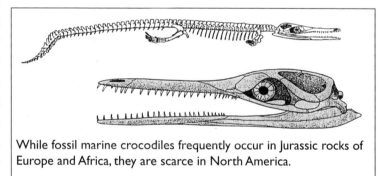

While fossil marine crocodiles frequently occur in Jurassic rocks of Europe and Africa, they are scarce in North America.

was present, including 35 bone fragments and 15 teeth, in the famous Nut Bed locality. Unlike its modern counterpart, the superb skull of the *Pristichampsus* had broad, bladed teeth, similar to those of carnivorous dinosaurs. At the time the alligator lived, the Clarno climate was warm and humid, favorable to its existence. As members of the crocodilian family, they do not tolerate cool weather, and as the climate became more temperate into the Oligocene and Miocene, alligators disappeared from Oregon.

A native Oregonian, Dave Taylor's paleontology interests developed when he participated as a student in summer field camps offered by the Oregon Museum of Science and Industry. Here he is excavating a tortoise (*Geochelone*) near Jordan Valley in eastern Oregon during an OMSI camp in 1995. (Photo courtesy D. Taylor.)

To date only a single fragmentary dinosaur bone has been collected in the state. Dave Taylor recovered a sacrum (the fused backbone pieces above the pelvis) of what appears to be a hadrosaur or duck-billed dinosaur. The bones were embedded in a very hard layer of the late Cretaceous Cape Sebastian Sandstone a few miles south of Gold Beach in Curry County. Since this formation is part of the Gold Beach terrane, which was displaced over 80 miles northward from California, the dinosaur may be one of the state's first immigrants.

As pervasive as dinosaurs are in Mesozoic rocks elsewhere in the world, it is curious that their local record is so meager. This may be because Triassic rocks, present in both the Klamath and Blue mountains, are almost exclusively volcanic or offshore continental shelf limestones and shales, unlikely to entomb terrestrial beasts. Similarly, much of the Oregon Jurassic and Cretaceous represents a marine deep-water setting inhabited by open ocean dwellers. Prime targets for locating dinosaurs in eastern Oregon would be sediments of the Coon Hollow Formation along the Snake River, the Hudspeth Formation in Wheeler County, as well as rocks in the Greenhorn area of Baker County. In the western part of the state, shallow-water nearshore sands of Jurassic plant locales in Douglas County and sands and shales of the Cretaceous Hornbrook Formation in and around Ashland may eventually yield the bones of dinosaurs. The virtual absence of dinosaurs here is undoubtedly due to the lack of a thorough search for them.

While dinosaurs inhabited the land, ichthyosaurs, fish-like swimming reptiles,

Even though duck-billed dinosaurs were common in North America 100 million years ago, just one small piece of these reptiles has been found locally.

reached their greatest diversity in the ocean during the Jurassic Period, 160 million years ago, before suffering extinction near the end of the Cretaceous. These reptiles, along with sharks, were frequently the top carnivore in the marine food chain. Their rows of sharp conical teeth, set in a beaked jaw, were adapted to feeding on fish and swimming mollusks such as squid and ammonites. With a sleek streamlined body, most of these creatures were similar in size to modern porpoises, but some giants attained a length of 60 feet.

Among the most striking of marine reptiles, ichthyosaurs have been reported from every continent except Africa and Antarctica ranging from the middle Triassic to the latest Cretaceous. In Oregon, the first remains of ichthyosaurs, found in 1895 by Othniel Marsh of Yale University, were vertebrae from the Blue Mountains. Collected in conjunction with the shallow-water clam *Trigonia*, the specimen couldn't be identified with certainty. More than 20 years after Marsh's find, Earl Packard extracted two ichthyosaur vertebrae (centra) from Cretaceous Hudspeth strata of the Mitchell Quadrangle in Wheeler County. The bones represented a species new to North America, but they also could not be identified even to the generic level.

It was to be another 30 years before the upper and lower jaws of an ichthyosaur from Sisters Rocks in Curry County were unearthed from dark-gray mudstones of the Jurassic Otter Point Formation along with shelled mollusks and the squid-like belemnites. Identified as comparable to *Ichthyosaurus*

Fascinating studies of the stomach contents of Jurassic ichthyosaurs from Great Britain have revealed thousands of tentacle hooks, sucker rings, and mouth parts of squid-like cephalopods. Intact ichthyosaur remains from Germany, which have premature or unborn young inside the body cavity, show that, unlike other reptiles, these reptiles were not egg-laying, making them fully independent of the land. In German legends, gnomes were practiced in excavating the skeletons of ichthyosaurs. (From Pouchet 1882.)

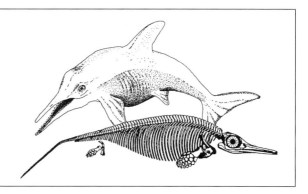

Ichthyosaurs, fish-like marine reptiles, were familiar inhabitants of Mesozoic oceans until vanishing forever shortly before the extinction of dinosaurs.

californicus, the jaw, with a dozen teeth, had been crushed and deformed. As with several other Mesozoic fossils from the Klamath Mountains, the ichthyosaur is notable since it was entombed in an exotic terrane that originated in California before being shifted to Oregon by faulting.

In 1978 Professor William Orr extracted scores of articulated vertebrae and ribs in the Triassic Martin Bridge limestone along Eagle Creek in the Wallowa Mountains. The bones were identified as those of the ichthyosaur *Shastasaurus* heretofore known only from northern California. A partial skull from the same site was discovered several years later. Although the specimen was badly distorted by metamorphic processes, it was intact enough to be assigned to the same genus. Quite unlike Triassic ichthyosaurs elsewhere in North America, Shastasaurids from Oregon and California provide additional evidence that these western margins of the continent are parts of exotic or accreted terranes.

Rivaling the ichthyosaurs, plesiosaurs were unknown in Oregon until 2004, when a nearly complete skull and much of a skeleton were excavated by NARG members (North American Research Group) from Cretaceous Hudspeth mudstones near Mitchell. These marine superpredators had a flat inflexible body and small tail, so propelled themselves along with well-developed paddles. The Oregon specimen was the short neck variety, and it is regrettable that this exciting discovery was sent to South Dakota.

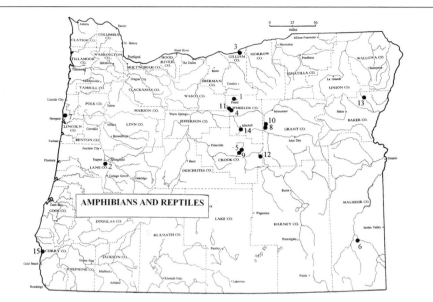

AMPHIBIAN AND REPTILE LOCALITIES

The following are Oregon fossil amphibian and reptile localities and a selection of authors who list faunas from the locales.

AMPHIBIANS

Salamanders

1. Fossil, Wheeler Co. John Day Fm. Oligocene. Naylor 1979.
2. Goshen, Lane Co. Fisher Fm. Eocene. Van Frank 1955.

REPTILES

Turtles

3. Arlington, Gilliam Co. The Dalles Fm. Miocene. Fry 1973.
4. Nut Beds, Wheeler Co. Clarno Fm. Eocene. Cope 1873; Hanson 1996.
5. Rattlesnake Creek, Crook Co. Mascall/Rattlesnake Fms. Miocene. Brattstrom and Sturn 1959; Downs 1956; Hay 1903, 1908.
6. Rome, Malheur Co. Rome Beds? Miocene. Brattstrom and Sturn 1959.
7. Spencer Creek, Lincoln Co. Astoria Fm. Miocene. Packard 1940, 1942.
8. Turtle Cove, Grant Co. John Day Fm. Oligocene. Hay 1902, 1908; Merriam and Sinclair 1907.

Lizards and Snakes

9. Camp Creek, Crook Co. John Day Fm. Oligocene. Berman 1976.
10. John Day River. John Day Fm. Oligocene. Gilmore 1938.

Crocodiles and Alligators

11. Nut Beds, Wheeler Co. Clarno Fm. Eocene. Hanson 1996; Retallack, Bestland, and Fremd 2000.
12. Suplee, Crook Co. Snowshoe Fm. Jurassic. Buffetaut 1979; Stricker and Taylor 1989; Bland, Rose, and Currier 2007.

Ichthyosaurs

13. Eagle Creek, Wallowa Co. Martin Bridge Fm. Triassic. Orr and Katsura 1985.
14. Mitchell, Wheeler Co. Hudspeth Fm. Cretaceous. Merriam and Gilmore 1928.
15. Sisters Rocks, Curry Co. Otter Point Fm. Jurassic. Koch and Camp 1966.

Birds are really reptiles, or, more specifically, they are related to dinosaurs. There are several known finds of Cretaceous dinosaurs with feathers, and the structural similarities between birds and reptiles are astonishing. Even though the two are alike, birds are still placed in their own avian classification because modern species have adapted so efficiently and thoroughly to their surroundings. Birds are among the most intensely studied of vertebrates, but the dismal avian fossil record continues to keep their ancestry and relationships unclear, in part because their fragile air frame of hollow bones seldom preserves well. Few identifiable remains can be found from the Mesozoic Era of Oregon, and most Tertiary specimens are of aquatic species. Archaic birds became extinct by the end of the Cretaceous, 66 million years ago, and most living avian families can be traced forward from the Eocene.

Oregon's 100-million-year-old *Pteranodon* from marine Hudspeth sediments near Mitchell in Wheeler County has significantly added to the scarce record of these reptiles in North America. This Cretaceous specimen is the only such find in the state. Pterosaurs had a wingspan of almost

An imaginary scene depicts what central Oregon might have looked like during the Ice Ages some 12,000 years ago. (From Pouchet 1882.)

Among the small number of early avian paleontologists, Robert Shufeldt's efforts stand out. Shufeldt's passion for birds enabled him to develop a unique knowledge of their characteristics and habitats. As a child in the 1860s he accompanied his father, a rear admiral in the navy, on his travels, enabling the young Robert to indulge his interest in natural history. Obtaining a medical degree from Columbia University, Shufeldt joined the army as an assistant surgeon, and, while serving on the western frontier, he continued his collecting. Shufeldt married Florence Audubon, granddaughter of the famous ornithologist. A long association with the Smithsonian brought an appointment as curator until his death in 1904.

Lloye Miller is one of the few scientists who devoted a lifetime to the field of paleo-ornithology. Miller was born in Louisiana in 1874, and, although interested in many aspects of nature, he focused on avian studies for a Ph.D. in 1912 from the University of California, Berkeley. Miller's first contact with Oregon was accompanying vertebrate paleontologist John Merriam on an expedition to the John Day basin in 1899, and most of his subsequent Oregon research focused on the eastern part of the state. Miller taught at the University of California, Los Angeles, retiring in 1943, but continuing research until his death in 1970 at 96 years of age. (Photo courtesy Page Museum Archives, Los Angeles.)

In determining that the humerus (leg) bone of a pigmy goose (*Anabernicula*) from Fossil Lake was a distinct species, the meticulous Hildegarde Howard examined and measured 300 skeletal parts from seven locations. Paleo-ornithologist Howard was born in Washington, D.C., in 1901 but spent her professional life in California. Receiving a Ph.D. in zoology in 1928 from the University of California, she was offered a curatorial position at the Los Angeles County Museum. There she processed thousands of birds from Rancho La Brea along with identifying and cataloging those from Oregon. Howard retired in 1990, when, as she was fond of saying, two men were hired to replace her. She died in 1998. (Photo courtesy Archives, Page Museum, Los Angeles.)

20 feet, a weight of around 35 pounds, and a skull longer than its trunk. Some are believed to be covered with hair, but in most cases this is a reference to parallel bundles of fibers attached to wings to aid in flight. Pterosaur arm bones typically display a flange for muscle attachment, and the combination of a very light air frame in conjunction with the skeletal and muscle structure suggests that these animals remained aloft by soaring rather than by flapping wings. Delicate limbs were highly adapted to gliding flight, and it is difficult to imagine them folded for locomotion on the ground. Three smaller toes positioned midway on the front of the wing may have assisted in walking or clutching onto trees or rocks.

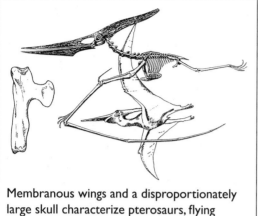

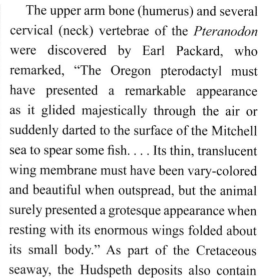

Membranous wings and a disproportionately large skull characterize pterosaurs, flying reptiles. Their humerus displays a characteristic flange for attachment of wing muscles.

The upper arm bone (humerus) and several cervical (neck) vertebrae of the *Pteranodon* were discovered by Earl Packard, who remarked, "The Oregon pterodactyl must have presented a remarkable appearance as it glided majestically through the air or suddenly darted to the surface of the Mitchell sea to spear some fish. . . . Its thin, translucent wing membrane must have been vary-colored and beautiful when outspread, but the animal surely presented a grotesque appearance when resting with its enormous wings folded about its small body." As part of the Cretaceous seaway, the Hudspeth deposits also contain mollusks, foraminifera, fish-like ichthyosaurs, and a reptilian plesiosaur.

Tertiary bird bones are randomly scattered at localities throughout Oregon. In Coos and Washington counties, remains are fragmentary, whereas east of the Cascades in the Blue Mountains province they are more abundant. Only at Fossil Lake can the bones be found in considerable numbers. In Coos County, an Eocene locality at Sunset Bay produced the leg bones from an auk. Embedded in a matrix with an abundance of marine mollusk shells, the remains of *Hydrotherikornis oregonus* were found in 1926. The long-legged *Hydrotherikornis* is regarded as a diving and swimming nearshore bird similar to living auks and murrelets.

Another single bird find, *Phocavis maritimus*, similar to a pelican, is represented by a leg bone from the Eocene Keasey Formation near Vernonia in Washington County. James Goedert, of the Los Angeles County Museum, reported the specimen in 1988. He notes that this is the oldest recorded member of the now extinct Plotopteridae family and the first known in Oregon. The Plotopteridae may have become extinct because of the drop in numbers of fish or because of the rise in pinnipeds concurrent with middle Miocene global warming. Since fossils of the large flightless bird were found with mollusks that lived in 1,500 feet of water, it evidently operated some distance from shore.

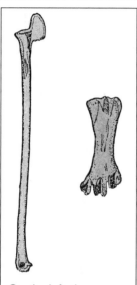

On the left, the upper leg bone (tibiotarsus) of an auk (*Hydrotherikornis*) from the Coaledo Formation in southwest Oregon contrasts in size to the lower leg bone (tarsometatarsus) of a flightless pelican (*Phocavis*) from the Keasey Formation in the northwest part of the state.

During the winter of 1995 a sensational fossil bird egg, also from Keasey strata, was recovered by James Leary of Cottage Grove.

The slightly crushed specimen was somewhat smaller than a chicken egg. The surface and a cross-section of its shell were examined under an electron microscope by William Orr at the University of Oregon. He confirmed that the details of the shell architecture compared most closely with that of members of the pelican family. In addition, a CAT scan of the intact egg revealed no internal embryo. Although known from several intervals in the geologic record, eggs are among the rarest of fossils because they are so fragile. When Orr announced the find,

The first clues that the Keasey egg was the real item were the shell thickness and the presence of pores in the surface. An unmistakable crystal pattern within the shell verifies the identification. (Photo courtesy W. Orr.)

he remarked ruefully that "he had been looking at 'eggs' brought to him over a span of 30 years," and all but one turned out to be rounded stream pebbles, nodules, or concretions.

Oligocene and Miocene sedimentary basins of eastern Oregon generated a quantity of skeletal bird parts. During these epochs, lakes and ponds expanded with heavy rainfall. Under such conditions birds would gather in the thousands around the waterways where today a handful of remains, preserved in volcanic ash and tuffs, might be the only evidence of their presence. Leg bones from the John Day valley were identified by collector Robert Shufeldt as those of a *Colymbus* (grebe), a *Larus* (gull), a *Limicolavis* (surfbird), a *Phalacrocorax* (cormorant), and a *Phasianus* (pheasant). Similar collections of Miocene *Phasianus* bones near Paulina Creek were found in Mascall sediments in Crook County, while Willow Creek in Malheur County yielded *Larus*, *Limicolavis*, and a *Phalacrocorax* from what is probably the John Day Formation.

Bird bones on the Owyhee plateau were unearthed by field parties from the California Institute of Technology and identified by ornithologist Lloye Miller. At ancient lakebeds in Malheur County, Miocene bird fossils lie near the surface but are so scattered that the parts of any one individual couldn't be associated. The bones near Rome were permineralized and the accumulation of flamingos (*Phalacrocorax*), geese (*Branta*), swans (*Cygnus*), pelicans (*Pelecanus*), and indeterminate eagles, along with those of beaver (*Dipoides*), antelope (*Sphenophalos*), horse (*Pliohippus*), and smaller rodents suggested to Miller a lake, peripheral marsh, and wetlands.

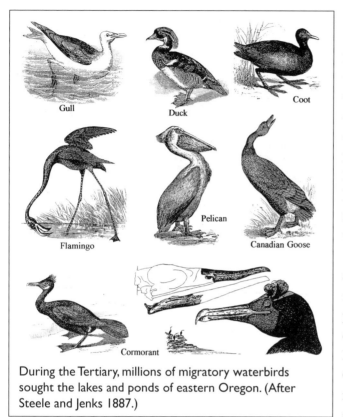

Gull

Duck

Coot

Flamingo

Pelican

Canadian Goose

Cormorant

During the Tertiary, millions of migratory waterbirds sought the lakes and ponds of eastern Oregon. (After Steele and Jenks 1887.)

A late Miocene locale at McKay Reservoir in Umatilla County yielded two ducks (*Nettion*), a quail (*Lophortyx*), and a sandpiper (*Bartramia*). The bird remains, along with a vertebrate fauna, dated between 5.8 and 8.5 million years ago, suggested to Pierce Brodkorb, of the University of Florida, a streamside setting at the reservoir, with grassy savannas in the distance.

By far the greatest number and assortment of birds were the inhabitants of the broad inland pluvial lakes of eastern Oregon during the cool moist Pleistocene Epoch some 12,000 years ago. Concentrations of shells and bones reflect their vast numbers living in and around wetlands and lakes. The diversity of birds at Fossil Lake shows that, as with Klamath and Malheur lakes today, these waterways were stopovers for migratory birds, although the juveniles present confirm that many of the birds nested here as well. The specific Fossil Lake locality became confused historically when the terms Christmas Lake, Silver Lake, and Fossil Lake were shown on official maps as only a single basin. Most of the vertebrate remains described from this vicinity at the turn of the twentieth century were probably from Fossil Lake.

Shiny jet black bones scattered across the expansive surface of dry Fossil Lake are strikingly evident. In northern Lake County an optimum environment for burial and fossilization existed. Soft, organic-rich mud of the lake bottom gently covered the carcasses, while subdued wave activity kept wear and abrasion to a minimum. Today drifting sands expose thousands of bones of vertebrates, birds, freshwater fish, and shells. Although fish fossils far outnumber those of birds, the careful observer working on hands and knees can uncover finds without difficulty. Unfortunately the skeletons have been separated and dispersed. By examining the patterns of wear (taphonomy), preservation, and breakage on an extensive collection of bird bones made in 2003, Jim Martin and students recently reinterpreted the Fossil Lake paleoenvironment. He suggests that the bone accumulations are the result of volcanic action and not from the scouring of wind or work of water as was concluded in 1966 by Ira Allison.

Fossil Lake became famous for its vertebrates in the 1860s, and it was the site of particular interest to early visitors such as Thomas Condon, Edward Cope, and Charles Sternberg. The postmaster at Silver Lake, George Duncan, frequently provided room and board and served as a guide to the "bone yard." Thomas Condon made the first excursions in 1877 when "the rarest and most valuable fossils . . . were the eighty bird bones which were unusually perfect and in a fine state of preservation." His explorations were followed shortly by those of Edward Cope of the Philadelphia Academy of Sciences and Charles Sternberg, a commercial collector. As soon as he came upon the fossils, Sternberg went down on his hands and knees, picking up bones and teeth and putting them in piles. In his autobiography he wrote "I

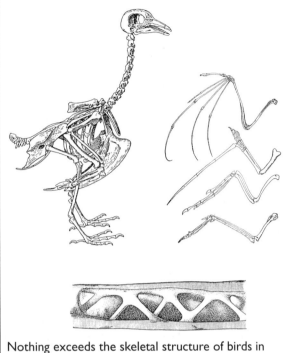

Nothing exceeds the skeletal structure of birds in ratio of strength to weight. The construction of the internal wing struts, shown here from a vulture, is known in engineering parlance as a Warren's truss (bottom). It is notable that the four vertebrate wing constructions, that of a bat, a pterodactyl, an archaeopteryx, and a modern bird all evolved independently. (After Lucas 1902.)

did not find a single bone or tooth in its original position . . . but that all were loose, detached, and scattered, and that Indian implements were lying about in the same way." Today Fossil Lake is closed to public collecting.

Robert Shufeldt identified 51 species, 13 of which were new, from a total of 1,500 bone fragments from Fossil Lake. Hildegarde Howard revised this list in 1946, finding 66 species, the most numerous by far being *Aechmophorus* (grebe), followed by *Colymbus* (grebe), *Phalacrocorax,* (cormorant), *Cygnus* (swan), and *Branta* (goose). *Larus californicus* (gull), *Phoenicopterus copei* (flamingo), *Ardea paloccidentalis* (heron), *Fulica minor* (coot)*, and *Pelecanus erythrohynchos* (pelican) once living here are now extinct.

Howard pointed out that Shufeldt's 1913 study of Fossil Lake bird faunas hadn't been revised until her work in 1946, and it was 25 years after this that Robert Storer used new methods of analyzing skeletal material to reach different conclusions. At the University of Michigan, Museum of Zoology, Storer compared

The wingspan of the Pleistocene teratorn from Woodburn could be estimated at 12 feet by measuring the length of the humerus (upper arm bone).

The bones of vultures (*Coragyps*) are among the remains at 8,000-year-old Indian sites near The Dalles. (From Pouchet 1882.)

the skeletons of recent *Aechmophorus occidentalis,* a Western grebe from Canada, to those from the Pleistocene of Fossil Lake to show that previous techniques of matching bone sizes between fossil and modern birds to determine new species might not be valid. Storer evaluated the measurements of skeletons of 119 Western grebes that had been frozen in Lake Newell, Alberta, in 1959 against those already housed in museums. He concluded that significant skeletal differences between the two groups could be attributed to geographic variations, postmortem wear, or preparation techniques, and not to genetic variations.

Avian paleontology was examined on a broad geographic scale in 1912 by Lloye Miller when he looked at bird fossils from nine west coast locations including those from Fossil Lake to provide species lists and draw general conclusions on distribution, environment, and correlation among the different faunas. Discussing the extinction of birds during the late Tertiary he proposed that while humans probably did not directly cause their disappearance, the demise of large mammals certainly contributed to the end of carrion-eating birds. Disease and climate changes are also suggested as causes for the extinction of certain species. Miller recounts that in 1908 and 1909 thousands of seabirds, dying over a relatively short interval, were found to be filled with intestinal worms. Diminished rainfall after the Ice Ages brought about a drying of lakes in the Great Basin, and temperature changes would have substantially altered the composition of the forests, both factors affecting bird populations. In more recent times, one of the consequences of El Niño weather systems has been seabird mortality in fantastic numbers due to lack of food at tropical and mid latitudes.

Throughout the 1990s, excavations of several Pleistocene bogs within the city of Woodburn in Marion County uncovered bird bones along with those of rodents, weasels, deer, amphibians, and bison. Carbon-14-dated at 12,500 years in the past, the most spectacular were skeletal elements from the extinct family Teratornithidae. With an estimated wingspan of 16 feet, *Teratornis* was the world's largest known flying bird, soaring much like a condor. Initially placed in the vulture family, teratorns have been pictured as sitting in a tree, waiting to pounce on their victims. However, the structure of the skull and mandible indicates that they didn't tear their prey apart but fed on small rodents, birds, or frogs, swallowing them whole.

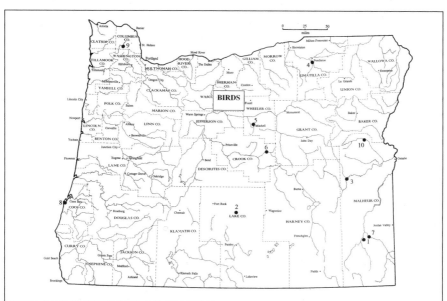

BIRD AND PTEROSAUR LOCALITIES
The following are fossil bird and pterosaur localities and a selection of authors who list faunas from the locales.

1. Dry Creek, Malheur Co. ?Juntura Fm. Miocene. Brodkorb 1961; Miller 1944.
2. Fossil Lake, Lake Co. No Fm. Pleistocene. Cope 1878, 1889; Howard 1946; Jehl 1967; Miller 1911; Shufeldt 1912, 1913, 1915; Storer 1989.
3. Juntura, Malheur Co. ?Juntura Fm. Miocene. Brodkorb 1961.
4. McKay Reservoir, Umatilla Co. McKay Fm. Miocene. Brodkorb 1958.
5. Mitchell (Pterosaur), Wheeler Co. No Fm. Cretaceous. Gilmore 1928.
6. Paulina Creek Valley, Crook Co. Mascall Fm. Miocene. Shufeldt 1915.
7. Rome, Malheur Co. No Fm. ?Oligocene. Brodkorb 1961; Miller 1944.
8. Sunset Bay, Coos Co. Coaledo Fm. Eocene. Miller 1931.
9. Vernonia, Columbia Co. Keasey Fm. Eocene. Goedert 1988.
10. Willow Creek, Malheur Co. ?John Day Fm. Miocene. Brodkorb 1961; Miller 1944; Schufeldt 1915 .

MARINE AND FRESHWATER VERTEBRATES

The peg-like cheek teeth and canines of a sea lion reflect its varied diet of fish and mollusks. (Photo courtesy Condon collection.)

Bony fish, sharks, and marine mammals are treated together for convenience. All occupy an aquatic habitat, and, with few exceptions, the remains of marine fish and open ocean mammals are found in the same Tertiary formations of western Oregon, whereas those of freshwater fish are primarily east of the Cascades. Aquatic vertebrates fall into three groups, the sharks and rays, the bony fish, and the marine mammals.

The evolutionary history of fishes, both bony fish and sharks, is shrouded in the past, but by the late Silurian and early Devonian, some 400 million years ago, numerous primitive fish-like vertebrates had evolved. Some were successful and became established, while others vanished, but two classes arose, sharks (Chondrichthyes) and bony fish (Osteichthyes). They have many contrasting characteristics. Bony fish evolve rapidly, while sharks change slowly and thus have limited use in geology. Fish can exist in both marine and nonmarine environments, and some, such as salmon, can migrate from salt to freshwater, an astonishing feat. Sharks, on the other hand, are primarily found in a marine setting. As scavengers, sharks are invariably drawn to carcasses, and it is not uncommon to find their broken teeth embedded in the bones of bigger animals. Shark teeth and skin armor (denticles) are important for identification, while bony fish are recognized by their scales and skeletal parts.

Marine mammals range from awkward-looking nearshore herbivores to torpedo-shaped open ocean predators, but despite their diversity, they all share characteristic mammalian features such as hair and warm blood physiology. Mammals fall into a wide range of classes, but both the carnivores, such as otters and sea lions, and cetaceans, or whales, split off in the Tertiary from primitive land-dwellers before returning to the ocean, where they adapted completely to the water environment. Several unique creatures, such as the hippopotamus-like desmostylids and the bear-like *Kolponomos*, lack modern representatives but flourished in Oregon during the Miocene Epoch.

Preservation

The ocean environment acts as something of a deterrent to the preservation of vertebrates. An impressive array of scavengers from sea urchins to sharks to crabs awaits the opportunity to eat any carrion. With smooth efficiency this cleanup crew strips off the flesh before dispersing the bones across the ocean floor. Carcasses washing ashore as well as the bodies of coastal dwellers carried into the ocean are subject to intensive sorting, scattering, and abrasion by the high energy of the surf. In the terrestrial realm, there are many specialized environments such as quicksand, tar pits, and bogs that trap, mire, and bury animals to preserve them, often intact, but these situations are lacking in marine waters.

Some bony fish remains consist of bits and pieces, some are single bones and scales, and others are intact skeletons, whereas almost all shark remains are isolated skin denticles and teeth. Sharks have a flexible cartilaginous skeleton, which usually deteriorates too rapidly for preservation, while their multiple teeth are replaced continuously over a lifetime. The same is true for skates and rays, bottom-feeders with a pavement-like dental surface in the mouth adapted for crushing shells. In addition to the production of copious numbers of teeth, the calcium phosphate of the tooth enamel is extremely resistant to wear, abrasion, and dissolution.

In spite of the comparatively massive body size of many marine vertebrates, their fossil record is remarkably poor when compared to that of terrestrial animals, and their limited remains tend to be separated and worn. Several factors contribute to this. One is that not many members of a marine population are living at any given time. Except for small fish, most ocean-dwelling vertebrates are at or near the top of food pyramids, and the environment can only support a few such individuals. Even if an animal is physically large, numbers are what really count with respect to being represented in the fossil record. If a whale is buried, it is much less probable that this single specimen will be discovered or collected than for remains of thousands of smaller animals of equal total weight to be preserved.

Paleoenvironments and Fish

Bony fish are among the most successful and adaptive of all vertebrates, and their skeletal debris is readily apparent in aquatic sediments. Where their scales and bones are abundant, they are useful in distinguishing ecological conditions during deposition of the strata. Variations in specific parameters such as temperature, depth, or habitat stability are reflected by alterations in the fish inhabitants themselves. In reviewing the history of fossil freshwater fish in North America, Ted Cavender of Ohio State University noted that climate cooling, regional uplift, and disruption of drainage systems near the end of the Eocene brought about the extinction of many warm-water fish species. This worldwide trend was also reflected in a decline in the diversity of mollusks and other invertebrates, in a loss of plant species, and in a similar diminishing of vertebrates. Cyprinidae (minnows) and Umbridae (mudminnows) first appeared following this event as part of the replacement (turnover) fauna. Cavender also surmised that increased rainfall and warmer temperatures in the early Miocene may have accounted for the spread of Centrarchidae (sunfish) in the Basin and Range of Oregon and northern California.

Sharks and Rays

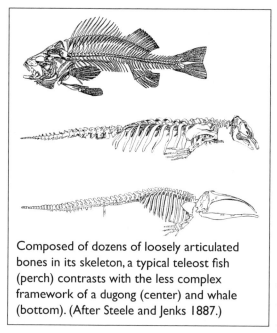

Composed of dozens of loosely articulated bones in its skeleton, a typical teleost fish (perch) contrasts with the less complex framework of a dugong (center) and whale (bottom). (After Steele and Jenks 1887.)

Fossil sharks were mentioned as early as 1849 by James Dana of the U.S. Exploring Expedition, but references to these fish are rare and generally included with accounts of invertebrates. Oregon's preserved shark teeth and denticles (armor and spines embedded in the skin) range from the Cretaceous through Tertiary periods. Of the 16 genera of sharks, skates, and rays inhabiting the coastal waters today, 10 are present in the fossil record. Reflecting shelf or intertidal habitats, the teeth of the dogfish shark *Squalus* are the most frequently encountered.

Some of the characteristic details used when identifying sharks and rays from the teeth alone include the shape of the crown and root, the position of the nutrient canals, the tooth size, and the serrations on the crown. Diversity in the teeth of an individual jaw must also be taken into account, and one of Bruce Welton's most successful techniques has been to reconstruct entire "tooth sets" to show all variations within a species.

Three teeth of the only reported Cretaceous shark are from the Oregon Hudspeth Formation near Mitchell, where the carcass might have washed up on what was then the shallow continental shelf open to the ocean. Of the three teeth, only one could be tentatively identified as that of a goblin shark *Scapanorhynchus*. Mollusks are not uncommon in the mudstones, and occasional marine reptiles such as ichthyosaurs and even a flying pterosaur are known from these sediments.

Bruce Welton, who specializes in the study of sharks, is one of the few researchers in that field. Born in Portland, Welton attended summer camps organized by the Oregon Museum of Science and Industry before going on to obtain a Ph.D. in 1979 from the University of California, Berkeley. His research focused on Cretaceous and Cenozoic Squalimorphii, primitive sharks of the northwest Pacific Ocean. Retiring from Mobil Oil, Welton lives in New Mexico. (Photo courtesy B. Welton.)

Complete tooth set of the shark *Hexanchus* showing the characteristic comb-like pattern of the lower teeth (After Welton 1972.)

Although the ocean shoreline fluctuated during the Tertiary, the overall trend was one of retreating seas. Eocene shark fossil localities mark the position of the sandy beaches and rocky headlands, which extended from Coos County northward to Lincoln, Columbia, and Washington counties. The few Oligocene locales on the west slope of the Willamette Valley and in Columbia County parallel that ancient shoreline, but by Miocene time the strand was restricted to the western margin of the Coast Range where fossil shark sites are limited to the beach cliffs at Cape Blanco, Newport, and Astoria. Some strata are richer than others in shark teeth, but at least 75 percent of all those recorded in Oregon are from the Eocene.

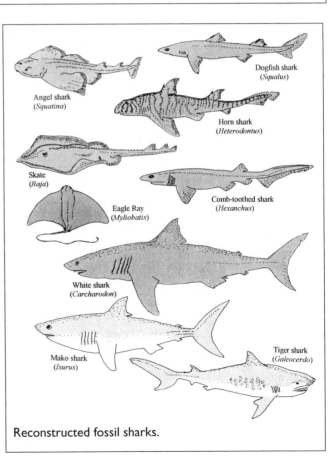

Reconstructed fossil sharks.

Among the Eocene formations on the coast, the Coaledo at Shore Acres State Park in Coos County frequently yields fragments and complete teeth of the sand shark *Odontaspis*, the dogfish shark *Squalus,* and the eagle ray *Myliobatis.* The Coaledo has the most *Myliobatis* material of any strata in the state and the highest variety of sharks and rays, including *Aetobatus, Carcharias, Galeorhinus, Heterodontus, Isurus, Lamna, Rhinoptera, Scyliorhinus, Squalus,* and *Squatina.* To the north near Toledo, poorly preserved prism-shaped teeth of the mako shark *Isurus,* the tiger shark *Galeocerdo,* and small teeth of the skate *Raja* accompany mammal bones and fish remains in an embayment setting of the Eocene Nestucca Formation.

Along Scoggins Creek in Washington County, dark-gray marine shales of the Eocene Yamhill Formation yielded 22 vertebrae, "patches of hollow cubes" of calcified cartilage, and the single tooth of a sand shark (*Odontaspis*). These rare specimens were collected in 1967 and examined by paleontologist Shelton Applegate of the Los Angeles County Museum, who identified the components from the skull, jaws, gills, and fin supports of *Odontaspis macrota.* The modern equivalent is the sand shark *Odontaspis taurus,* a somewhat sluggish predator, capable of short bursts of speed. Today it is found on shallow ocean bottoms along the coastline of the eastern United States and in Europe and the Mediterranean Sea.

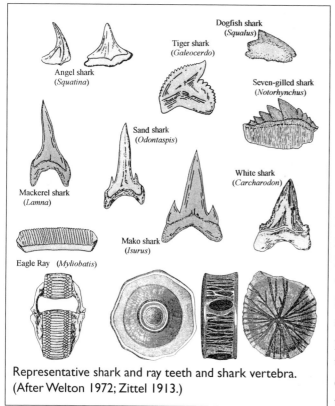

Representative shark and ray teeth and shark vertebra.
(After Welton 1972; Zittel 1913.)

This interval in northwest Oregon near Mist produced several sharks such as *Squatina, Odontaspis,* the great white shark *Carcarodon,* as well as the seven-gilled *Notorhynchus* and the six-gilled *Heptranchias.* These were inhabitants of shelf and deep basin environments during the Cowlitz and Keasey intervals.

In the Willamette Valley, Bruce Welton documented over 2,000 teeth, 95 percent of which belong to *Odontaspis macrota* within a thin lens of the Eocene Spencer Formation. Spencer strata at Helmick Hill in Polk County provide one of the few localities where the teeth are affected by detrimental conditions. Here beach wear and leaching by groundwater have smoothed and rounded the previously sharp edges.

Oligocene and Miocene shark remains are less frequent, reflecting a diminishing ocean across the state. Near Newport, *Isurus*, *Galeocerdo*, *Carcharodon* (great white shark), *Echinorhinus* (bramble shark), *Hexanchus* (cow shark), *Raja* (skate), and *Myliobatis* (spotted eagle ray) lived along the continental shelf at moderate depths in the Oligocene Yaquina and Nye and Miocene Astoria periods. *Squalus*, one of the few Miocene sharks from the Empire Formation at Coos Bay, is associated with mollusks from what was a large coastal embayment. It is similarly found with invertebrates in the Pleistocene Port Orford Formation. In 2009 Bruce Welton finalized an analysis of Pleistocene Port Orford and Elk River fishes at Cape Blanco based on collections primarily from the *Psephidia* (clam) beds.

In Lane County, shark teeth are occasionally found in the mid Tertiary Eugene Formation. Saw-like lower teeth of *Hexanchus* and the single cusp tooth of galeoid sharks have been noted. Shark teeth from this formation show beach wear as well as the darkened enamel typical of phosphatic composition. (Photo courtesy Condon collection.)

Bony Fish

Extremely diverse, bony fish have dominated both marine and fresh waters since the Devonian, and modern finned fish such as haddock, cod, trout, and perch number more than 25,000 species. Not only are their skeletons made of bone, but these fish have incorporated dermal bone into their jaws, gills, and palate as well.

Fossils of marine fish are not uncommon in coastal Tertiary formations marking the Pacific shoreline, whereas freshwater fish are found in ancient lake and stream deposits east of the Cascades. Research in the fields of marine and freshwater fossil fishes has been very sporadic. Edward Cope described several freshwater minnows and suckers in the Great Basin in 1883. After that effort, a gap of around 30 years lapsed before David Starr Jordan, mostly working in California, revised the list in 1907. Jordan also published on fossil sharks and rays of the Pacific slope. Carl Hubbs in the 1940s, Lore Rose David in the 1950s, and more recently Ted Cavender, Robert Miller, Jerry Smith, and Peter Kimmel have all worked with bony fish. Of these, only David focused on marine fish.

For her marine deepwater fish studies, Lore Rose David examined fish scales from Eocene through Pliocene intervals of California, Oregon, and Washington. Most of these belonged to the Gadiformes, living hake and cod, a group known as

Landmark papers by paleoichthyologist, Lore Rose David, which specifically treat fossil marine fish, are unique even to this day. Born and educated in Germany, she received a Ph.D. in biology at the University of Berlin in 1931. From there she moved to Belgium as ichthyologic curator in the Museum at Tervuren before immigrating to the United States. Working in laboratories at California Institute of Technology and Richfield Oil Company, David changed careers in 1954, to serve as a librarian until retirement from Santa Clara County. She died in San Jose in 1985. (Photo courtesy J. Jorgensen.)

Both paleoichthyologists at the University of Michigan, Gerald "Jerry" Smith and Bob Miller (holding the 13-pound squawfish) have traced the ancestry of freshwater fish in North America. (Photo courtesy G. Smith.)

During his early years, Peter Kimmel lived with his family at Monument, Oregon, allowing him to explore the John Day fossil beds as an OMSI student. Work on fossil sharks from Millican led to a graduate degree at Oregon State University and a subsequent Ph.D. from the University of Michigan on fossil fish in eastern Oregon and Idaho. He currently lives in Virginia. (Photo with his daughter; courtesy P. Kimmel.)

"herrings of the deep sea" because of their abundance in bathyal ocean waters. She set up a precise stratigraphic sequence using the brown to amber-colored scales, which are similar to a fingerprint, showing that the surface pattern of swirls and lines on the face of fish scales was a reliable tool for identification of species. By classifying the patterns, which she compared to those of modern fishes, and by familiarizing herself with the environments at which present-day species lived, Lore Rose David was able to extrapolate paleodepths for Pacific Coast Tertiary formations, in which the fossil fish were found.

David glued chips of rocks containing the flat scales to one end of wooden tongue depressors and then recorded data on taxonomy and stratigraphy of the scale on the other end of the stick. Ultimately she built a sizeable catalogue useful for exploring deep-water intervals. In oil well cores from Oregon, David found that the dolphin fish *Calilelpidus, Probathygadus*, and *Promarcrurus* were residents of bathyal depths in the Eocene Keasey at Mist as well as in the Oligocene Yaquina and Miocene Nye formations at Alsea Bay, Yaquina Bay, and Waldport. *Pyknolepidus*, an indicator of deep ocean waters, was recovered at Alsea Bay.

Individual ear bones, or otoliths, of fish are as distinctive as the scales. Otoliths belonging to Congridae (conger eels) and to the Macruridae (rat tails) have been reported in the Pittsburg Bluff Formation where they are associated with molluscan fossils. Bruce Welton's study of otoliths, teeth, scales, and other skeletal elements of sharks, rays, and bony fish from the Pleistocene Port Orford and Elk River formations makes

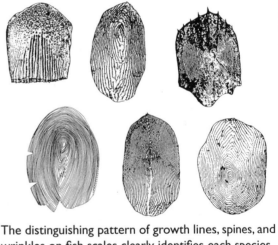

The distinguishing pattern of growth lines, spines, and wrinkles on fish scales clearly identifies each species. (After David 1956.)

several comparisons with modern species and environments. Welton identified 25 species of marine fish, most of which are teleosts. The assemblage is dominated by bottom-dwellers such as *Gadus* (cod), *Icelinus* (sculpin), *Merluccius* (hake), and *Sebastes* (rockfish). The sharks and rays *Cetorhinus*, *Squalus*, and *Raja* constitute a smaller percentage of the group.

Other than scales, marine fish parts are not uncommon in coastal Tertiary deposits, but they are generally poorly preserved and too fragmentary for identification. However, there are exceptions. The partial skull and scales of a billfish were recovered in 2001 from the Oligocene Yaquina Formation by Harry Fierstine of California Polytechnic State University, who described it as an *Aglyptorhynchus*

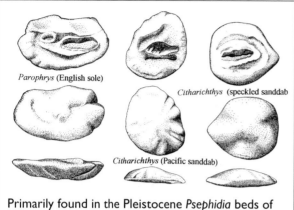

Primarily found in the Pleistocene *Psephidia* beds of southwest Oregon, otoliths, or ear bones, of fish are individually differentiated. (Illustration courtesy B. Welton.)

a member of the Xiphiidae family. The ancestral billfish, among the fastest of swimmers, made its way from the warm Gulf Stream of the Atlantic Ocean to the Pacific through the Panama seaway some 30 million years ago. The scant remains of marine fish teeth, spines, scales, vertebrae, and skulls from the Miocene Astoria Formation were supplemented in 2006 by the skull of a spectacular swordfish, weighing over 100 pounds. The specimen is retained by a private collector.

Freshwater Tertiary environments of the Pacific Northwest were summarized by paleoichthyologists Carl Hubbs and Robert Miller: "Hundreds of lakes have come and gone throughout the arid portion of the western United States. Over a long period of time—Miocene, Pliocene, and the many millennia of early and middle Pleistocene—lakes formed, filled their depressions with deposits, and disappeared." These lakes were home to thousands of fishes whose remains in Oregon are found from the Eocene, over 40 million years ago, and into the Pleistocene.

Eocene fish scales, vertebrae, and skulls, long noted in Clarno sediments at Ochoco Pass, were collected in 1966 by Ted Cavender, who taught at Ohio State University. Steep roadcuts at Ochoco Summit in Wheeler County yielded a variety of fishes, which are younger than the famous middle Eocene Green River fish of Wyoming but older than several other middle Tertiary faunas in western North America. Bowfins (*Amia*), mooneyes (*Hiodon*), catfish (Siluriformes), and sucker fish (*Catostomus*, *Amyzon*) are the most abundant from these dark laminated shales that have been interpreted as lake sediments. A rich accumulation of plant impressions accompanies the fish, as do occasional beetle wings. Thick elongate bowfin scales are distinctive and are usually preserved intact. In their

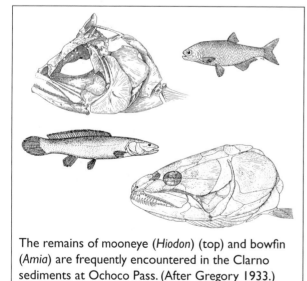

The remains of mooneye (*Hiodon*) (top) and bowfin (*Amia*) are frequently encountered in the Clarno sediments at Ochoco Pass. (After Gregory 1933.)

evolutionary history, bowfins are a particularly conservative group, having changed very little in 70 million years from their first appearance during the Cretaceous. Of the four fossil groups, only the Catostomidae live in the same region today. Mooneyes, catfish, and bowfins no longer inhabit the Columbia River basin, but catfish are known from late Tertiary sediments of ancient Lake Idaho in the western Snake River valley.

Calm waters of ponds and lakes, that characterized the John Day Formation of central Oregon over 33 million years ago, provided for the accumulation and preservation of freshwater fish long after the region had become dry and arid. Several virtually intact specimens of an Oligocene mudminnow, *Novumbra oregonensis* were found at two different locations of this formation in Wheeler County. Around four inches in length, living mudminnows inhabit the muddy bottoms and dense aquatic vegetation of quiet lowland streams. They can endure considerable variation in temperature and dissolved oxygen but have a somewhat limited tolerance for strong currents and changes in salinity. The skull and jaw fragments of *Novumbra* from Knox Ranch and Allen Ranch were associated with plant fossils belonging to the Bridge Creek flora. Additional freshwater fish from the same strata were identified by Cavender as pikes or pickerels (Esocidae). Previously, in 1927, paleobotanist Ralph Chaney uncovered fragmentary fish remains from the Bridge Creek shales at Gray Ranch in Crook County, but these were too poorly preserved to name them with any certainty. One suggestion was that they might either be related to a primitive perch or to *Pholidophorus*, an ancient fish that had eyes surrounded by bony plates and the large teeth of an aggressive predator.

The Miocene of eastern Oregon was ushered in by renewed volcanic activity, impacting the landscape and impounding numerous lakes. One of the largest of these was ancient Lake Idaho that backed up in the Snake River Plain from Vale and Baker, Oregon, all the way to Twin Falls, Idaho, 15 to 5 million years ago. Fish and mammal remains in the lake deposits were abundant, and after filling and emptying several times, drainage of this huge lake in the late Pleistocene time brought about the extinction of almost half of the species.

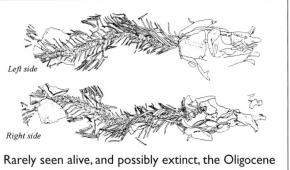

Rarely seen alive, and possibly extinct, the Oligocene mudminnow *Novumbra* is even more scarce in a preserved state. (After Cavender 1969.)

Disarticulated, broken fish bones occur in lakeshore and floodplain sands of the Deer Butte Formation at a number of localities near Quartz Basin in Malheur

County. Peter Kimmel listed 24 species of salmon, sunfish, sculpin, and whitefish, and showed high percentages of the catfish (*Ictalurus*), huge salmon-like predators (*Paleolox, Rhabdofario*), bottom-dwelling omnivorous suckers (*Catostomus*), and mollusk-eating minnows (*Idadon, Mylocheilus, Ptychocheilus*). The dimensions of some of the predatory fish, which include several that are more than three feet in length, suggest a lake of considerable size. An *Esox* (pike family), with a streamlined body enhancing its ability to capture moving prey, was also a lake resident.

Compiling a detailed paleoenvironmental picture, Kimmel pointed out that the massive sands and siltstones interbedded with pure volcanic ash could be evidence of either slow deposition or rapid deposition during very intermittent volcanic activity. However, he concluded that the presence of beautifully articulated fish and mammal bones and the incidence of conglomerate beds supports rapid deposition typical of a floodplain or the edge of a lake disrupted by swiftly flowing streams.

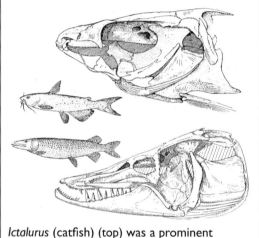

Ictalurus (catfish) (top) was a prominent resident in the Miocene lakes of eastern Oregon, as was the long-jawed predaceous pike *Esox*. Related to salmon, the Esox from Quartz Basin in Malheur County was the first recorded west of the Continental Divide. (After Cavender, Lundberg, and Wilson 1970; Gregory 1933.)

By the Pliocene Epoch, the warm tropical climate of eastern Oregon had disappeared along with the dense thick forests. Across the irregular topography, floods of lava created barriers behind which cool-water lakes were inhabited by fishes such as salmonoids and sculpins. Over 30 species of salmon occupied western North America waters during the late Tertiary, but fossils of these fish are astonishingly rare. Occasional teeth, vertebrae, and skull fragments of an enormous extinct salmon turned up in California as early as 1917, but it wasn't until the 1950s to 1960s that more complete specimens of a new species of Chinook Salmon *Smilodonichthys rastrosus* were described from Oregon.

In 1950 Ann Brownhill, a student at the University of Oregon, discovered numerous skeletal parts of a single large salmon near Gateway in Jefferson County from the 5-million-year-old Deschutes strata. The fossil was later collected by Arnold Shotwell, the director of the University of Oregon Museum of Natural History. During the summer of 1964 the skull, jaws, and gill arches of a second specimen from the same area were recovered. Further searching at Gateway over the past 40 years has yielded a steady stream of additional fragments, totaling more than 100 skeletal pieces and a complete skull 17 inches in length, producing a truly spectacular specimen.

Comparing the many pieces of the salmon to others collected in the Pacific Northwest, Ted Cavender and Robert Miller were able to develop something of an ancestry for this giant. Estimated to reach eight feet in length, the size of this fish and the details of its bone structure indicate that, like modern salmon, it was anadromous, splitting its life cycle between freshwater and the ocean. In addition to its massive size, there is evidence that the Oregon specimen was a breeding male and as such sported fang-like fighting teeth roughly 1.5 inches in length, hence the name *Smilodonichthys*, after the sabre-toothed cat. In spite of its fierce appearance and name, this great fish fed on plankton aided by a delicate and elaborate system of over 100 gill rakers in the skull that functioned as a sieve. *Rastrosus* is the Latin word for rakers.

At the Always Welcome Inn in Baker City, freshwater lake deposits were excavated by Jay Van Tassell of Eastern Oregon University. The fine-grained sediments yielded a fauna of minnows (*Achrocheilus*) and sunfish (*Archopites*), along with snails and clams, frogs, salamanders, and pond turtles. The site also has the bones of small vertebrates that indicate a possible Pliocene date. Nearby in the Powder River valley, sunfish (*Archoplites*), whitefish (*Prosopium*), and catfish (*Ameiurus*) populated a deep lake around 4 million years in the past before conditions became more arid.

Jaws of large minnows (*Mylocheilus*), found in lake deposits of Malheur County, have characteristic blunt or rounded shell-crushing teeth which, at first glance, resemble those of a primate. The sharp teeth of *Ptychocheilus* (middle) represent the most abundant and largest of the minnows found in the Deer Butte Formation, reaching a length of four feet. (After Gregory 1933; Smith 1975.)

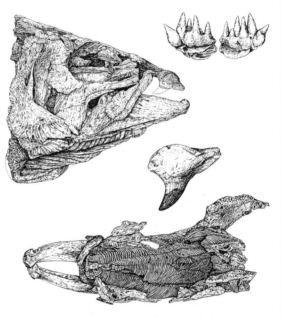

Views of the skull of the salmon *Smilodonichthys rastrosus*, showing gill rakers (bottom) and a breeding tooth (middle), along with the breeding teeth of the salmon-like *Onchorynchus* (top right), occur at Fossil Lake. (After Cavender and Miller 1972.)

The jaw bones of *Oncorhynchus*, a close relative to the *Smilodonichthys,* were unearthed in 1968 from an abandoned railroad cut near Worden in Klamath County. In probable Pliocene Yonna Formation strata, the salmon was accompanied by the bones of Cyprinidae (minnows) and Catostomidae (suckers), as well as fragments of plant and mammal remains. The Klamath basin periodically filled with pluvial lake waters, which, during the Pliocene and Pleistocene, may have flowed south into the Klamath River and on to the Pacific Ocean.

The state's most famous Pleistocene freshwater fish locality lies in northern Lake County at Fossil Lake. Even though terrestrial vertebrate remains are also known from this site, fish are by far the most common fossil. Jet black fish bones litter the sands of what was a large body of water during the late Tertiary and Pleistocene. The lack of fine-grained clays across much of the old lake surface has largely precluded the preservation of fragile scales, but skull bones, vertebrae, and ribs abound and are easily collected. The turbulent, shallow environment of the lake thoroughly scattered individual fish skeletons, and no articulated remains are reported. Some of the fish skulls occur in concretions, and bones were even encountered in pits dug into the lakebeds. In his review of the fauna of Fossil Lake, Ira Allison listed five species of carp and suckers (*Chasmistes, Cliola, Leucus, Myloleucus*) and the king salmon (*Oncorhynchus tschawytscha*), represented by skull bones, jaw, teeth, and vertebrae. The salmon bones indicate that former Fort Rock Lake was once much larger, encompassing the present-day areas of Fossil Lake, Christmas Lake, and Silver Lake. It may have overflowed through an outlet to the Pacific Ocean as late as Pleistocene time. Even after it became an isolated body of water, some fish were able to adapt and live there. The Fossil Lake site is almost entirely public land, owned by the U.S. Bureau of Land Management, and is closed to private collecting.

Marine Mammals

Historically, Oregon's marine mammals have received sporadic attention, in part, because of their fragmentary nature. Since the 1970s renewed interest can be attributed to the description and close scrutiny of new material by several researchers such as Lawrence Barnes and Ed Mitchell of the Los Angeles County Museum, by Clayton Ray at the Smithsonian Institution, Richard Tedford of the American Museum of Natural History, Charles Repenning at the U.S. Geological Survey, and Annalisa Berta and Thomas Demere of San Diego State University. The amount of marine vertebrate bones available was significantly enlarged with the fieldwork of two collectors along the Oregon coast, Douglas Emlong and Guy Pierson. Today knowledge of the evolution and taxonomy of marine vertebrates has reached the stage where significant advances are being made.

Born in Portsmouth, Virginia, Lawrence Barnes received most of his education in California, completing a Ph.D. from Berkeley in 1972 in vertebrate paleontology and mammalogy. Holding concurrent appointments at the Los Angeles County Museum and the Berkeley Museum of Paleontology, Barnes specializes in marine mammals of the Tertiary. Barnes is responsible for a large volume of recent work on marine vertebrates of the Oregon coast. (Photo courtesy Archives, Page Museum, Los Angeles.)

Living on the central Oregon coast, Douglas Emlong became fascinated with fossils from an early age, collecting his first marine vertebrate at Fogarty Creek in Lincoln County when he was 14 years old. After high school, Emlong opened a private museum at Depoe Bay where admission was 50 cents. Originally working under the supervision of Arnold Shotwell at the University of Oregon, Emlong contracted with the Smithsonian Institution, eventually selling them his collection in 1967. Something of a recluse, Emlong fell to his death at Devils Punch Bowl in 1980, while still a young man of 38. (Photo courtesy of Oregon Department of Geology and Mineral Industries.)

 Guy Pierson's interest in fossils began when he would gather dinosaur bones as a young boy visiting relatives in Montana. Although trained as an economist, Pierson stated in a newspaper interview that he "fell back into my old ways" when he moved to Moolack Beach in Lincoln County, where he was "sitting right on the fossil beds." For many years Pierson gave talks about fossils to Portland schools and displayed his collection to visitors. (Photo c. 1980; courtesy S. Pierson.)

Marine mammals come with a variety of common names, but they can be broadly grouped into the otters, seals, sea lions, and walruses (Carnivora), the whales and porpoises (Cetacea), and the manatees, dugongs, and sea cows (Sirenia). Several unusual creatures, such as the desmostylids (Desmostylia) and the bear-like *Kolponomos* lack modern representatives. Coastal exposures are the richest source for marine mammals in the state, reflecting the position of the ancient shoreline just east of where it lies today.

Carnivores constitute the families Mustelidae (sea otters), Phocidae (true seals), Otariidae (sea lions and fur seals), Odobenidae (walruses), and the Amphicynodontidae (*Kolponomos*). As predators that once thrived on land, carnivores inevitably invaded, adapted, and diversified in the marine environment as well. Modern carnivorous mustelids are familiar as weasels, skunks, minks, and river otters, but less well known as sea otters. These animals are generally slender, with powerful muscles, and an irritable disposition. The only fossil mustelid from the marine environment of Oregon is that of a sea otter *Enhydra*. Two *Enhydra* femurs from the Pleistocene Elk River Formation at Cape Blanco are identical to the

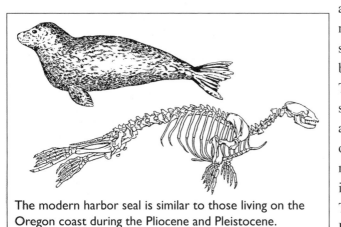

The modern harbor seal is similar to those living on the Oregon coast during the Pliocene and Pleistocene.

modern *Enhydra lutris* and are the earliest record of sea otters in North America. After death, this animal apparently washed into a quiet embayment where it was preserved along with the remains of an eared seal and a tapir. An assortment of foraminifera, molds of shells, fragmentary fish and birds, bryozoans and sponges, as well as crustaceans are also represented in this formation.

Seals (Phocids) and sea lions (Otariids) first appeared in the Atlantic, Mediterranean, and Pacific basin during the middle Miocene. True seals in the Phocidae family and eared seals and sea lions in the Otariidae are collectively known as pinnipeds. Although somewhat dated, the term is still useful in describing carnivores with webbed or pinnate feet. During the late 1800s these animals were further divided into the "walkers" and the "wrigglers." Seals which are able to flex their back legs and walk on shore are the otariids, while the phocids awkwardly wriggle on their stomach with their hind legs splayed behind. Charles Repenning and Richard Tedford attribute the comparatively poor fossil record of pinnipeds more to sampling than to actual frequency. "As much as anything this seems to be the result of mammalian paleontology stopping at the water's edge . . . pinniped remains frequently have been ignored or described as more or less isolated curiosities."

At Cape Blanco, the Pleistocene Port Orford Formation has produced Oregon's only true fossil seal, a phocid. This marginal record is consistent with that for phocids elsewhere. Lawrence Barnes and Edward Mitchell compared the skull fragments from a small young female seal *Phoca vitulina,* collected about 1960, to the bones of living seals inhabiting the Pacific Northwest coastline today. Both are similar, and they speculated that *Phoca vitulina* became extinct during the Pleistocene after being isolated by advancing ice. At the same time modern phocids, by contrast, were becoming established.

Three exceptional otariids (sea lions) from the Oligocene and Miocene, *Enaliarctos, Pteronarctos,* and *Desmatophoca,* in additional to the Pleistocene *Eumetopias,* are important parts of marine mammal history. The extinct *Enaliarctos* and *Pteronarctos*, aquatic bear-like creatures, are the most primitive of the pinnipeds. The new species *Enaliarctos barnesi, E. emlongi,* and *E. tedfordi,* and one previously described, *E. mitchelli* are all from Lincoln County between Newport and Waldport. The skulls were from three successive formations, the Oligocene Yaquina Formation and Miocene Nye Mudstone and Astoria strata. A careful comparison of the characteristics of the well-preserved skulls was unable to resolve the relationship between the three enlaliarctid species, although they may have shared a common ancestry.

The skull of *Pteronarctos,* a primitive bear-like aquatic animal (After Barnes 1990.)

Pteronarctos piersoni and *Pteronarctos goedertae*, from the Astoria Formation of Lincoln County, share several characteristics with modern true sea lions, fur seals, and primitive fur seals. A complete skull with canine teeth, lying among loose boulders on Moolack Creek, was found in April, 1983, by Guy Pierson and later described by Lawrence Barnes. The second species *Pteronarctos goedertae* was found at Nye Beach, and Barnes noted that *Pteronarctos* was the first new pinniped genus from the Pacific Northwest

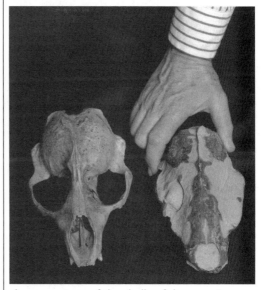

A comparison of the skulls of the aquatic mammal *Pteronarctos* (right) to that of a modern seal. (Photo courtesy Condon collection.)

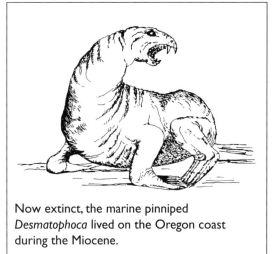

Now extinct, the marine pinniped *Desmatophoca* lived on the Oregon coast during the Miocene.

coast since Thomas Condon's description of *Desmatophoca*.

Astoria strata near Newport also yielded the skull and humerus (upper arm bone) of an extinct sea lion, *Desmatophoca oregonensis*, discovered and described by Thomas Condon in 1906. Even though the detailed physical characteristics of *Desmatophoca* are not clear, it may have been the size of living sea lions, with a long slender head and sharp teeth. The discovery of a second skull, *Desmatophoca brachycephala*, from near Knappton, Washington, as well as a partial skull and canine collected north of Newport, enabled Barnes to speculate on the animal's habitat. Known only from the northeast Pacific margin, *Desmatophoca*, along with *Pteronarctos*, inhabited a warm-temperate ocean on an open shelf at moderately shallow depths. Both pinnipeds occupied coastal Oregon during the Miocene, but more specimens are needed to determine whether they lived here during different seasons or migrated simultaneously to the same beaches.

Remains of only one sea lion (otariid) *Eumetopias* have been reported in the state. Ewart Baldwin found the radius (lower arm bone) of a sea lion in Pleistocene sediments of the Port Orford Formation near Cape Blanco. The ancestors of living *Eumetopias* first appeared as far back as two million years ago near Japan, and this sea lion was similar to those present on the coast today.

Yet another unique carnivore, *Kolponomos newportensis,* with shared characteristics between pinnipeds and bears, was recovered in 1969 from the Nye Formation. Similar to that of a sea otter, the partial skull, encased in a concretion, was found north of Newport by collector Douglas Emlong. Six years later he found a second part of the same concretion containing the remainder of the skull, mandible, and teeth. *Kolponomos* has been described as a partially aquatic, near-shore animal whose molar teeth were suited to crushing the hard shells of mussels, limpets, and abalone after prying them off rocks with sharp incisors and canines. It's possible the creature had a large upper lip, as in a modern-day walrus, that was sensitive to the touch, enabling the *Kolponomos* to feel its prey in turbulent seawater when visibility was poor. The lifestyle and habitat of *Kolponomos* probably closely resembled that of modern sea otters, although, in size, it was huge by comparison. Since no remains of this mammal have been found in deposits younger than middle Miocene, *Kolponomos* may represent the end of a lineage.

Walruses of the Odobenidae family are among the largest of the carnivorous marine mammals, attaining a weight of 1.5 tons on a diet of shrimp, starfish, and mollusks. Just one fossil walrus has been found in Oregon, and the identification of the skull was not confirmed until approximately 70 years after discovery. A skull, vertebrae, and rib fragments were extracted from a sandstone bluff at Fossil Point in Coos County by a local merchant and then sold to paleontologist William Dall. The late Miocene Empire Formation here produces a wealth of invertebrates as well as the bones of fish and marine mammals. The vertebrae and rib were identified as from a humpback whale *Megaptera*. At that time the skull was thought to be from a new species of sea lion *Pontolis magnus*. However, as additional bones appeared sporadically, a closer examination years later placed them tentatively among the walruses. Near Newport in Lincoln County, Douglas Emlong collected five skulls, along with teeth, vertebrae, and other skeletal pieces from the Astoria Formation. Identified with the Odobenidae family and named *Proneotherium*

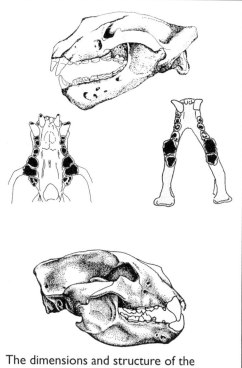

The dimensions and structure of the *Kolponomos* skull (top) are strikingly similar to that of modern bears (bottom). Flared broad cheek teeth characterize this marine mammal from the Miocene. (After Tedford, Barnes, and Ray 1994.)

repenningi by Thomas Demere and Annalisa Berta in 2001, these specimens possessed pointed piercing teeth enabling them to seize and hold their prey rather than shearing and chewing it.

Cetacea, or whales and porpoises, have fully adapted to the marine environment with a thick insulating cover of blubber and streamlined bodies, yet they retain a range of unmistakable mammalian features in their flesh and bone structure. Recently, they have been grouped with living artiodactyls such as the hippopotamus. Cetacea, the most abundant and varied of the marine mammals, are divided into the Archaeoceti or extinct toothed whales, the Odontoceti or modern toothed whales and porpoises, and the Mysticeti or baleen "whalebone" whales. The first two are carnivorous predators occupying top positions in food pyramids. The Mysticeti, by contrast, have essentially regressed by eliminating several rungs in the food pyramid to feed directly on small planktonic arthropods. Almost all modern great whales are Mysticeti, and their body shape readily distinguishes their lifestyle and habitats. The frequency of Mysticete fossils suggests that these whales lived in abundant populations off the West Coast during the Miocene epoch, much like their closest living relative, the gray whale. The modern gray whale closely resembles the fossil baleen whale *Cophocetus*.

It is noteworthy that whale remains in the Pacific Northwest are recovered from Oligocene sediments, an epoch when there was a pronounced drop in the number of cetaceans worldwide. Although two new groups, Mysticeti and Odontoceti, appear for the first time during the Oligocene, this interval is marked by limited populations and low diversity of cetaceans. This lull correlates well with a simultaneous global waning in plankton. Even though the baleen whales (Mysticeti) consumed enormous quantities of plankton, there can be little doubt of the key role of plankton in the food pyramids of toothed whales as well. Further evidence suggests that the diminished plankton populations may have been set off by slow but profound changes in global sea temperatures and ocean circulation. It is not uncommon for an environmental crisis to trigger the appearance of new and innovative life-forms even as it extinguished others on a large scale.

Primitive toothed archaeocete whales first appeared more than 50 million years ago during the Eocene epoch and then disappeared before the end of the Oligocene after branching into modern toothed and baleen whales. There are gaps in the ancestry of whales, and their remains have yet to be located in several formations such as the Pittsburg Bluff where sharks and bony fishes occur. However, in view of the wide range of paleoenvironments favorable for the preservation of marine vertebrates, the eventual discovery of additional cetacean material is inevitable.

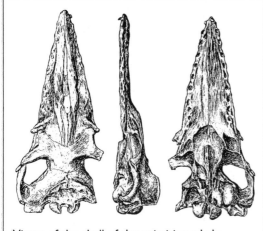

Views of the skull of the primitive whale *Aetiocetus* (After Emlong 1966.)

Bones of whales from the Eocene through the Pleistocene are scattered on the coast from Newport to the mouth of the Columbia River and even in the foothills of the Western Cascades. The Miocene Astoria Formation produces the richest numbers of fossil whales, but the oldest cetacean remains in the state are as yet undescribed vertebrae from Keasey sediments.

One of the most noteworthy finds is that of a primitive 12-foot-long toothed whale from the Oligocene Yaquina Formation in Lincoln County. *Aetiocetus* was initially discovered in rocks just north of Seal Rock State Park by Douglas Emlong in 1964. After excavating the 30-million-year-old specimen, he described the new genus as an archaeocete of the species *Aetiocetus cotylalveus*. The partially crushed,

complete skull, teeth, and most of the vertebrae of the
Aetiocetus were so distinctive that a new family name
Aetiocetidae was proposed for this unique creature, in
spite of the fact that the skeleton of the animal displays
many characteristics of baleen whales (Mysticeti).
Wear on the tooth crowns along with numerous crab
shells and fish remains at the same spot attest to the
diverse diet of *Aetiocetus*. The new classification has
since been disputed, and some paleontologists have
assigned the whale to the mysticetes in spite of its
multiple sharp teeth.

Other skeletal remains of *Aetiocetus* have turned
up on the coast. Two well-preserved skulls and
associated bones have come from the Yaquina and
Alsea formations. Another nearly complete skull, also
from near Seal Rock State Park, was found in 1969,
while the remainder of the skeleton, uncovered a few
years later, was named *Aetiocetus weltoni* by Larry
Barnes.

Part of Guy Pierson's collection, the
skull of a long-nosed porpoise *Ziphius*,
with tiny, slightly hooked teeth, is
dwarfed by that of a modern toothed
whale. (Photo courtesy Condon
collection.)

Fragmentary whale bones of *Aetiocetus* from the
Scotts Mills Formation have come to light along Butte
Creek in Marion County. This is the most easterly marine cetacean in the state, and
several vertebrae were discovered in the creek bed by local residents. A large slab
containing the fossils was removed intact from the late Oligocene siltstones by
William Orr during the summer of 1972. The presence of both juveniles and adults
of *Aetiocetus* indicates a developing coastal population of these predators, which are
similar to modern Orca killer whales. With the recent discovery of a well-preserved
complete specimen of *Aetiocetus* from
middle Tertiary sediments in Japan, a
geographic distribution of this primitive
whale is beginning to emerge.

Toothed Odontoceti whales have only
recently been reported from Oregon,
and most of the skeletons and fragments
of baleen whales (Mysticeti) along
the coast are from Miocene Astoria
strata, long known for its richness of
vertebrates. Indeed, the first such whale
to be recorded from Oregon was from

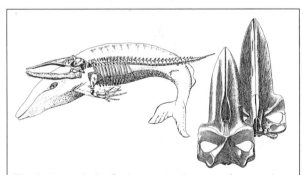

The baleen whale *Cophocetus* is the most frequently
recovered fossil whale in Oregon. The spade-shaped
skull is typical of cetaceans. (From Orr and Orr
1999b; Packard and Kellogg 1934.)

the Astoria by James Dana, a geologist accompanying an expedition sent out by the U.S. government to explore the west. In 1841 Dana collected vertebrae and ribs of a whale along with fish, crabs, mollusk shells, and plants near the entrance to the Columbia River. Another explorer to the region from 1848 to 1860 was naturalist and geologist George Gibbs, who sent cetacean fossils from this same formation to the Academy of Natural Sciences in Philadelphia.

Following this, a new genus of a primitive baleen whale, *Cophocetus oregonensis,* from the Astoria was named by Professors Earl Packard of the University of Oregon and Remington Kellogg of the University of California, Berkeley. The skeleton, lying in a nearly horizontal position with part of the jaw and skull exposed, was found at Newport in 1920. A few years later additional baleen material was recovered, while a third skull was collected about 1932 by a storekeeper at Otter Rock, who turned it over to Packard. As is typical of Astoria fossils, several of the bones were encased inside large hard concretions. As state paleontologist after the death of Thomas Condon, Packard was routinely notified of fossils picked up on Oregon beaches, so that over the years he examined well over a hundred occurrences of whale fragments. Similarly, in ensuing years, museum curators have identified dozens of whale remains from the coast, most of which turned out to be *Cophocetus.* Several porpoises and needle-nosed ziphids (Odontocetes) from the Astoria were collected by both Pierson and Emlong, but they have yet to be described.

Eroding from the cliffs at Cape Blanco, Miocene and Pleistocene whale bones include mandibles, vertebrae, ribs, limbs, and skulls from the Empire and Port Orford formation. Many were too broken to be identified, but a unique specimen showing a carbonized impression of baleen plates of a mysticete whale caught Packard's attention in 1947. Stiff at one end and fringe-like at the other, baleen is suspended in the mouth of a whale to screen and filter its food. Unlike bone, baleen is of chitinous organic composition unlikely to be preserved without rapid burial to protect it from bacteria. The irregular water-worn chitinous block was 18

Remains of an enormous Ice Age baleen whale were excavated from near Cape Blanco. (Photo by D. Taylor, courtesy Oregon Department of Geology and Mineral Industries.)

inches long, 11 inches wide, 4 inches thick, with a pattern of bristle-like fibers regularly spaced across the surface. The specimen was not collected in place, but rock matrix still adhering to the block confirmed it was from Empire strata.

One of the most recent whale finds, projecting from sediments of the Pleistocene Port Orford Formation, was recovered in 1988. Bones of an enormous baleen whale, on state park land at Cape Blanco, were brought to the attention of paleontologist David Taylor. With a team of volunteers from the Oregon Museum of Science and Industry, Taylor excavated vertebrae, ribs, and portions of a fin, which now reside at the Northwest Museum of Natural History in Portland.

Sirenians, or modern-day dugongs and manatees, first appear during the Eocene, but the only evidence of their presence in Oregon is from remains found in early Miocene sediments of the Nye Mudstone. A partial skull and jaw of the dugong, *Halitheriinae*, were found in 1984 by Guy Pierson. Enclosed in a concretion that had been drilled through by clams, the fragments came from Lost Creek south of Newport in Lincoln County. Researchers concluded that *Halitheriinae* reached the West Coast from the Caribbean when the ocean waters were at their warmest.

The desmostylids were yet another fascinating mammal inhabiting coastal waters. With a long skull and forward-projecting tusks, these creatures were something of a marine hippopotamus wading in the surf. Very few skulls of this peculiar extinct mammal have been found, and *Desmostylus* is largely represented in the fossil record by its frequent and distinctive cheek teeth. The generic name *Desmostylus,* derived from the Greek *Desmos* (bundle) and *Stylos* (pillar/column), refers to the structure of these odd molars as a tightly packed group of tapering cylinders.

The ancestry of desmostylids, which inhabited the Pacific Ocean between 10 to 30 million years ago, was examined in 1937 by Vertress L. Vanderhoof, at the Museum of Paleontology, Berkeley. Vanderhoof studied all of the available desmostylian remains. He assigned them to the sea cow (Sirenia) order of marine mammals. In attempting to come to some conclusions about the niche of desmostylids, Vanderhoof noted that all known fossils occur in shallow-water marine sediments and speculated that the animals were omnivorous, living on a diet of mollusks, seaweed, and shore grasses. He further pointed out that the California locations where *Desmostylus* turns up are invariably shell beds.

A complete skull from Spencer Creek in Lincoln County was chiseled out of the Miocene Astoria sandstone in 1914, but it wasn't until several years had passed that the beast was named *Desmostylus hesperus* by Oliver Hay of the Smithsonian Institution. The distinctive molar teeth of *Desmostylus* are multiple cylinders. (Photo courtesy Condon collection.)

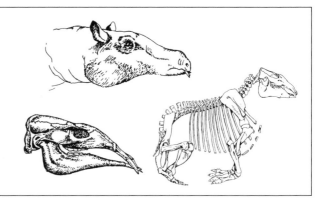

A reconstructed *Desmostylus* skull and skeleton. (From Orr and Orr 1999b.)

In the 1980s, the relationship among these creatures was reconsidered. Examining several lines of evidence, researchers concluded that while there was a common ancestor between elephants (Proboscidea) and desmostylids (Desmostylia), sea cow (Sirenia) was a separate group. In addition, the resemblance of desmostylid body size, build, and jaw to the hippopotamus suggests similar lifestyles. Never straying far from saltwater, these creatures may have fed on algae and other marine plants. Whereas Vanderhoof felt the projecting tusks were ideal for digging up mollusks, Daryl Domning of Howard University notes the "incisors and canines . . . seem well suited to forking up masses of vegetation, detaching plants from rocks . . . or uprooting mats of rhizomes."

A jaw, teeth, and tusk of what would be described as a new species of desmostylid, *Behemotops emlongi,* from Lincoln County were explained in a letter of 1977 by Douglas Emlong: "I stopped at Seal Rock . . . and found the most interesting thing of all—a giant desmostylian-like mandible. . . . The Oligocene specimen is far larger and heavier and I am sure it is a great find. . . . It may be related to that giant tusk . . . from the Yaquina Formation, and is not far from that area." Both Emlong and Pierson collected a substantial inventory of desmostylian material from the Yaquina, Nye, and Astoria formations. While skulls and mandibles of the *Desmostylus* are well known in Oregon, postcranial material is incomplete. By contrast, whole articulated skeletons of desmostylians have been recovered from mid-Tertiary shallow-water in both California and Japan.

As work proceeds on Oregon's fossil aquatic vertebrates, a more complete picture will emerge to complement the present knowledge of these fascinating animals.

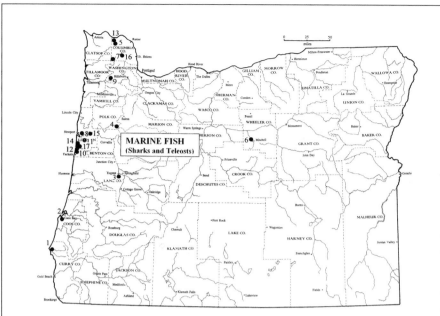

SHARK AND MARINE FISH LOCALITIES

The following are fossil shark and marine fish localities and a selection of authors who list faunas from the locales.

SHARKS

1. Cape Blanco, Curry Co. Empire/Port Orford Fms. Miocene-Pleistocene. Welton 1972, 1979.
2. Coos Bay, Coos Co. Coaledo Fm. Eocene. Welton 1972.
3. Eugene, Lane Co. Eugene Fm. Oligocene. Hickman 1969; Steere 1958.
4. Helmick Hill, Polk Co. Spencer Fm. Eocene. Welton 1972, 1979.
5. Mist, Columbia Co. Keasey Fm. Eocene. Welton 1972, 1979.
6. Mitchell, Wheeler Co. Hudspeth Fm. Cretaceous. Welton 1972.
7. Nehalem River Valley, Columbia Co. Cowlitz/Pittsburg Bluff Fms. Eocene-Oligocene. Steere 1957; Welton 1972.
8. Newport, Lincoln Co. Yaquina/Nye/Astoria Fms. Oligocene-Miocene. Welton 1972, 1979.
9. Scoggins Creek, Washington Co. Yamhill Fm. Eocene. Applegate 1968; Welton 1979.
10. Spencer Creek, Lincoln Co. Astoria Fm. Miocene. Welton 1979.
11. Toledo, Lincoln Co. Nestucca Fm. Eocene. Welton 1972.

MARINE (BONY) FISH

12. Alsea Bay, Lincoln Co. Yaquina Fm. Oligocene. David 1956.
13. Mist, Columbia Co. Keasey Fm. Eocene. David 1956.
14. Seal Rock, Lincoln Co. Yaquina Fm. Oligocene. Fierstine 2001.
15. Toledo, Lincoln Co. Nestucca Fm. Eocene. David 1956.
16. Vernonia, Columbia Co. Pittsburg Bluff Fm. Oligocene. David 1956; Moore 1976.
17. Yaquina Bay, Lincoln Co. Nye Mudstone. Miocene. David 1956.

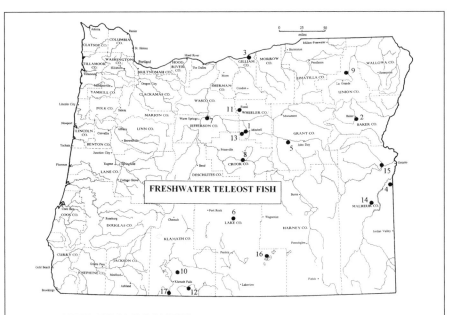

FRESHWATER TELEOST FISH

FRESHWATER FISH LOCALITIES

The following are fossil freshwater fish localities and a selection of authors who list faunas from the locales.

1. Allen Ranch (Mitchell), Wheeler Co. John Day Fm. Oligocene. Cavender 1969.
2. Always Welcome Inn (Baker City), Baker Co. No Fm. Pliocene. Van Tassell et al. 2007.
3. Arlington, Gilliam Co. The Dalles Fm. Miocene. Cavender and Miller 1972.
4. Blackjack Butte, Malheur Co. Deer Butte Fm. Miocene. Kimmel 1975, 1982.
5. Dayville, Grant Co. ?Mascall Fm. Miocene. Cope 1878, 1883d; Uyeno and Miller 1963.
6. Fossil Lake, Lake Co. No Fm. Pleistocene. Allison 1966; Allison and Bond 1983; Cope 1883d, 1889b; Hubbs and Miller 1948; Uyeno and Miller 1963.
7. Gateway, Jefferson Co. Deschutes Fm. Pliocene. Cavender and Miller 1972; Uyeno and Miller 1963.
8. Gray Ranch (Post), Crook Co. John Day Fm. Oligocene. Chaney 1927.
9. Imbler, Union Co., No Fm. Pliocene Van Tassell, McConnell, and Smith 2001.
10. Klamath Falls, Klamath Co. Yonna Fm. Pliocene. Newcomb 1958.
11. Knox Ranch, Wheeler Co. John Day Fm. Oligocene. Cavender 1969.
12. Lost River, Klamath Co. No. Fm. Pleistocene. Hubbs and Miller 1948; Jordan 1907; Uyeno and Miller 1963.
13. Ochoco Pass., Wheeler Co. Clarno Fm. Eocene. Cavender 1968.
14. Quartz Basin, Malheur Co. Deer Butte Fm. Miocene. Cavender, Lundberg, and Wilson 1970; Kimmel 1975.
15. Vale, Malheur Co. Chalk Butte Fm. Miocene. Kimmel 1985; Smith et al. 1982.
16. Warner Lake, Lake Co. Pleistocene. Cope 1883d.
17. Worden, Klamath Co. ?Yonna Fm. Pliocene. Cavender and Miller 1972.

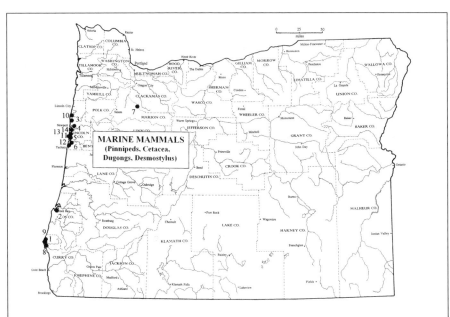

MARINE MAMMAL LOCALITIES

The following are fossil marine mammal localities and a selection of authors who list faunas from the locales.

OTTERS, SEALS, SEA LIONS, WALRUSES

1. Cape Blanco, Curry Co. Elk River/Port Orford Fms. Pleistocene. Barnes, Domning, and Ray 1985; Barnes and Mitchell 1975; Leffler 1964; Mitchell 1966; Packard 1947 .

2. Fossil Point, Coos Co. Empire Fm. Miocene. Dall 1909; Ray 1976; Repenning and Tedford 1977; Shotwell 1951.

3. Moolack Beach, Lincoln Co. Astoria Fm. Miocene. Barnes 1989, 1990; Demere and Berta 2001; Packard 1947; Ray 1976 .

4. Newport, Lincoln Co. Nye/Astoria Fms. Oligocene-Miocene. Barnes 1987, 1989; Berta 1991; Condon 1906; Demere and Berta 2002; Mitchell 1966, 1975; Tedford, Barnes, and Ray 1994.

5. Seal Rock, Lincoln Co. Yaquina Fm. Oligocene. Berta 1991.

WHALES

6. Alsea Bay, Lincoln Co. Alsea Fm. Oligocene. Ray 1976; Whitmore and Sanders 1976.

7. Butte Creek, Marion/Clackamas Cos. Scotts Mills Fm. Oligocene. Orr and Faulhaber 1975; Orr and Miller 1983.

8. Cape Blanco, Curry Co. Elk River Beds. Pliocene. Packard 1940.

9. Cape Blanco, Curry Co. Empire/Port Orford Fms. Miocene-Pleistocene. Packard 1947.

10. Otter Rock, Lincoln Co. Astoria Fm. Miocene. Packard 1935.

11. Seal Rock, Lincoln Co. Yaquina Fm. Oligocene. Barnes 1976; Barnes, Domning, and Ray 1985; Barnes et al. 1994; Emlong 1966.

12. Spencer Creek, Lincoln Co. Astoria Fm. Miocene. Hay 1916; Moore 1963; Packard and Kellogg 1934; Ray 1976.

DUGONGS, DESMOSTYLIDS

13. Seal Rock, Lincoln Co. Yaquina/Nye Fms. Oligocene. Barnes, Domning, and Ray 1985; Barnes et al. 1994; Domning and Ray 1986; Domning, Ray, and McKenna 1986; Packard 1921; Ray, Domning, and McKenna 1994; Vanderhoof 1937.
14. Yaquina Bay, Lincoln Co. Astoria Fm. Miocene. Vanderhoof 1937.

LAND MAMMALS

Land mammals first appeared during the Triassic period more than 200 million years in the past, about the same time that the dinosaurs emerged. These early mammals were small rat-sized creatures that fed on insects and co-existed with dinosaurs for several million years before the demise of those large reptiles at the end of the Cretaceous Period. During the subsequent Cenozoic Era, beginning about 66 million years ago, the smaller animals rapidly diversified to dominate terrestrial habitats.

A 1925 fossil collecting expedition to the John Day region from the University of California, crossing the Columbia River by ferry between Roosevelt and Arlington. (Photo courtesy Archives, Page Museum, Los Angeles.)

Fossils of mammals typically are found in nonmarine sedimentary rocks such as alluvial fan deposits, wind-blown dune sands, fine-grained lake sediments, or the coarse sands and gravels of streams and rivers. While ocean strata are not a likely place for preservation of extensive land-dwelling vertebrates, occasional mammals other than whales or seals are found in marine rocks. Since most of Oregon's Mesozoic rocks, including the Triassic, Jurassic, and Cretaceous periods are marine, the state's oldest land mammals occur in 44-million-year-old terrestrial Eocene exposures in Wheeler County. Virtually all of the Eocene through Miocene nonmarine intervals are in eastern Oregon, whereas the Pliocene is limited to Klamath and Baker counties, while Pleistocene locations are scattered throughout the state.

Dating Rocks Using Land Mammals

Initially it was difficult to resolve the age of nonmarine rocks in North America with vertebrate fossils because there was no standardized biostratigraphic succession to use as a guide. When the geologic record was being delineated and refined in the 1800s, research was carried out in the marine basins of Great Britain and western Europe, where terms like Eocene, Miocene, and Pliocene were based on characteristic marine mollusk shells occurring in those regions. In rare areas

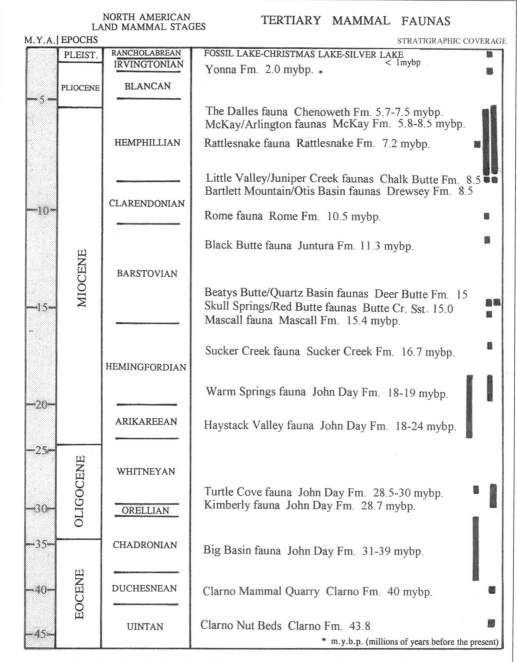

	NORTH AMERICAN LAND MAMMAL STAGES	TERTIARY MAMMAL FAUNAS	
M.Y.A.| EPOCHS			STRATIGRAPHIC COVERAGE

Stratigraphic distribution of Oregon terrestrial fossil mammals (After Armentrout 1981; Fisher and Rensberger 1972; Hunt and Stepleton 2004; Orr and Orr 1999b; Prothero, ed. 2001; Retallack, Bestland, and Fremd 2000.)

where marine rocks could be traced laterally directly into nonmarine strata of the same age, a prehistoric strand or shoreline between the old ocean and land was crossed. In these few cases, in which the marine and nonmarine layers were the same age, the terrestrial rocks could be dated by using mollusks from the adjacent ocean deposits. But for the most part, correlating or dating terrestrial strata by using mammal fossils was difficult and imprecise.

To remedy this problem, North American paleontologists established a separate chronology for mammal stages based on regional vertebrate faunas. One characteristic of mammals is that they evolve rapidly and thus tend to have very short stratigraphic ranges. The duration, then, of mammal species in the geologic record from first appearance to disappearance is notably short compared to that of fossil plants and invertebrates, and the limited stratigraphic ranges of mammals pinpoint the age of the entombing rock particularly well. Because of their generous populations in contrast to that of the larger mammals, rodents are of considerable practical value in biostratigraphy. By comparing rodent skeletal material, stratigraphic chronologies can be drawn and matched with those in other western regions. One example can be seen in the abundant rodent fauna in the John Day Formation that makes it possible to define three biostratigraphic zones based on genera of *Meniscomys* (mountain beaver), *Entoptychus* (pocket gopher), and *Mylagaulodon* (horned rodent). Recognized as far east as South Dakota, the three zones succeed one another from oldest to youngest.

Using very precise dating techniques such as radioactive decay (absolute dating) and magnetic stratigraphy, the mammal-based spans of time or stages were tied in with the established European epochs. One of the newest methods of magnetic reversal stratigraphy is based on changes in polarity of the earth's magnetic field. Integrating these new timelines with the fossil records of vertebrates, mollusks, foraminifera, and plants into a single chronologic framework has provided a standard for comparison with global time charts.

Land Mammal Preservation

In the fossil record not all animals are created equal. Because of their sizable populations and prolific reproduction rate, smaller mammals such as gophers, rats, mice, moles, and shrews produce far more fossil material than do the less numerous ungulates or hoofed mammals. Ungulates, in turn, invariably leave a better record than the more rare carnivores. A single vertebrate animal can produce over one hundred separate fossil bones, but the complexity of the skeleton is such that the various parts can readily be associated with a given species. Mammals are represented in the fossil record by their bones and teeth, but the teeth are most commonly preserved since they are the most durable part of the skeleton. The

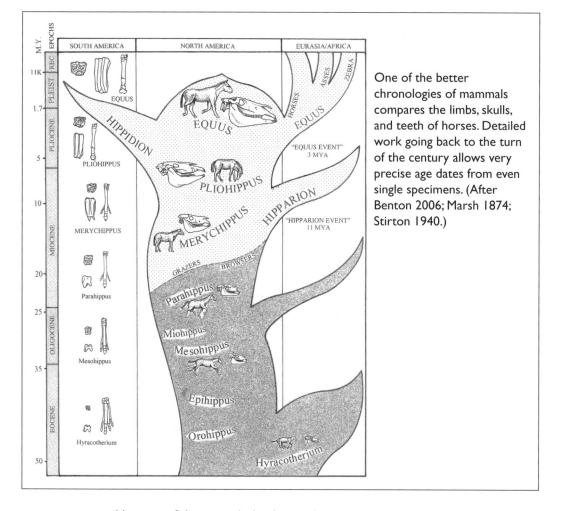

One of the better chronologies of mammals compares the limbs, skulls, and teeth of horses. Detailed work going back to the turn of the century allows very precise age dates from even single specimens. (After Benton 2006; Marsh 1874; Stirton 1940.)

architecture of the enamel, dentine, and cement, along with normal wear on the teeth, produces a characteristic pattern for most species much like a fingerprint, and the appearance of the crown surface of the cheek or jaw teeth is distinctive for identification purposes.

Habitats also play a role. Fossil bones of animals that lived in a lowland or plains, where sedimentary deposition takes place, are invariably more numerous than those living in a hilly or mountainous region, where erosion predominates. If animal carcasses are transported some distance by streams or scattered by scavengers before being deposited, much of the skeleton becomes worn, dispersed, or destroyed entirely. In some cases, the calcium makeup of the bone material may be soluble in acidic soils, thus a meager fossil record of vertebrates might reflect poor conditions for preservation rather than a limited population. To address these problems, an entire area of paleontology called taphonomy has emerged.

Taphonomic studies consider all manner of impacts on animals from their death to burial and preservation, and then to discovery.

Paleoenvironments and Land Mammals

In addition to aiding in the identification of a given mammal, aspects of the skeleton reflect the animal's niche in the environment. For example, the high-crowned teeth of the horse *Equus* indicate a diet of abrasive grass. The condition and frequency of skeletal parts also may provide insight into conditions at the time and place of burial and entombment. If a generous number of bones of a specimen are found at one site, and if they show minimal wear, death and burial probably occurred at that location.

Early work on the relationships of Tertiary mammals to their local environment was undertaken by J. Arnold Shotwell, at the University of Oregon. Shotwell focused on the interaction of community groups, their composition, structure, and habitat during the Tertiary. He realized that community members shared a particular habitat, and as changes took place in the surrounding environment, the creatures themselves adjusted accordingly. By counting fossil material and computing the numbers of genera, he was able to determine paleoenvironments. For example, grasslands encouraged the abundance of one community type, forested areas another. Determining that populations adapted to modifications in the surroundings, Shotwell then concluded that climate changes, in which the vegetation was altered or replaced, appeared to have the most significant effect on mammalian development. Ultimately, extreme variations in climate resulted in the evolution of the group or in complete destruction of the members themselves. In such a case, a new community could develop.

Eocene Land Mammals

Eocene mammals are only sparsely known in Oregon, and most are from the John Day River basin, where they have been entombed in a complex succession of lava, mudflows, and stream deposits of the Clarno Formation. Located a few miles from the abandoned village of Clarno in Wheeler County, the strata not only contains vertebrates of the widely known Mammal Quarry, but fruits, leaves, seeds, and wood of the famed Nut Beds as well. In contrast to the younger 40-million-year-old quarry, the 43.8-million-year-old Nut Beds have ample representatives of plant life but limited animal bones. The lush heavy vegetation, tropical rains, and wetlands of the Oregon Eocene produced a habitat favorable for such animals as tiny horses, camels, rhinoceroses, brontotheres, and carnivores. Fish, aquatic turtles, and alligators of formidable size completed the picture.

The Clarno Formation at Clarno Ferry is made up of layers of mudflows and lava along with multicolored fossil soils. (Photo courtesy Condon collection.)

Even though both are exposed in the John Day basin, Eocene Clarno vertebrates were discovered only in 1942, almost a century after discovery of fossils in the younger John Day Formation. Lorene Bones, whose husband amassed a collection of plant remains, found a mammal tooth, which proved to be that of *Hyrachyus*, a rhinoceros about the size of a Great Dane. The premolar came from the Clarno Nut Beds, and it would be another ten years before Lon Hancock found bones in what came to be known as the Hancock Mammal Quarry, about one-half mile to the northeast. Hancock, who had been exploring the region for some time, began working in the quarry over the summer of 1954. Steady excavations by Hancock and a team of student volunteers from the Oregon Museum of Science and Industry (OMSI) turned up a spectacular 3,000 fragments of bone and teeth of Eocene animals. These finds had an electrifying effect on the paleontology community and focused international attention on eastern Oregon. Professor Earl Packard, then at Palo Alto, California, called it "one of the most important vertebrate discoveries in the Oregon Territory since the early (18)80s." A formal contract was drawn up between OMSI and the Museum of Natural History at the University of Oregon for the site to be excavated jointly by Hancock and J. Arnold Shotwell at Eugene, where the material was to be housed. OMSI sporadically quarried the beds until 1972, and Shotwell's oversight assured that the resource wasn't publicly exploited in the same manner as happened with the John Day vertebrates. This Clarno locality is now part of the John Day National Monument and is closed to collecting.

The Clarno Nut Beds in Wheeler County, recognized for their diverse vegetation, also preserve a fauna of some 100 land-dwelling animals. In addition to *Hyrachyus, Hadrianus* (land turtle), *Orohippus* (four-toed horse), *Patriofelis* (cat), *Telmatherium* (brontothere), and *Pristichampsus* (alligator) were highly unusual in appearance and are represented by single samples of teeth, bones, or portions of jaws. These creatures are preserved in what was probably a stream-channel deposit, although some researchers have suggested a calm, shallow-lake environment. The presence of crocodilians and large land tortoises indicates a frost-free subtropical climate.

Alonzo "Lon" Hancock was born in Arkansas in 1884. His career with the U.S. Postal Service in Portland didn't hamper his interest in paleontology and geology. As far back as 1940, when Lon and his wife Berrie worked in eastern Oregon, they were accompanied by young students, who combed the region for fossils as they learned the geologic history. Many of these summer students went on to become professional paleontologists. After retirement, the Hancocks organized and operated a permanent camp that was formalized in 1951 as Camp Hancock under the auspices of the Portland-based Oregon Museum of Science and Industry. The few tents and open-air facilities were replaced by permanent buildings some ten years later. Before his death in 1961, Hancock donated his personal collection to OMSI. (Photo courtesy Condon collection.)

The clay/conglomerate layer of the Mammal Quarry, which preserves the vertebrates, is restricted to a small section in the upper part of the Clarno Formation. A number of large skulls were found intact, supporting the conclusion that the animal remains accumulated on the inside meander of a stream, where they decomposed before burial. Since the skeletons were scattered but showed little signs of wear, the carcasses may have been carried along in the stream while still covered with flesh. Overall, the quarry has yielded around 2,000 specimens of large animals, dominated by the elephant-sized brontotheres. An aquatic turtle is the only recorded North American specimen of the Chelydridae family from the interval 40 to 30 million years ago. Only partial skeletal elements of horses (*Haplohippus* and *Epihippus*), rhinoceroses (*Procadurcodon* and *Teletaceras*), tapirs (*Protapirus* and *Plesiocolopirus*), brontothere (*Protitanops*), the pig-like *Heptacodon*, and a saber-tooth cat (Nimravidae family) were unearthed.

Skulls and other bones of the small Clarno rhinoceros *Teletaceras radinskyi*, found in articulated condition, show that the animal may have evolved from an Asian migrant. Displaying affinities with faunas from Japan, Korea, and Siberia, these creatures might have crossed the Bering land bridge during the Eocene to displace mammals living in North America.

Oligocene Land Mammals

Of all the rock formations in the state, none approach the John Day for the quality of preservation and the richness of mammal remains. As noted by Chester Stock of the University of California, "Perhaps nowhere in North America . . . does a representative portion of the past history of mammalian life unfold so clearly and impressively as in the John Day region of north-central Oregon." The John Day

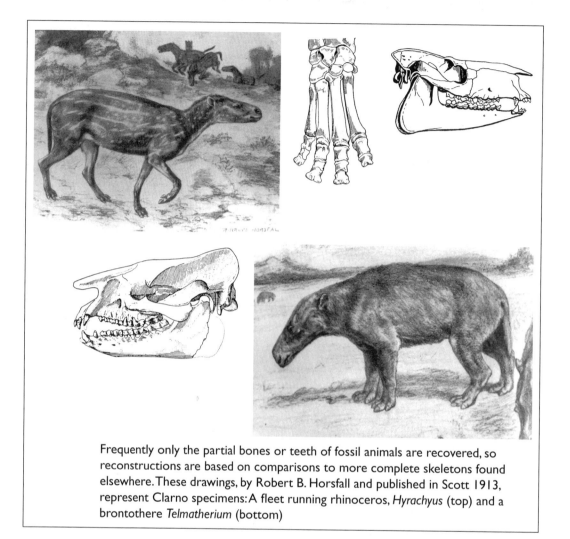

Frequently only the partial bones or teeth of fossil animals are recovered, so reconstructions are based on comparisons to more complete skeletons found elsewhere. These drawings, by Robert B. Horsfall and published in Scott 1913, represent Clarno specimens: A fleet running rhinoceros, *Hyrachyus* (top) and a brontothere *Telmatherium* (bottom)

environment, so conducive to vertebrate preservation, was one of slow-moving streams spreading sediments across low-lying areas. Ash falls blanketed the area, contributing to water-laden mudflows. Today the colorfully eroded sediments are plainly visible in the Painted Hills, at Foree (4-E) and Logan Butte, near Kimberly, and at Sheep Rock.

The John Day Formation spans the late Eocene, Oligocene, and early Miocene epochs. The strata represent the warm, humid conditions of the lower Big Basin Eocene, giving way to a drier, cooler climate during the Turtle Cove interval, which fostered wooded grasslands, seasonal flooding, and well-developed stream networks. In the Miocene, as temperatures and rainfall continued to decline, herds of horses, expanding grasslands, and flowing streams marked the Kimberly and Haystack Valley environments.

The discovery of fossil beds in the John Day occurred in the mid 1800s, long before those of the Clarno were investigated. Thomas Condon is credited with first bringing the presence of fossils here to light. In 1865 or 1866 Condon had journeyed south from The Dalles with a cavalry company under Captain John Drake, at which time he collected fossil bones, teeth, and leaves on the John Day River and at Bridge Creek. The bones were sent to Joseph Leidy, a professor of anatomy at the University of Pennsylvania, who identified the specimens as two new forms of rhinoceros, a *Leptomeryx* (modern pecora), *Lophiodon* (tapir-like creature), and *Elotherium* (pig-like animal), several Oreodontidae (sheep-like creature), and *Anchitherium* (small horse).

Othniel C. Marsh and Edward Cope continued in Leidy's footsteps. Unlike Leidy, who had to work with mere fragments sent to him, in many cases without precise locations or stratigraphic data, Cope and Marsh took to the field themselves to inspect specimens *in situ*. A fierce competition ensued between these masters of paleontology to see who could amass the largest and most complete collections. Both employed professional agents in the race to triumph, a life-long rivalry that tainted much of their scientific contributions.

Clarno mammals include the sleek cat, *Patriofelis* (top), small browsing four-toed horses *Haplohippus* (middle), and *Orohippus* (bottom). All drawings by R.B. Horsfall, published in Scott 1913.

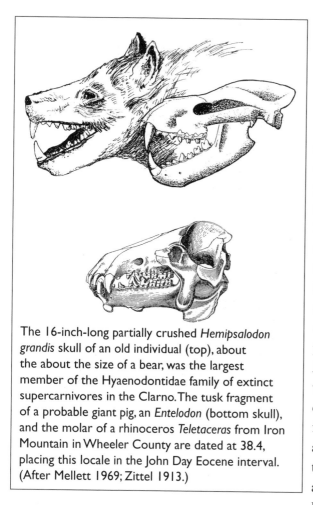

The 16-inch-long partially crushed *Hemipsalodon grandis* skull of an old individual (top), about the about the size of a bear, was the largest member of the Hyaenodontidae family of extinct supercarnivores in the Clarno. The tusk fragment of a probable giant pig, an *Entelodon* (bottom skull), and the molar of a rhinoceros *Teletaceras* from Iron Mountain in Wheeler County are dated at 38.4, placing this locale in the John Day Eocene interval. (After Mellett 1969; Zittel 1913.)

The search for the remains of fossil mammals was the stimulus that brought paleontologists and museum representatives to central Oregon once the scent of unique and ample finds became known. Wagonloads of material moved eastward, but the bulk of it languished in storage for many years; much was never examined, and some was subsequently destroyed. By November 1870 Marsh had become aware of Thomas Condon's John Day finds, and his letters display his anxiety that he had not yet seen them. Condon did send him a small package of specimens, and Marsh, who was greatly pleased, named a new species of peccary *Platygonus condoni*. Marsh's appetite was merely whetted, however, and he decided to visit Oregon himself in the fall of 1871. Arranging to meet Condon at Canyon City, his party consisted of the local rancher Sam Snook, the guide and commercial collector Leander Davis, horses, wagons, supplies, and 15 or so young college men from Yale. Rushing past many sites such as the fossil-rich Turtle Cove, they examined Condon's collection at The Dalles, then left for home. Marsh identified the horses as *Anchitherium, Anchippus, Miohippus*, and *Protohippus*, but later expanded the list to include *Diceratherium* (rhinoceros), *Eporeodon* (sheep-like animal), and two peccaries *Thinohyus* and *Dicotyles*. Key John Day specimens allowed Marsh to demonstrate a complete lineage of the evolution of horses from the Eocene *Orohippus* to the modern *Equus*.

In 1871 Condon began to exchange letters with Edward Cope, who requested material, sent scientific journals, and offered to purchase the fossils. As with others, Condon shared his duplicates, but refused to sell any of his "first-class specimens." Because he lived in Oregon, Condon's initial role was to provide information, localities, and fossils to the better-known eastern paleontologists, which he generously did. Later, as he began to build up his own collection, Condon became more reluctant to part with his material. In 1870 he forwarded

Although he was from a family of modest means, Othniel Charles Marsh (back center) was subsidized by a well-to-do uncle until he graduated from Yale in 1862, where he subsequently took a position on the faculty. From 1870 to 1875 he led expeditions across the western territories, paying for many of these trips with his own funds. Appointed as the official United States vertebrate paleontologist in 1882, he worked in conjunction with the U.S. Geological Survey. Amassing fossils, which form the nucleus of the vertebrate sections at the Smithsonian Museum and at Yale University, Marsh himself was completely overwhelmed by the amount of material, and much was left for later paleontologists to examine and describe after he died in 1899. In the photograph, Marsh is shown with members of an expedition in 1872. While the display of firearms may seem theatrical, it was just four years later that General Custer was defeated at Little Big Horn. (Photo courtesy Yale University, Archives.)

specimens to Marsh, marking those to be returned, and five years later he loaned a number of prized bird bones from Fossil Lake to Cope, some of which had no duplicates. Both men were negligent in returning what they had borrowed. Even though Condon wrote to inquire about the status of the bird remains, it wasn't until 1926 that they were recovered. Marsh's loans weren't sent back until 1906, years after his death and shortly before Condon himself died.

Scientists continued to exploit the west. Princeton University sponsored expeditions of professors and students from 1877 through 1889. Led by William Berryman Scott and assisted by the professional collector Leander Davis, they were camping at Turtle Cove during mid July, 1889, when the remains of rhinoceroses, the three-toed horse, camels, deer, oreodonts, turtles, several carnivores, and

Edward Drinker Cope, who was Marsh's competitor, was born of a wealthy Quaker family in 1840. A fascination with the unexamined western regions took him to those parts of the United States, where he uncovered remarkable numbers of fossils while working with state geologic surveys. An abrasive personality combined with scientific errors, which were inevitable when covering such vast areas so quickly, led Cope into acrimonious disputes with Othniel Marsh. Cope is remembered for unraveling the complexities of Tertiary fossils throughout the west and for his numerous eclectic publications. Cope contributed close to 40 major works on western fossils. His most remarkable work was the 1009-page, ponderous, five-and-one-half-inch-thick book *Vertebrata of the Tertiary Formations of the West,* which includes 100 plates. Cope taught at the University of Pennsylvania until his death in 1897. (Photo courtesy Archives, Smithsonian Institution.)

rodents were removed from John Day strata. At the end of the field season, several tons of bones were taken by wagon to Dayville and there packed for shipment to Princeton.

Paleontologists alone weren't responsible for attrition to the fossil beds, and many residents collected to one degree or another. As noted in *The Oregonian* of November 1946, "Nearly everyone in the vicinity is a fossil hunter of sorts." Locals were hired as guides, while farm and ranch families provided room and board. The Days and Mascalls were two such families. Collector Charles Sternberg described William Day as knowing "every inch of the fossil beds and all the best camping grounds; his services were invaluable." The superior value of a good skull over skeletal parts, which were somewhat disparagingly termed "joints," was brought home to Sternberg, who found that the fossil beds had already been picked over. "Here and there we would run across a pile of broken bones and a hole from which a skull had been taken." Day replied, "We were only looking for heads, though we sometimes saved knucks and jints."

Tourists were another problem, and in 1933 concerns were expressed by Thomas Large, publisher of *Northwest Science,* about "tourists . . . who for souvenirs did not hesitate to break up and carry off parts. . . . [S]ome form of Federal administration should be established." Others felt roadside exhibits should be set up to call attention to the fossil resource, and State Park Superintendent Sam Boardman intended to put markers at mileposts to point out the geologic story.

Expressing similar reservations in 1943, John Merriam viewed the fossils as a great heritage, "which we must understand, interpret, and use as in some measure a guide for the future." To that end he proposed that a detailed study of the John Day area be undertaken, that educational displays be set up at the university in Eugene, and that a book be written with chapters on individual aspects of the region. Notable outcomes were professors Warren Smith's

Overlooking the Columbia River near The Dalles, this 1901 expedition may have been part of John Merriam's group from the University of California. (Photo courtesy Condon collection.)

A University of California collecting trip down the Deschutes River to The Dalles at the end of May 1899 consisted of paleontologist John Merriam, ornithologist Lloye Miller, photographer Frank Calkins, mineralogist George Hatch, a fisherman, and guide Leander Davis. Encountering many difficulties during the journey, they managed to load over 300 pounds of specimens on wagons in addition to extracting the jaw of an oreodont, two teeth of an *Anchitherium* (horse), and the skulls of the pig-like *Entelodon*. Left to right, unknown, Lloye Miller, John Merriam, and Leander Davis. (Photo courtesy Condon collection.)

Scenic Values in the John Day and Chester Stock's *Oregon's Wonderland of the Past—The John Day*.

A committee of paleontologists called The Associates was to oversee the John Day Park; but the notion of setting aside land as a national park wasn't met with enthusiasm by the locals. Ultimately, however, The Associates was responsible for much of the John Day Park acquisition. In order to protect the fossils in the basin, 14,402 acres was set aside as the federal John Day Fossil Beds National Monument on October 26, 1974. The Monument takes in three separate tracts: the Clarno region, the Painted Hills, and Sheep Rock in Grant and Wheeler Counties. Just recently, new displays opened at the Thomas Condon Paleontology Center at the Sheep Rock station.

The pioneer William Mascall not only provided a bed and home-cooked meals to visitors, but he offered a temporary storeroom for fossils as well. He settled near Dayville in 1864, and today the ranch still remains in the family. Mascall standing atop the fossiliferous rock formation named for him. (Photo courtesy Mascall family.)

Due to professional interactions and physical proximity, West Coast geologists began to play a dominant role in central Oregon by the early 1900s. From the University of California at Berkeley, John Merriam (left) was one of the leading practitioners of paleontology, who spoke of "my very deep interest in Oregon and all it represents." Born in Iowa in 1869, Merriam moved with his family to Berkeley where he studied botany and geology. In 1893 he received a Ph.D. in vertebrate paleontology from the renowned Karl von Zittel at Munich University, after which he joined the staff at Berkeley. He lived his final years on the West Coast, traveling often to Eugene, before dying in Oakland in 1945. Merriam sponsored many students, and his influence on the course of West Coast paleontology is immeasurable. John Merriam and Lloye Miller extract fossils from the massive greenish ash of the middle John Day Formation in 1899. (Photo courtesy Condon collection.)

Thomas Jefferson Weatherford, an old-time cowboy, trapper, and stage driver, began to amass a sizable number of vertebrates from the John day area in the late 1800s. Even though the precise localities weren't always known, Weatherford hoped to sell his collection to the government. But it wasn't until after his death that his sons received $10,000 from the Grant County Chamber of Commerce for the fossils. (Photo courtesy E. Baldwin.)

At the University of California, some members of The Associates. Left to right, back row: Earl Packard, W.S.W. Kew, and J.D. Nomland; front row: John Buwalda, John Merriam, Chester Stock, and Bruce Clark. (Photo courtesy David K. Smith, Museum of Paleontology, University of California, Berkeley.)

In 1901 John Merriam divided the John Day Formation into three members based on color: the lower reddish beds, the middle greenish beds, and the uppermost cream to white layer. Since Merriam's original classification, the lithostratigraphic and biostratigraphic designations have been refined several times, and four separate intervals or members were established in 1972. The oldest, the deeply oxidized red claystone of the Eocene-Oligocene Big Basin Member has abundant plant remains. Above that, the greenish tuffs of the Oligocene Turtle Cove Member have well-preserved bone material, as do the buff Miocene Kimberly and uppermost tuffs and conglomerates of the Haystack Valley sequence. Of these, the Turtle Cove has by far the most vertebrate fossils. On the basis of fossil soils (paleosols), the Big Basin Member, which extends from the upper Eocene into the Oligocene, was further separated into three units by Greg Retallack at the University of Oregon, while the Haystack Valley was split into four by Ellen Stepleton and Robert Hunt from the University of Nebraska.

Additionally, the John Day Formation was broken into three main geographic areas: an eastern facies along the John Day River, a southern section on the Crooked River, and western exposures on the Deschutes River. The eastern and southern regions have yielded the most mammalian remains, which become scarce toward the west near Warm Springs in Jefferson County.

Even though there are minor gaps in the record of plants and animals from the John Day valley, the stratigraphy still offers one of the most complete sequences of Tertiary fossils anywhere. An extraordinary diversity of more than 100 genera of vertebrate remains from the John Day Formation includes many varieties of cats and dogs, weasels, rabbits, and numerous rodents. Hoofed mammals such as tapirs, horses, pigs, elephants, rhinoceroses, camels, and oreodonts are extinct or only vaguely resemble their present-day relatives. The Turtle Cove Member had the most abundant and varied fauna with at least 61 species represented. A subtropical rather than temperate climate is evidenced by the presence of warmth-loving tortoises (*Stylemys capax*), whereas a landscape of mixed vegetation can be seen in the presence of grazing rhinoceroses (*Diceratherium*), browsing horses (*Miohippus*), and squirrel-like rodents.

The hoofed mammals, for which the John Day beds are most famous, are the extinct oreodonts, similar to sheep in appearance, but otherwise unrelated. Belonging to two families, Agriochoeridae and Merycoidontidae, their numerous remains suggest that large herds roamed in eastern Oregon. Ranging in size from the size of a dog up to that of a small horse, they displayed a wedge-shaped skull, like that of a pig, which prompted their early designation as ruminant hogs; however, the tooth structure is vastly different. The seledont molar cusps of the browsing oreodonts form pairs of crescents in crown view, whereas those of pigs, with an omnivorous diet, possess low cusp rounded molars or bunodont teeth. Although related to the Merycoidontidae, *Agriochoerus* had claws instead of hoofs, and it has

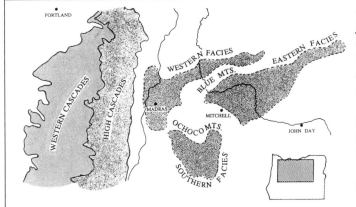

Geographic divisions of the John Day Formation are separated from each other by the highlands of the Blue and Ochoco mountains. (After Fisher and Rensberger 1972; Hanson 1996; Hunt and Stepleton 2004; Retallack, Bestland, and Fremd 2000; Robinson, Brem, and McKee 1984; Woodburne and Robinson 1977.)

been suggested that they climbed trees like raccoons. Other John Day oreodonts include *Desmatochoerus, Eporeodon, Eucrotaphus, Oreodontoides, Hypsilops, Merycochoerus, Paroreodon, Promerycochoerus*, and *Ticholeptus.*

Marsupialia (opossum): Opossums are routinely known from strata as old as late Cretaceous in North America, but only one such fossil has been reported in Oregon, and that was in the southern facies of the John Day beds near Logan Butte in Crook County. The fragmentary skull and lower jaw of a new species *Peratherium merriami* were similar in size to the modern Virginia *Didelphis.*

Insectivora (shrews and moles): One remarkable anomaly in the John Day faunas is the paucity of insectivores. Insectivores, or insect-eaters, are known from rocks as old as Cretaceous, and elsewhere in the world they have a persistent though not overly spectacular Tertiary fossil record. Found near Courtrock in Grant County, the skull and left jaw of *Micropternodus morgani* represent a new species, which has no modern equivalent, and even its relationship to other groups is uncertain.

Chiroptera (bat) and Dermoptera (lemur): The skeletal elements of a bat are among the exceptionally rare finds from the John Day. One of the very few fossil bats known from Oregon is represented by only a jaw and shoulder bones recovered from upper Oligocene shales behind the schoolhouse at Fossil in Wheeler County. Because the fragments were in such poor condition, the bat could not be identified, and, indeed, the fossil record of Chiroptera before the Pleistocene is sparse.

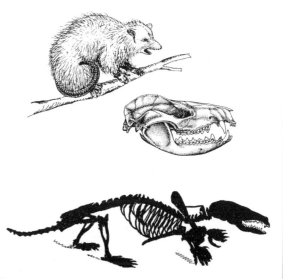

Opossum skull (top) and mole skeleton (bottom) (After Stock and Furlong 1922; Steele and Jenks 1887.)

The other extraordinary discovery is that of an upper tooth fragment of *Ekgmowechashala*, similar to a lemur and as obscure as its peculiar name. The initial specimen was found by John Rensberger in the pale green Turtle Cove tuffs near Picture Gorge in 1961. In the Sioux language, "Ekgmo" means cat, "wechasha" means man, and "la" means little. Taken together, the name means little cat man. Originally thought to be a primate, it has now been placed in the order Dermoptera, which has living equivalents in Asia, Indonesia, and the Philippines. Existing on an herbivorous diet, modern members are nearly helpless on the ground but are quite adept in the trees, where they fly by means of a gliding membrane between their limbs.

After the initial tooth fragment of the primitive
Ekgmowechashala was found in the pale green
Turtle Cove tuffs in 1961, additional teeth
were uncovered near Sheep Rock in the 1990s.

The skull and mandible of *Nimravus* (top) and
Dinictis, rapacious cats from the John Day.
(From B. Horsfall, in Scott 1913; after Eaton
1922; Merriam 1906.)

The skull and mandible of the primitive canids
Cormocyon (bottom) and *Osbornodon* (top).
(From B. Horsfall, in Scott 1913; after Matthew
1899; Merriam 1906.)

Carnivora (dogs, cats, bears, raccoons): The list of carnivores or flesh-eating mammals is small, and of the unusual finds the cats *Dinaelurus* and *Nimravus*, the dogs *Cormocyon, Leptocyon, Mesocyon, Nothocyon, Osbornodon*, the bear-dogs *Cynodictis* and *Temnocyon*, and the bears *Allocyon, Parictis*, and *Enhydrocyon* are the most frequent. Sleek lightweight predators such as *Eusmilus* and *Dinictis*, which were somewhat larger than a lynx, lack the sharp extremely elongate canine teeth common to those of later periods. *Cormocyon*, about the size of a fox and almost weasel-like in the proportions of its long tail and body, was more of a primitive dog, but *Nothocyon* is considered to be a raccoon-dog. Raccoons, pandas, and related animals of the Procyonidae family along with bears of the Ursidae family evolved from dogs during the middle Tertiary, and both share similar characteristics. The bear-dogs (*Daphaenodon* and *Pericyon*), the dogs (*Cynarctoides, Desmocyon*), and mustelids (*Hesperocyon, Oligobunis*) are Miocene species from the Haystack Valley and Kimberly members of the John Day.

Artiodactyls (even-toed hoofed pigs, oreodonts, camels, and deer): Even-toed hoofed mammals known from the John Day basin include giant pigs, peccaries, oreodonts, camels, and deer, many of which exhibit a basic resemblance to their modern counterparts. A wealth of pigs and peccaries (*Cynorca, Thinohyus*) from this stratum includes the giant pig-like scavengers, entelodonts. Entelodonts had a distinctive flat bone shelf that projected just below the eyes, a pronounced high ridge or sagittal crest atop the skull, which attests to their powerful jaws, and the low, rounded bunodont cheek teeth of an omnivore.

Resembling a pig, the powerfully built predator-scavenger, *Entelodon* is customarily found in Tertiary faunas. (From B. Horsfall, in Scott 1913; after Pearson 1927.)

North American camels are restricted to the Tertiary period, but as plains dwellers, their fossil record is extensive and relatively complete. Their delicate small hoofs contrast to the flat wide pads of the modern camels, specifying a late adaptation to arid or desert environments. The camel *Protomeryx* occurs in the Oligocene interval of the John Day Formation, whereas *Gentilicamelus* and *Paratylopus* are in both the Oligocene and Miocene. Deer likely appeared in the Oligocene, and their fossil remains are less frequently encountered than those of other cervids. A rare upper molar of the primitive cervid *Blastomeryx* from the Warm Springs Miocene is smaller in size than that of its living relatives.

Oreodonts became quite diverse in John Day time. (Reconstruction from B. Horsfall, in Scott 1913; photos from the Condon collection; after Schultz and Falkenbach 1968.)

Fossils of the small mouse-deer *Hypertragulus hesperius* are the most common artiodactyl in John Day sediments. Only distantly related to deer, they are mostly found as broken fragments. About the size of a jackrabbit, these animals had an upper horny pad in their mouth, much like that of a cow, which cropped off their food.

Large, clawed, browsing perissodactyls, Chalicotheriidae were never numerous, and their fossil remains from sites in Oregon and on the Gulf Coast of Florida are even more exceptional. Muscular, but relatively small in size, *Moropus oregonensis* frequented well-watered coastal habitats rich in plant life. These curious beasts first appeared during the Eocene Epoch and persisted into the Pleistocene when they fell to extinction. It is easy to see why early workers failed to link the teeth, skull, and great claws of *Moropus*, initially placing the animal in the horse family. Although close relatives, chalicotheres had

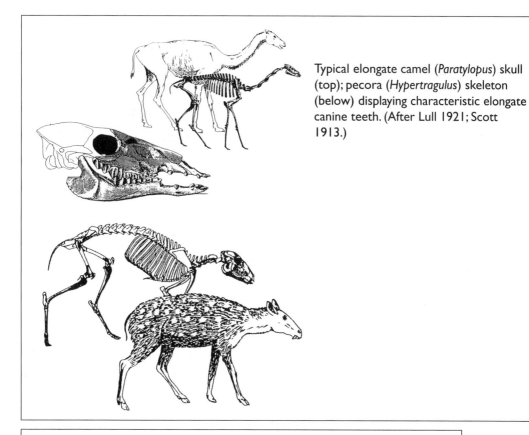

Typical elongate camel (*Paratylopus*) skull (top); pecora (*Hypertragulus*) skeleton (below) displaying characteristic elongate canine teeth. (After Lull 1921; Scott 1913.)

Perissodactyls (odd-toed hoofed horses, rhinoceros, and tapirs): As a domestic animal long exploited for agriculture and transport, horses have a special place in human history. Explorers found no horses when they arrived in North America, but once introduced, the animals spread widely across the continent. Horses, along with camels and elephants, had been abundant here since the Eocene, but became extinct as recently as 11,000 years ago. Since horse fossils are plentiful in the North American Tertiary, they provide a continuous evolutionary picture, and those from the John Day valley are particularly outstanding in quality and quantity. The three-toed Oligocene horses *Mesohippus* and *Miohippus* were small, with low-crowned teeth suited for browsing. *Miohippus* may have given rise to all later species of horses. Miocene John Day strata include the genera *Archaeohippus* and *Merychippus*.

elongate claws instead of hoofs, and the family has since been reevaluated. The pronounced claws may have been digging aids, as is the case with later sloths. The teeth and foot bones of several *Moropus* were collected in the John Day valley in the late nineteenth century by Thomas Condon, Othniel Marsh, and others. These specimens were reexamined, reevaluated, and augmented considerably by teeth and bone fragments collected in the 1990s.

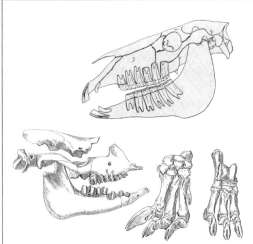

Horse skull showing high-crowned (hypsodont) teeth and the deep elongate jaw (top); skull and claws of *Moropus*. (After Holland and Peterson 1913.)

Rodentia (beavers, mice, squirrels, gophers, rats); Lagomorpha (rabbits): Rodents probably shared a common ancestor with rabbits in the distant past, but the groups separated sometime in the early Cenozoic into two orders, Lagomorpha and Rodentia. In the Oregon fossil record, only three genera of rabbits are known, *Archaeolagus*, *Hypolagus*, and *Palaeolagus*. This is on par with a modest diversity of 12 Tertiary genera worldwide. Rodents, on the other hand, virtually exploded into 400 genera, and fossil beaver, gophers, and squirrels are common locally. One of the more cosmopolitan rodent families, the Sciuridae, includes squirrels, chipmunks, prairie dogs, and marmots. Some 14 species of fossil squirrels alone have been recorded from the John Day and later formations. Other Oligocene and Miocene John Day rodents include the beavers (*Allomys, Meniscomys, Palaeocastor, Stenofiber*), the aplodontid (*Sewellelodon*), the gophers (*Entoptychus, Geomys, Pleurolicus Schizodontomys*), mice and rats (*Leidymys, Palustrimus*), the squirrels (*Sciurus*), and the horned gophers (*Mesogaulus, Mylagaulodon*).

Relatively rare in North America, tapirs (*Protapirus*) can be found in the Miocene interval of the John Day. (From B. Horsfall, in Scott 1913; after Hunt and Stepleton 2004.)

The skull of the aquatic rhinoceros *Metamynodon* displays the attachments, which enable the powerful jaw muscles to operate the anvil-like grinding molars. (From B. Horsfall, in Scott 1913; after Cope 1887; Hatcher 1901.)

Tapirs are comparatively rare in the Eocene and early Oligocene of North America, and their abrupt appearance here may reflect a migration from Europe. The postcranial skeleton of these odd animals bears a striking resemblance to that of rhinoceros, and, indeed, tapirs, rhinoceroses, and horses are closely related. A small but stocky John Day *Protapirus* was just about half the size of the living South American tapir. *Nexuotapirus* and *Miotapirus* are found in the Haystack Valley and Kimberly members of the Miocene John Day.

Rhinoceroses have consistently been one of the most diverse and widespread perissodactyls, well adapted to an herbivorous lifestyle and far outnumbering horses, tapirs, and brontotheres. Once living in great herds in North America, they are represented in the John Day Formation by *Caenopus, Diceratherium*, and *Metamynodon.*

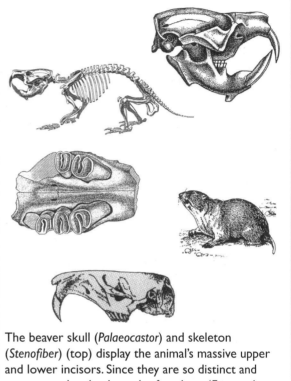

The beaver skull (*Palaeocastor*) and skeleton (*Stenofiber*) (top) display the animal's massive upper and lower incisors. Since they are so distinct and numerous, the cheek teeth of gophers (*Entoptychus, Geomys*, bottom) are particularly useful for age-dating intervals of the John Day Formation. (After Peterson 1905.)

Caenopus was large and hornless, approximately the size of a modern white rhinoceros. *Diceratherium* was somewhat smaller with paired horns placed side-by-side on the tip of the snout, and *Metamynodon* was a large aquatic animal, similar to an hippopotamus in appearance.

Miocene Mammals

From the Oligocene into the Miocene, ocean waters receded with steady uplift of the Coast Range. In western Oregon, deposition continued in the shallow marine basins, which formed on the narrow coastal plain, while the eastern part of the state was covered with thick layers of volcanic debris. Building a plateau across northeast and central Oregon as well as into Washington and Idaho, over 50,000 cubic miles of accumulated flows of Columbia River basalts had the most impact on the topography, flora, and fauna. Volcanic events demolished whole habitats even as new plant and animal communities flourished between disturbances. The uneven basalt surface provided depressions where animal carcasses, carried by streams, accumulated and were buried under conditions optimum for preservation.

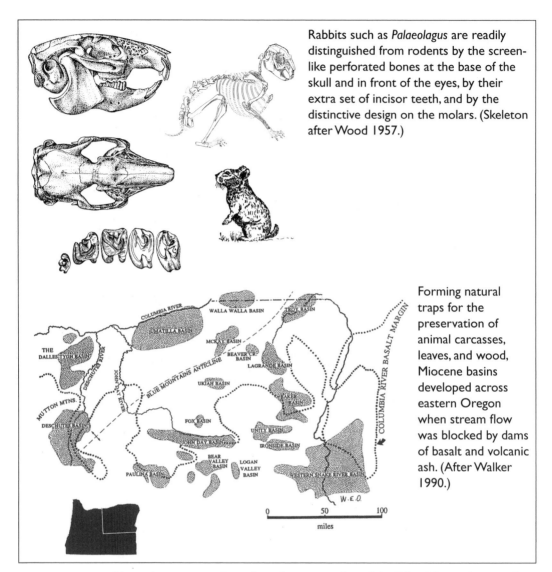

Rabbits such as *Palaeolagus* are readily distinguished from rodents by the screen-like perforated bones at the base of the skull and in front of the eyes, by their extra set of incisor teeth, and by the distinctive design on the molars. (Skeleton after Wood 1957.)

Forming natural traps for the preservation of animal carcasses, leaves, and wood, Miocene basins developed across eastern Oregon when stream flow was blocked by dams of basalt and volcanic ash. (After Walker 1990.)

Along the west, south, and southeast margins of the Blue Mountains Province and merging with the Owyhee Plateau to the south, separate depositional lake basins characterized the Miocene. All were ephemeral, the sizes varied, and most were geographically restricted so that separate formational names were assigned to each one of the sediments. Early day paleontologists collected the individual sites, but there were few attempts to relate the prehistoric animals, plants, and environments of the isolated basins to each other. Specific vertebrate localities were investigated by C. Lewis Gazin, who described the Skull Springs fauna in 1932, by David W. Scharf, who listed vertebrates from Succor Creek in 1935, by Robert Wallace, who collected Beatys Butte in 1946, and by Theodore Downs, who worked on the Mascall fauna in 1956. Ralph Chaney and Arnold Shotwell both

Chester Stock took a Ph.D. from the University of California in 1917, where he had been so enthused by John Merriam's lectures that he majored in geology. Stock was born in San Francisco in 1892, and spent most of his career as a professor at California Institute of Technology and as a staff member of the Los Angeles County Museum. A popular lecturer, Stock sponsored many students who went on to outstanding careers in paleontology. His early studies emphasized Pleistocene mammals, but he branched out into the Eocene and Miocene of the Great Basin, Pacific Coast, and Northern Mexico. Devoting much of the summer of 1916 to fieldwork in Oregon, he helped to secure larger collections from both the Rattlesnake and Mascall formations in Grant County. Stock died in Pasadena in 1950. In the picture Chester Stock is measuring dire wolf skulls from the LaBrea tar pits. (Photo courtesy Archives, Page Museum, Los Angeles.)

In the collection, preparation, and cataloging of material from newly discovered sites, Stock was aided by Eustace Furlong, who followed him from Berkeley to the California Institute of Technology. A meticulous fieldworker, Furlong contributed significantly to unraveling the Tertiary history of the Great Basin and Mexico by leading and supervising parties of graduate students and by his careful curation of specimens. At the invitation of the University of Oregon, Furlong traveled to Eugene in 1945 to prepare and study oreodonts from Thomas Condon's collection. While there he was struck by an automobile and severely injured. Furlong never fully recovered, and, returning to Davis, California, he died in January 1950. When Arnold Shotwell began his career at the Museum in Eugene, the oreodonts, still untouched, were laid out in the preparation laboratory with Furlong's notations. Eustace Furlong is inspecting a rhinoceros skull. (Photo courtesy Condon collection.)

assessed the Miocene setting of eastern Oregon. In 1959 Chaney compared floras, time relationships of the deposits, and climates producing the vegetation, while about the same time Arnold Shotwell evaluated the composition and environment of each community.

On the southern edge of the Blue Mountains, the Owyhee Plateau remained virtually unexplored until field reconnaissance began in 1924. Paleontologists recovered and itemized new vertebrate faunas at Succor Creek, Skull Springs, Quartz Basin, and Red Basin in Malheur County, and at Beatys Butte in Harney

In 1951 Theodore Downs produced a detailed study of the Mascall fauna for his Ph.D. from the University of California under the guidance of Reuben Stirton. Accepting the hospitality of the Mascalls and other local ranching families, Downs collected at numerous sites in the area as well as reviewed specimens stored in various museums. Downs pursued a career at the Los Angeles County Museum where he was curator until his death in 1997. (Photo courtesy Archives, Page Museum, Los Angeles.)

The most significant additions to understanding the geologic history of the eastern Oregon Miocene were made by J. Arnold Shotwell, who worked in the Museum of Natural History in Eugene as curator from 1947 to 1953 and as director from then until his retirement in 1971. Born in North Bend in 1923, he completed his Ph.D. in paleontology from the University of California, Berkeley, in 1953. Specializing in the paleoecology of fossil mammals and the effects of extinction, climate, and migration on populations, he spent his summers in eastern Oregon where he located and collected many new sites of Miocene vertebrates. At present the Shotwells live in Bay Center, Washington. (Photo courtesy Archives, University of Oregon.)

In spite of the massive elephant leg bone in his hands in this photograph, Howard Hutchison is known for his work on Tertiary shrews and moles, including many species from eastern Oregon. Hutchinson, who worked with Shotwell, completed a Master's degree from the University of Oregon in 1964 on late Tertiary insectivores of the Great Basin and a Ph.D. in paleontology at the University of California, Berkeley, in 1976, where he stayed on at the museum. His eclectic interests included Cretaceous and Tertiary turtles, vertebrates from Burma, and alligators from Texas, along with his focus on the systematics, morphology, and ecology of recent and fossil Talpidae (moles). Hutchison retired in 1993 and moved to Utah where he continues his research. (Photo courtesy Archives, University of California, Berkeley.)

County. These middle Miocene depositional sites, marked by varying amounts of fossil mammals, were part of the same landscape around 15 million years ago when the preservation of animal remains was linked to thick accumulations of volcanic debris in lake bottoms.

Derived from a nearby caldera complex, ash of the Sucker Creek Formation is nearly white in color but weathers to green, brown, or yellow in badlands topography. More famous for its beautifully preserved fossil plants such as oak, pine, willow, and maple than for its fragmentary vertebrates, the Sucker Creek represents forested hills surrounding basins filled by lake waters. Browsing horses (*Hypohippus*, *Parahippus*), a chalicothere (*Moropus*), primitive cervids (*Dromomeryx*), rhinoceroses, peccaries (*Prosthennops*), and a dog-like carnivore were first collected from this formation during the 1920s.

In 1992 Kevin Downing of the University of Arizona inventoried small mammal populations from Sucker Creek layers, washing and sieving more than a thousand pounds of matrix. Downing found almost 30 separate species, including several that were new at Devils Gate south and east of the Owyhee Dam in Malheur County and at a second locality immediately across the border at Stagestop, Idaho. Shrews and moles (Insectivora), bats (Chiroptera), rabbits (Lagomorpha), and rodents (Rodentia) could be identified from 2,700 skeletal fragments of teeth and limb bones, many encased in concretions. New genera, recorded for the first time, include a bat (Phyllostomatidae family), a flying squirrel (*Petauristodon*), and a rodent (*Leptodontomys*). Interestingly, vegetative successions from woodlands to grasslands, brought about by volcanic disturbances, were not reflected by parallel changes in the composition of small mammal assemblages. Citing observances after the 1980 eruption of Mt. St. Helens, Downing notes that rodents occupy environmentally diverse habitats and can therefore recover more quickly from overwhelming disasters than can plants.

Mammal faunas at Skull Springs and Beatys Butte have many elements in common with those from Succor Creek, and only *Archaeohippus* (horse) and an unidentified insectivore from Beatys Butte don't occur in all three areas. The presence of *Merycodus*, a grazing antelope, *Liodontia*, a large beaver, along with bones of the horses *Hypohippus, Parahippus*, and *Merychippus*, and *Dromomeryx*, a cervid, point to a paleolandscape of open woodlands adjacent to grassy plains and ponds. A surface assortment at Skull Springs, within the Butte Creek Volcanic sandstone layer, produced teeth and bones encased in nodules. To the southwest at Beatys Butte, Chester Stock made collections over several summers, beginning in 1938. Water-deposited tuffs near the base of an old volcanic cone, where uniform layers and stratification convey burial in calm water, are particularly productive. Larger bone fragments regularly show gnaw marks made by rodents' teeth, and

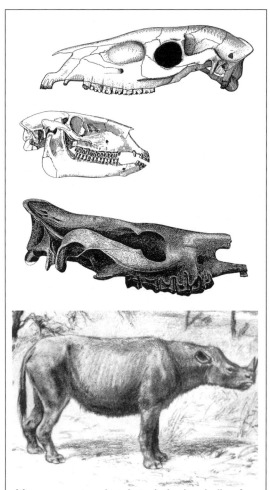

Miocene perissodactyls include the skulls of the small three-toed *Merychippus* (top) and *Parahippus* (middle), horses with relatively short (low-crowned) jaw teeth, which were adapted to browsing on brush and trees. The horned *Diceratherium* rhinoceros (bottom) enjoyed a remarkable variety of habitats from forest to plains and rivers. (After Cope 1887; Downs 1956; Peterson 1920; from B. Horsfall, in Scott 1913.)

traces of rodent burrows, now filled with ash, appear as irregular elongate cylinders.

Vertebrate remains at Quartz Basin in the Deer Butte Formation and at Red Basin in the Butte Creek Volcanic Sandstone were the focus of a study by Arnold Shotwell with a field crew from the University of Oregon during the summers of 1960 and 1961. By perusing aerial photos and locating areas that looked promising, Shotwell successfully uncovered remarkably rich pockets of bones in the sandstones and conglomerates. All of the key or index species, which delineate the specific Miocene stages of 16 to 5.5 million years ago, were present, enabling him to establish a Miocene geologic time sequence, which was applicable elsewhere in the Northern Great Basin.

Sites at Quartz and Red basins yield a large and diverse assortment of insectivores, rodents, carnivores, and hoofed mammals; however, contrasting groups of animals occurred at the two areas. Larger species were more abundant at Red Basin, while smaller animals such as insectivores and rodents characterize the Quartz Basin region. A total of 70 species were reported from both areas, but only eight species of small mammals were common to both. Differences between the two faunas prompted Shotwell to conclude that they were nearly identical in age but from markedly different environments. Red Basin represented a placid, shallow lake at a high elevation, whereas Quartz Basin was a streamside basin at a lower setting.

Theodore Downs speculated that periodic floods and suffocating volcanic clouds of ash and gases from erupting Cascade and local cones were carried by winds and spread over a broad shallow basin in Grant County on the southwestern border of the Blue Mountains province. The rain of falling ash particles would have overcome and destroyed all life. Extensive stream systems carried the bloated carcasses into lakes where they settled into muddy layers on the bottom. Successive intervals of the Mascall Formation, deposited in the lakes and ponds of the John Day Valley, accumulated both vertebrate and plant remains during this period some 15.4 million years ago. Fossil leaves, wood, and seeds, scattered throughout the buff to white tuffs of the Mascall, suggest open upland forests, grasslands, and wetlands in and around the basin.

The concentration of predators, hoofed mammals, birds, rodents, fish, and turtles in the Mascall Formation is not as high as in the older John Day strata. Large animals, such as elephants, would have been a familiar sight on the Miocene landscape of eastern Oregon, but the sheep-like oreodonts, so prevalent during the older John Day, are markedly reduced in numbers. Plains-dwelling horses, antelope, and primitive deer were more frequent, especially the graceful pony-sized, three-toed horse *Merychippus*. Superbly preserved skulls of *Merychippus* display short, broad, grinding

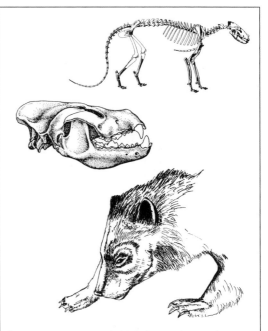

Bears, raccoons, and dogs, living during the Miocene, descended from primitive, heavy, bear-sized canids such as *Amphicyon* and *Hemicyon*, which had long thick tails. The skunk-sized mustelid *Leptarctus* (bottom) sported characteristic twin saggital crests on the skull. (After Peterson 1910; Scott 1913; Stock 1930.)

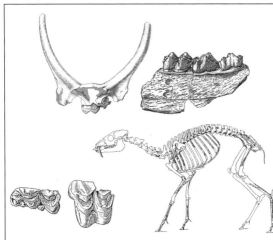

Artiodactyls (even-toed mammals) of the Miocene include the deer-like *Dromomeryx* (top) that sported fluted curved horns and the *Blastomeryx* (bottom) with long canines. (From B. Horsfall, in Scott 1913; after Downs 1956; Gazin 1932; Matthew 1909.)

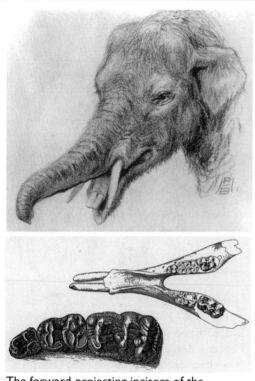

The forward-projecting incisors of the *Gomphotherium* may have been used for plowing up plants and roots from the forest floor. (From B. Horsfall, in Scott 1913; after Gaudry 1867.)

teeth adapted for browsing on tough shrubs. *Dromomeryx*, a primitive cervid with curved blunt horns, is represented by numerous skeletal parts, and a partial jaw of the horned gopher *Mylagaulus* from Paulina Creek in Crook County displays the characteristic complex pattern on the crown surface of the molar teeth. The nearly complete skull and mandible of *Tomarctus*, a large flesh-eating bear-dog, and a sloth (Megalonychidae family), two exotic creatures, were found at the Camp Creek and Crooked River locales. Several mandibles of the mustelid *Leptarctus oregonensis* came from Dayville in 1930 and again in 1990.

Near Baker several primitive Miocene elephants were exhumed by local ranchers in 1924. Parts of a distinctive mastodon, identified as *Gomphotherium*, were found 15 miles east of Baker in volcanic ash and tuffs mapped as Mascall strata. The relatively complete mandible was eventually examined by Downs, who described the massive creature and conjectured that the body had been carried by floods and deposited before much decomposition could take place. The most striking feature of the mastodon was the forward projecting elongate incisor teeth. The extension of the entire lower jaw was pronounced, and the family Gomphotheriidae is called the "long-jawed mastodons."

Yet another elephant of Mascall age was uncovered at Ironside in Malheur County in 1941. Lon Hancock, a private collector, and Chester Arnold, a paleobotanist from the University of Michigan, were searching for fossil leaves in Miocene lake sediments when Hancock broke open a piece of sandstone to reveal mammalian bones. Many hours of chiseling freed the enormous 450-pound skull of a *Miomastodon*. Mislabeled for years as the similar *Tetrabelodon*, the significance of the jaw wasn't realized until George Gaylord Simpson, a paleontologist from the American Museum of Natural History in New York, examined the specimen, which he recognized as *Miomastodon merriami*. Twelve years later, Hancock returned to the site where he located and excavated the lower jaw, teeth, and short tusks about 15 feet from the spot that the skull had been discovered earlier. In the evacuation process the jaw was broken, and the entire skull and jaw were transported to the Museum of Natural History at Eugene to be pieced back together over several years. Around 1960 the specimen was moved to the Oregon Museum of Science and Industry, Portland, where the identification is under review.

On the Owyhee Plateau, just beyond the southern margin of the Blue Mountains, lake and stream locations continued as sites of optimum vertebrate preservation into late Miocene time. In Malheur County the Juntura basin is overlapped and obscured by younger lavas, volcanic tuffs, mudflows, and sandstones of the older Juntura and younger Drewsey formations. Thick Juntura ash beds reflect voluminous volcanic activity, whereas the significantly smaller amount of ash in the succeeding Drewsey lakebed deposits signals waning activity. Layers of diatoms in the central basin, along with the presence of freshwater snails, record a large placid water body, which drained during the latest Juntura interval. Erosion of higher elevations accompanied by some volcanic activity is evident in deposition of the Drewsey Formation.

The older Juntura strata contain the temperate Stinking Water leaf flora but no vertebrate remains, while the younger Juntura yields diatoms and a wealth of mammals. The uppermost Drewsey Formation also produces a wide variety of bones. At Black Butte, a peak south of the town of Juntura in Malheur County, mammal remains occur in the latest Juntura sediments, which have been dated at 11.3 million years before the present. The fauna, geology, and paleoenvironment of the basin were described by Arnold Shotwell, who based the age on the presence

A partial listing of middle Miocene land vertebrates:

Marsupalia (opossum): Didelphid

Insectivora (moles and shrews): *Alluvisorex, Heterosorex, Ingentisorex, Paradomnina, Scalopoides, Scapanoscapter, Scapanus,*

Carnivora (dogs, cats, bear, raccoons): *Amphicyon, Bassariscus, Hemicyon, Leptarctus, Martes, Mustela, Osteoborus, Pliocyon, Pliotaxidea, Pseudaelurus, Tomarctus, Vulpes*

Artiodactyla (pigs, camels, antelope): *Blastomeryx, Cynorca, Dromomeryx, Megatylopus, Merycodus, Miolabis, Paratylopus, Procamelus, Prosthennops, Ticholeptus*

Perissodactyla (horses, oreodonts, rhinoceros): *Aphelops, Archaeohippus, Diceratherium, Hipparion, Hypohippus, Merychippus, Moropus, Parahippus, Pliohippus, Teleoceras, Ustatochoerus*

Proboscidea (elephant): *Ambylodon, Gomphotherium, Mammut, Mastodont, Miomastodon, Platybelodon*

Rodentia (mice, beaver, gopher, squirrel): *Adjidaumo, Citellus, Diprionomys, Dipoides, Epigaulus, Eucastor, Eutamias, Hystricops, Leptodontomys, Liodontia, Microtoscoptes, Monosaulax, Mylagaulus, Peridiomys, Perognathus, Peromyscus, Petauristodon, Pliosaccomys, Prodipodomys, Sciurus, Spermophilus, Tardontia*

Lagomorpha (rabbit): *Hypolagus, Oreolagus*

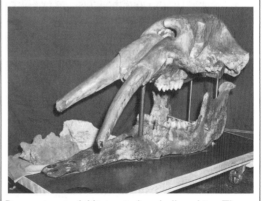

Reconstructed *Miomastodon* skull and jaw. The lower jaw was discovered a dozen years after the skull. (Photo courtesy Condon collection.)

of the beaver *Eucastor malheurensis* and the absence of the horse *Hypohippus*. Shrews, small rodents such as mice, the carnivorous hyena-like *Aelurodon*, the bone-eating dog *Osteoborus*, and rare tooth remains of the mustelid *Sthenictis* were also uncovered here. A comparison of the Oregon *Sthenictis* to one from Mongolia, suggests that this carnivore migrated from North America to China in the middle Miocene. Black Butte herbivores include elephants (*Mammut*) and the shovel-tusk mastodon *Platybelodon*, camels (*Procamelus* and *Megatylopus*), and a horse (*Hipparion*). Years later Shotwell recalled his amazement when he first saw the large fragments of the elephant *Platybelodon* lying on the surface during a reconnaissance trip. Tracing the pieces up slope he found a mandible and tusks and proceeded immediately to excavate.

During the 1950s and 1960s many new and exciting Miocene faunas were discovered in eastern Oregon by Arnold Shotwell and his field teams from the University of Oregon when they delved into unknown strata. (Photo courtesy Condon collection.)

Dated at 8.5 million years ago, sediments of the Drewsey Formation lie immediately above the Juntura. Deposited during the late Miocene in several areas close to Drinkwater Pass, Otis, and Bartlett Mountain in Harney and Malheur counties, tuffs of the Drewsey yield scattered vertebrates, which were initially prospected by Chester Stock and Eustace Furlong during the 1920s and 1930s and by Arnold Shotwell during the middle 1950s. Most collections were made from surface finds, and many of the species are similar to those at Rome, McKay Reservoir, and in the Rattlesnake Formation. The characteristic elongate molars of the grazing horse *Pliohippus spectans*, along with the remains of *Microtoscoptes disjunctus* (mouse), *Mylagaulus* (horned gopher), and *Dipoides stirtoni* (beaver), common at Bartlett Mountain, are also found at Rome and in the Rattlesnake. The only known rodents with horns, which belong to the Mylagaulidae family, may have developed this feature for digging, and not for sexual prowess, as has been suggested. *Liodontia furlongi*, a mountain beaver, occurs at the Rome locale but not in the Drewsey Formation, probably an indication of different environments.

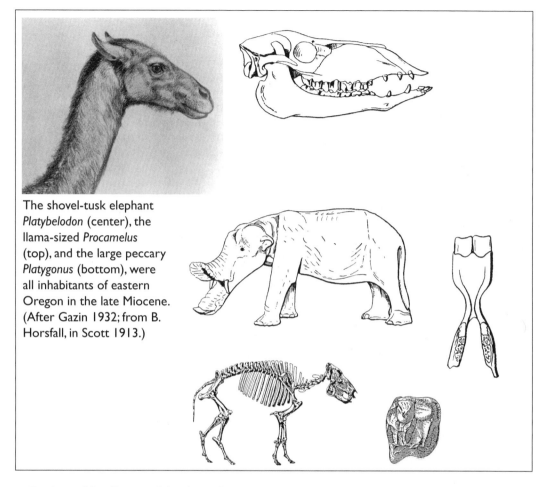

The shovel-tusk elephant *Platybelodon* (center), the llama-sized *Procamelus* (top), and the large peccary *Platygonus* (bottom), were all inhabitants of eastern Oregon in the late Miocene. (After Gazin 1932; from B. Horsfall, in Scott 1913.)

Pockets of fossils east of the town of Juntura near Little Valley and Juniper Creek are found in stream, marsh, and lake sediments of the Chalk Butte Formation. Accumulating in and around ancient Lake Idaho, which stretched across east central Oregon to present-day Twin Falls, Idaho, Chalk Butte sediments were deposited when stream drainage systems of the western Snake River Plain were periodically blocked by lava flows. Several successive lakes rose and then drained during the middle Miocene through the Pliocene. The obvious wear and fragmentary condition of the bones reflect postmortem water transportation and redeposition. Little Valley had more of the smaller rodents and shrews, while Juniper Creek had more carnivores, horses, camels, and rhinoceroses. A beaver (*Dipoides*), rabbit (*Hypolagus*), dog (*Canis*), camel (*Megatylopus*), horse (*Pliohippus*), and the rodent *Diprionolys* were present in both localities.

In the John Day Valley, episodes of tilting, folding, and erosion at the end of Mascall time separate that formation from the overlying gravels, tuffs, and silts of the younger Rattlesnake, dated at 7.2 million years before the present. Fossil

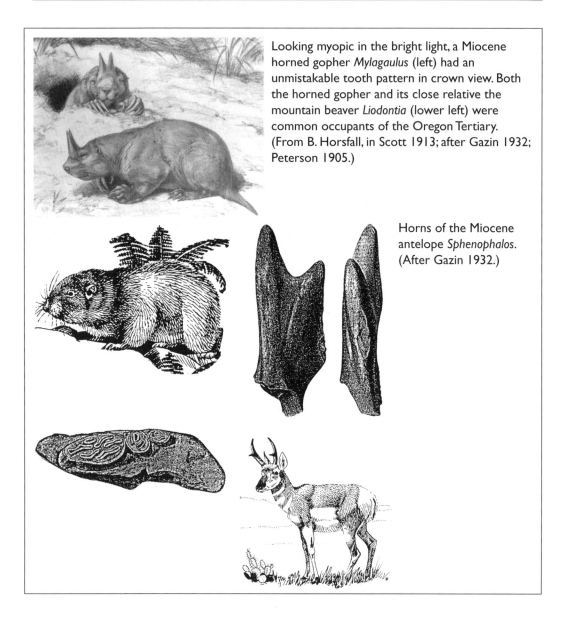

Looking myopic in the bright light, a Miocene horned gopher *Mylagaulus* (left) had an unmistakable tooth pattern in crown view. Both the horned gopher and its close relative the mountain beaver *Liodontia* (lower left) were common occupants of the Oregon Tertiary. (From B. Horsfall, in Scott 1913; after Gazin 1932; Peterson 1905.)

Horns of the Miocene antelope *Sphenophalos*. (After Gazin 1932.)

mammals and plants from the Rattlesnake Formation reflect a climate significantly drier than that of earlier intervals. Continuing expansion of grasslands accompanied the steady loss of the older, forested Mascall habitats. This environmental shift is most profoundly demonstrated by the evolution of horses, which shows a progressive increase in body size, elongation of the limbs, a reduction in the number of toes, and an adaptation of molar teeth from browsing to grazing. An important browsing animal, the rhinoceros became extinct in North America during the Pliocene, shortly after Rattlesnake time. A sharp increase in the populations of grazing horses (*Pliohippus*) during the late Miocene coincided nicely with a diminished occurrence of *Hipparion* in the northern Great Basin. These two distinctive horses attest to the profound response of the fauna to environmental change.

Although both occur along the John Day River in Grant County, the Rattlesnake beds have even fewer fossils than the moderately fossiliferous Mascall below, and both have considerably less than the wealth of vertebrate remains found in the earlier John Day Formation. Much of the Rattlesnake fauna is modern in aspect, but the presence of such exotic creatures as elephants and sloths gives it an unmistakable late Tertiary complexion. Cat (felid) and dog predators (*Canis, Osteoborus, Simocyon*), a bear (*Indarctos*), a pronghorn antelope (*Sphenophalos*), numerous horses (*Nannippus, Neohipparion*), mustelids (*Lutravus, Martes*), a rabbit (*Hypolagus*), rodents (*Peromyscus, Spermophilus*), and a very large peccary (*Prosthennops*) make up the fauna.

Several unusual finds from Rattlesnake strata include the three molar teeth and jaw of a distinctive grizzly-sized bear *Indarctos*,

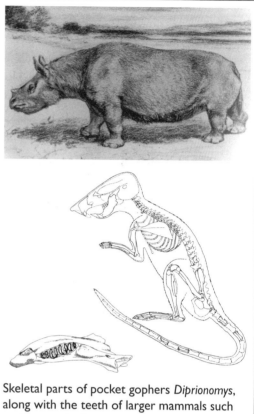

Skeletal parts of pocket gophers *Diprionomys*, along with the teeth of larger mammals such as the aquatic rhinoceros *Teleoceras*, are frequently found in screen washings of late Miocene sediments. (After Wood 1935.)

much like *Indarctos oregonesis*. These fragments were collected near the Mascall ranch in 1916, and a decade later, while searching at the same spot, John Merriam located the remainder of the jaw. The claw and partial left mandible of an immense ground sloth *Megalonyx* contrasts in size to the tiny humerus (front upper limb bone) of a *Condylura*, a star-nosed mole, both out of the Rattlesnake. This is the earliest known occurrence of the genus *Condylura* in western North America.

A study of Rattlesnake faunas by James Martin evaluated and described four new species from Picture Gorge in 1983. His listing includes the bones of a camel (*Hemiauchenia*), the tooth of a rodent (*Peromyscus*), that of a pig (*Prosthennops*), and the rare partial limbs of a bat (*Myotis*). The leg bones of a frog (*Rana*), a rhinoceros molar (*Teleoceras*), beaver (*Dipoides*), and the shell fragment of an aquatic turtle (*Clemmys*) point to the presence of a flowing stream.

Across northeast Oregon, small isolated depressions south and east of Arlington, south of Boardman and Hermiston, and at McKay Reservoir have a profusion of

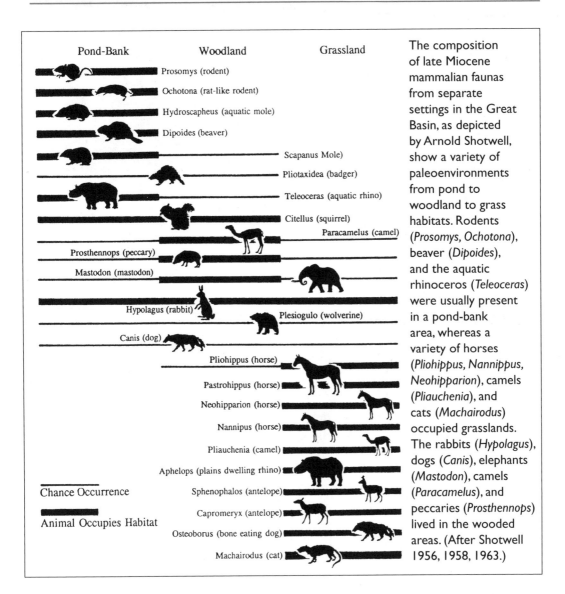

Pond-Bank   Woodland   Grassland

Prosomys (rodent)
Ochotona (rat-like rodent)
Hydroscapheus (aquatic mole)
Dipoides (beaver)
Scapanus Mole)
Pliotaxidea (badger)
Teleoceras (aquatic rhino)
Citellus (squirrel)
Paracamelus (camel)
Prosthennops (peccary)
Mastodon (mastodon)
Hypolagus (rabbit)
Plesiogulo (wolverine)
Canis (dog)
Pliohippus (horse)
Pastrohippus (horse)
Neohipparion (horse)
Nannipus (horse)
Pliauchenia (camel)
Aphelops (plains dwelling rhino)
Sphenophalos (antelope)
Capromeryx (antelope)
Osteoborus (bone eating dog)
Machairodus (cat)

Chance Occurrence

Animal Occupies Habitat

The composition of late Miocene mammalian faunas from separate settings in the Great Basin, as depicted by Arnold Shotwell, show a variety of paleoenvironments from pond to woodland to grass habitats. Rodents (*Prosomys, Ochotona*), beaver (*Dipoides*), and the aquatic rhinoceros (*Teleoceras*) were usually present in a pond-bank area, whereas a variety of horses (*Pliohippus, Nannippus, Neohipparion*), camels (*Pliauchenia*), and cats (*Machairodus*) occupied grasslands. The rabbits (*Hypolagus*), dogs (*Canis*), elephants (*Mastodon*), camels (*Paracamelus*), and peccaries (*Prosthennops*) lived in the wooded areas. (After Shotwell 1956, 1958, 1963.)

the late Miocene fossils. Carried by ancient rivers, the mammal remains were deposited in shallow basins atop the Columbia basalt plateau. Strong winds eroded the sands, producing long hollows or blowouts that have exposed bones of a rich varied fauna.

Mammal material here was so abundant that Arnold Shotwell was able to enumerate the variety of elements, which he could then use to reach broad paleoenvironmental deductions. Distinguishing three major settings, a pond-bank, grassland, and woodland, he concluded that the dominant local habitat near McKay Reservoir was a body of water, bordered by grassy prairies and forests at Ordnance and Arlington. By counting the statistical frequency of the separate bones of the skeleton, he could also differentiate between animals that had been carried into the basin (allochthonous) and those that had died *in situ* (autochthonous). In-place or local faunas are typically represented by a large array of skeletal elements in excellent condition, whereas transported material is often fragmented, incomplete, worn, and abraded. Overall, most of the smaller bones showed little abrasion. The outstanding preservation of the fossils at the federally managed McKay Reservoir site and the relative ease with which they could be separated from the soft matrix of white ash contributed to the success and accuracy of the study. Shotwell's counts also led him to conclude that certain elements of McKay Reservoir faunas are remarkably like those from the same time interval in north China, reinforcing the notion that there were migrations from that continent to North America which took place during the late Miocene. The carnivores, or flesh eaters, displayed the highest degree of similarity, whereas there was little in common between the rodents and hoofed mammals from both continents.

After he retired from the Museum of Geology at the South Dakota School of Mines, James Martin's research in northeast Oregon followed that of Shotwell. Martin's 1979 Ph.D. from the University of Washington focused on Miocene rodents of Ordnance, Arlington, and McKay Reservoir and their relationship to others from North America. At present, Martin and his students are reexamining faunas from Fossil Lake and reinterpreting the accumulations and preservations of bird remains. (Photo courtesy J. Martin.)

James Martin has examined a number of the puzzling concentrations of small vertebrate bones found in discrete pockets in Miocene exposures at three sites in Umatilla County, interpreting them as the fecal droppings of carnivorous mammals. Such accumulations had previously been considered as spread by water across a floodplain, but Martin concluded that many of the smaller bones had been in coprolites (fossil dung), while the larger mammal elements could not have been fecal in origin. Most of the bones had fresh breaks, and many were of juveniles. Beaver remains, parts of a young dog (Canidae), as well as the bones of the rats *Oregonomys, Parapliosaccomys*, and *Perognathus* were abundant in coprolites from Arlington, McKay, and Ordnance. The most numerous were predictably of squirrels (Sciuridae) and pocket gophers (Geomyidae) from Ordnance, the former Army depot south of Hermiston, those of fish, snakes, squirrels, and pocket gophers from Arlington, and bones of squirrels and rabbits from McKay Reservoir. Martin goes on to suggest that many paleoecological deductions might be reevaluated to consider the effects of contamination by animal fecal debris.

A partial listing of late Miocene vertebrates:

Amphibia (frog): *Rana*

Reptilia (turtle): *Clemmys*

Insectivora (moles and shrews): *Condylura, Hydroscapheus, Mystipterus, Scalopoides, Scapanus*

Chiroptera (bat): *Myotis*

Carnivora (dogs, cats, bears, raccoons): *Aelurodon, Canis, Felis, Indarctos, Lutravus, Machairodus, Martes, Osteoborus, Plesiogulo, Pliotaxidea, Simocyon, Sthenictis, Vulpes*

Proboscidea (elephants): *Amebelodon, Mammut, Platybelodon*

Artiodactyla (pigs, camels): *Capromeryx, Hemiauchenia, Megatylopus, Paracamelus, Platygonus, Pliauchenia, Procamelus, Prosthennops, Sphenophalos, Ustatochoerus*

Perissodactyla (horses, rhinoceros, sloth): *Aphelops, Hipparion, Megalonyx, Nannippus, Neohipparion, Pliohippus, Teleoceras*

Rodentia (mice, beaver, gophers): *Castor, Citellus, Dipoides, Diprionolys, Diprionomys, Epigaulus, Eucastor, Eutamias, Leptodontomys, Liodontia, Marmota, Microtoscoptes, Mylagaulus, Ochotona, Oregonomys, Parapliosaccomys, Perognathus, Peromyscus, Prosomys, Spermophilus, Tardontia*

Lagomorpha (rabbit): *Hypolagus*

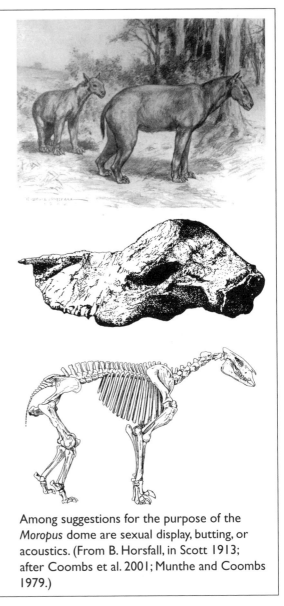

Among suggestions for the purpose of the *Moropus* dome are sexual display, butting, or acoustics. (From B. Horsfall, in Scott 1913; after Coombs et al. 2001; Munthe and Coombs 1979.)

The most northerly of the late Miocene basins at The Dalles has produced a handful of vertebrate remains. In volcanic debris redeposited by streams in eastward spreading fans, fragments of camel leg bones were found in an old stone quarry in the Chenoweth Formation by Thomas Condon around 1868. Few additional fossils have turned up since. A horse tooth and fragments of an elephant bone, the partial jaw of a primitive, short-faced dog (*Aelurodon*), and scant camel bones near a well-known Miocene plant locale on Chenoweth Creek are the few scattered fossils from this location. Comparing new canid teeth finds from the Chenoweth Formation with others in the northwest, James Martin outlined a possible biostratigraphic succession of borophagines (bone-eating dogs) during the Miocene.

By the middle Miocene, the ocean shoreline lay just slightly to the east of its present position, with a wide coastal plain stretching toward the Cascade peaks. Even with the potential for land mammals to be preserved in this wide area, their remains are remarkably limited. The only two known specimens found to date are from the Astoria Formation approximately four miles north of Newport in Lincoln County. One has been identified as the skull of a *Moropus* (chalicothere), a clawed, browsing horse-like animal, and the other is a tooth of an *Aphelops*, a plains-dwelling rhinoceros. The skull of the chalicothere has a very peculiar pronounced dome above the eyes, giving rise to speculation as to its significance. Among considerations are that the dome may have been used as an aquatic snorkel, as an attachment for jaw muscles, for water retention (like a camel's hump), for filtering or humidifying the air, or for increasing brain capacity.

Pliocene Mammals

There are very few fossil vertebrates in Oregon that are unquestionably from the Pliocene Epoch. A single mole *Scapanus* was described by Howard Hutchinson from unnamed Pliocene sediments at Enrico Ranch near the southern Oregon border. Because most late Tertiary moles are "fossorial" or burrowing animals, the limb bones are quite distinctive. The

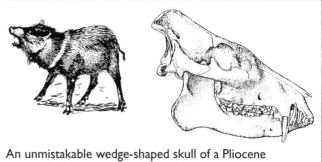

An unmistakable wedge-shaped skull of a Pliocene peccary from the Klamaths is similar to that of its modern counterpart. (After Newcomb 1958.)

upper front leg bone or humerus, in particular, is greatly enlarged and flattened as a support for the powerful digging muscles.

Pliocene sediments of the Yonna Formation in the Klamath River basin contain freshwater diatoms, leaf fragments, and invertebrates as well as the skull of a peccary *Prosthennops oregonensis*. Even though the stratigraphic position of the peccary, found in a quarry southeast of Klamath Falls, was not precisely established, the matrix proved to be similar to the light tan tuffs of the Yonna, placing it in the Pliocene.

What has been recently identified as an early Pliocene site, Always Welcome Inn in Baker City provides a rich faunal variety of shrews (*Paracyptotis*), rodents (rabbits, beavers), a vole (*Ophiomys*), a llama (Camelidae), and a mustelid (*Trigonictis*). Birds, fish, salamanders, frogs, and freshwater invertebrates add to the collection. The fish, diatoms, and gastropods suggest a warm lake and pond setting with streams flowing in from the surrounding forested mountains. Jay Van Tassell from Eastern Oregon University began working the site in 2004, concluding that the sediments and fossils show an environmental change from that of a shallow lake to a streamside. He suggests that the lack of large animal remains, such as those of horses, camels, and rhinoceroses, indicates that the bones of bigger mammals were not carried in to be deposited, whereas the local population was preserved at the site.

Pleistocene Land Mammals

Ending less than 12,000 years ago, the Pleistocene was a brief epoch that saw dropping temperatures and sheets of ice shaping the topography. Ice Age flooding brought devastation, but between flood events herds of exotic-looking elephants, bison, and clawed sloths, along with large carnivores such as cats and wolves, inhabited the state. Shallow, pluvial lakes in south and central Oregon provided a habitat for flocks of migratory birds along with fish and other animals, many of which became extinct around 11,000 years ago. With continued elevation of the Coast Range and retreat of glaciers, Oregon acquired its modern physiographic complexion.

The Pleistocene, or Ice Ages, of Oregon was a time of bleak and harsh cold weather at higher altitudes and heavy rainfall with vast wetlands at lower elevations. Although continental glaciers never reached this far south, smaller sheets of ice on mountain crests extended down into the valleys. The severe climate didn't restrict the range of mammals that were scattered throughout the state. Most of the reported remains, however, are concentrated in the Willamette Valley and in the south-central region of the Great Basin where swamps and pluvial lakes were conducive to the miring, entrapment, burial, and preservation of both large and small animals, birds, amphibians, and reptiles.

Over 10 feet high at the shoulder, the Columbian mammoth was the most common elephant in the Northwest during the Ice Ages. (From B. Horsfall, in Scott 1913.)

As with other fossil finds, and perhaps even more so with those of the Pleistocene, their discovery is often indirect, the consequence of human projects. Many teeth and bones of elephants from Baker County turned up in gold-bearing gravels of placer mining operations. Others, along the Columbia River, were discovered by engineers excavating for dam projects. But the highest frequency of known Pleistocene vertebrate sites correlates with the most densely populated parts of the northern Willamette Valley. Digging for wells and sewers, septic systems, and foundations for buildings or excavating for highways have uncovered many fossils. The extreme concentration of surface remains in a small area at Fossil Lake, famous for its mammal bones, makes this locale the exception.

Anticipating Harlan Bretz's theory of Ice Age floods, Thomas Condon drew attention to Pleistocene deposits in the Willamette Valley when he wrote in 1902 that terraces of what he called "Willamette Sound" were proof of waters reaching depths of 165 feet at Salem and as high as 325 feet over Portland. Doubtless the sudden floods drowned and then buried and preserved many mammals. Between the multiple events, forests, grasslands, and marshes encouraged herds of the ponderous mammoths, mastodons, and giant ground sloths. Bison, camels, horses, tapirs, deer, muskrats, weasels, and beavers rounded out the picture.

Several varieties of elephants (Proboscideans) wandered across the northwest, as suggested by their numbers preserved in Ice Age wetlands. Even though they fell to extinction during the late Pleistocene, the woolly mammoth, seen in European cave drawings, clearly coexisted at the same time as humans. *Elephas boreus, Elephas columbi*, and *Elephas imperator* lived in Oregon, and, of these *Elephas columbi,* the Columbian mammoth was by far the most prevalent. Found scattered throughout the state, this creature measured 11 feet at the shoulder, and its well-developed grinding molars attest to a diet of grasses.

In the Willamette Valley, remains of *Elephas columbi* have been documented from Newberg, McMinnville, Dayton, St. Paul, Silverton, Harrisburg, and Eugene, just to name a few sites. Fairly complete fossils from a bog near Silverton in Marion County generated much interest,

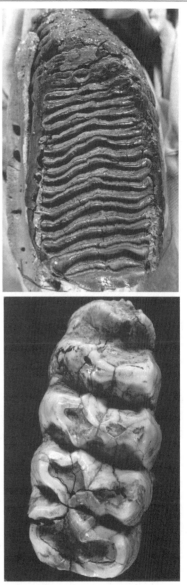

Crown views of elephant teeth clearly distinguish the broad low cusps of the browsing mastodon (below) from those of the grazing mammoth (above). (Photo courtesy Condon collection.)

but their fate was similar to many such finds. A rich deposit of bones, a partial skull, and tusks were found in 1946 during excavation for a spring in Evans Valley. Unfortunately, several weeks later, when finally examined by Oregon

State University paleontologist Earl Packard, the fossil material, which had been washed and then stored untreated in a barn, had dried and crumbled to fragments.

Occasionally fossil bones are encountered projecting from surface deposits. In a publicized case near Tualatin, the skeleton of a mastodon was discovered in 1960 half embedded in the mud near where the Safeway parking lot is now located. The exposed bones were in poor shape, but below the ground they were intact except for the skull. Excavated by geology students from Portland State University, the tusk and teeth were kept by a private collector, but the remainder of the bones were packed in wooden crates and variously stored at PSU, at the Portland Zoo, and even at a commercial storage unit. When brought back together thirty years later, the animal was reassembled in 1992. Since the mastodon had been lying on its left side, it was placed in bas-relief along with a silhouette of the flesh, and today the articulated beast may be seen adjoining the wall of the Tualatin city library.

The fossil remains of hoofed mammals with an even number of toes (artiodactyls) such as bison, camels, and deer often wash out of river gravels. Credit for discovery of the Oregon's first fossil bison goes to Ewing Young, a trapper and pioneer who arrived at Fort Vancouver on the Columbia in 1834. Young found a metatarsal bone along with elephant teeth and parts of a ground sloth on the bank of the "Walhaumet" River downstream from Eugene. In 1923, channel dredging uncovered remains of a bison above the falls at Oregon City. John Horner, a professor of history at Oregon Agricultural College (now Oregon State University), who founded the Horner Museum there, recovered the skull and horns of a superb *Bison occidentalis* on Lick Creek in Wallowa County. Horner wrote in February 1924 that when a charge of TNT, blasting for road construction by the U.S. Forest Service, was "put under a fir tree 2.5 feet in diameter (it) blew the tree out of the ground and left a hole 5 feet deep, at the bottom of which was found the skull."

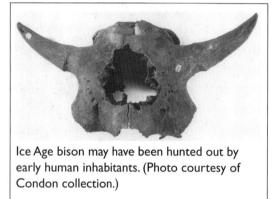

Ice Age bison may have been hunted out by early human inhabitants. (Photo courtesy of Condon collection.)

The Horner Museum officially opened as the College Museum in Corvallis on February 20, 1925, and consisted of collections assembled by several campus departments. The exhibits were eclectic and included natural history items, embroideries, historical weapons, and other artifacts. When "Jackie" Horner died in 1933, OSU President George W. Peavy recommended that the College Museum be named to honor him. Financial difficulty forced closure in 1995, when much of the collection was moved to the University Archives, stored, and ultimately sent to Benton County facilities in the 2000s.

A flurry of telegrams, letters, and postcards in the Spring of 1938 announced recovery of the "huge head of prehistoric animal" from near Medford. The fossil turned out to be that of an Ice Age bison *Bison antiquus*. The skull was inspected by John Allen of the Department of Geology and Mineral Industries, who advised sending the specimen to the University of California. To that end he encased it in plaster, crated it, and left it for shipment. However, Jackson County officials wanted to place it in a local museum, but, until arrangements could be made, the specimen was to be sent to the Condon collection in the Museum of Natural History at Eugene. But, as often happens, the ultimate destiny of the bison isn't known.

Other exotic creatures, such as camels and sloths roamed the state from Umatilla to Portland. Forty feet below the surface at Portland, a worn camel tooth was recovered while workers were excavating for the City Park Reservoir Number Three. The majority of Pleistocene camels are from the Ice Age lakes of south-central Oregon, where numerous bones have been identified as species of *Camelops* and the later *Camelus*.

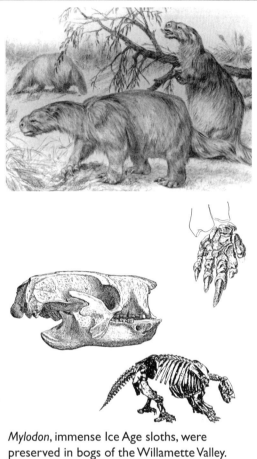

Mylodon, immense Ice Age sloths, were preserved in bogs of the Willamette Valley. (From B. Horsfall, in Scott 1913; after Stock 1925.)

Odd-toed hoofed mammals (perissodactyls) comprise the horses, which, except for those from Fossil Lake, appear mainly in the Willamette Valley. Many of the Oregon horse teeth have been identified as those of *Equus major*, *Equus occidentalis*, and *Equus excelsus*. Probably one of the most remarkable Pleistocene animals was the ground sloth, clawed member of the *Edentata*, which used the elongated digits for digging. Sloth remains are frequently found alongside those of Ice Age elephants, horses, and bison. Bones of the ox-sized *Megalonyx* and *Nothrotherium*, as well as those of *Mylodon*, which was twice as large, were recovered from river gravels in the Willamette Valley, in Klamath County, and from surface debris at Fossil Lake. The most colossal Pleistocene sloth, the 20-foot-high *Megatherium*, has not been reported in Oregon. Professor Earl Packard described several bones, housed at the Horner Museum, as *Mylodon harlani*, a form characteristic of western Oregon. Some species of sloths have hundreds of

Excavations at Fossil Lake in Lake County revealed a fossil elephant leg bone with only one toe digit missing. The small white "dot" near the lower leg is a snail shell. The name of the happy gentleman is also missing. (Photo by Allison 1966; courtesy Condon collection.)

Beaver were regular visitors in the ancient wetlands around Fossil Lake. (From Orr, Orr, and Baldwin 1992.)

button-shaped disks of bony armor imbedded in their thick leathery hides. The distinctive rounded pieces frequently turn up in muddy bogs.

Mylodon remains are a major component in the Mill Creek deposits at Woodburn in Marion County. This locale has been the focus of excavations since 1996 after bison bones were discovered during horizontal drilling to place a sewer pipe. During the 1990s bogs in several adjacent Mill Creek areas provided mammals, beetles, seeds, and birds identified by Professor William Orr from the Museum of Natural History and Cultural History in Eugene. A carbon-14 date of 12,500 years old was obtained from the rich accumulation of preserved wood.

Just two tapirs from the Pleistocene of the Pacific Coast have been reported. One came from Tuolumne County, California, and the other from Cape Blanco. The well-preserved Oregon jaw and teeth were collected from the Elk River beds in 1911 for the California Academy of Sciences. These were examined by John Merriam, who pronounced the creature similar to *Tapirus haysii*. Mollusk shells and microfossils associated with the tapir suggest a shallow cold-water embayment of the ocean. Like its modern counterpart, this tapir probably lived in the luxuriant forest, and, after dying, the carcass was washed offshore into the bay.

Major Pleistocene fossil concentrations in several wide, dry lakebeds of south-central Oregon have yielded the bones of vertebrates of all sizes. During this period of increased rainfall, vast pluvial lakes, bogs, and marshy wetlands filled shallow depressions across the Great Basin. Of these, two of the largest in Lake County were Lake Chewaucan at 461 square miles and Lake Coleman, now Warner lakes,

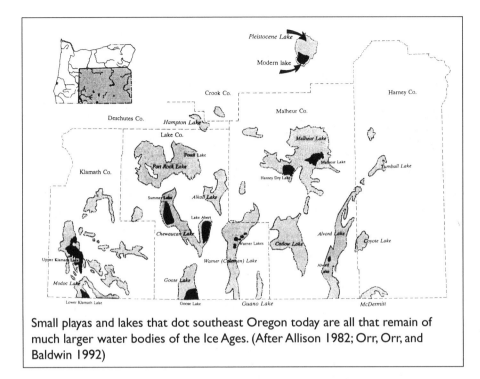

Small playas and lakes that dot southeast Oregon today are all that remain of much larger water bodies of the Ice Ages. (After Allison 1982; Orr, Orr, and Baldwin 1992)

at 483 square miles. Remnants of Lake Chewaucan are the current Summer Lake, Lake Abert, and the Chewaucan Marshes. Fort Rock Lake, now the dry Silver and Fossil lakes, reached 200 feet in depth. Cores from a delta in Summer Lake, drilled in 2004, helped to fill out the paleoenvironmental picture of the basin. Pollen and ostracods (arthropods) from the cores showed that several major water level changes took place in the middle and late Pleistocene. This conclusion is in agreement with that of Ira Allison, whose examination of ostracods in 1966 showed multiple stages of waxing and waning of lake levels.

At the height of the Ice Ages, these bodies of water attracted animals from the surrounding countryside, where today the greatest variety and multitude of Pleistocene mammals in the state are scattered on the surface at Fossil Lake. Among the beach sands and gravels are bones, tusks, and teeth of elephants, horses, camels, sloths, peccaries, antelopes, and numerous rodents. Carnivores such as wolves, bears, and cougar are also present along with birds and fish. Skeletal material is preserved by permineralization, where mineral fills in the pore spaces, creating a jet black polished surface. The bones are so thoroughly petrified that they ring like a bell when tapped.

Because fossils are so plentiful at these old lakebeds, they have been the focus of many interested parties since their discovery in the middle 1800s. Bones were picked up by cattlemen and brought to the attention of Governor John Whiteaker,

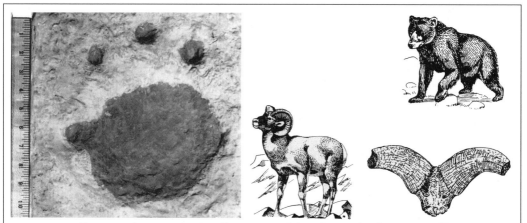

Bear remains are occasionally uncovered, but the track of an extinct Pleistocene bear from Lake County, measuring 10 inches in length, is exceptionally rare. Fossils of Pleistocene sheep are similarly scarce. Because the animal lived in an upland environment, the bones tended to be scattered or destroyed. Once locally extinct, the modern equivalent of the Pleistocene bighorn sheep has since been reintroduced in the Steens Mountains. (After Thoms and Smith 1973; photo courtesy Oregon Department of Geology and Mineral Industries.)

Charles Repenning started his partnership with bones when he brought home dead animals, usually tucked inside his shirt. A pilot captured during World War II, "Rep," as he was known, remained a prisoner until that conflict ended. Completing a Ph.D. at Berkeley, he began a career with the U.S. Geological Survey specializing in rodent and pinniped populations. Retiring in 1992, Repenning moved to Colorado where he was killed 13 years later by burglars, who wanted his fossil specimens. In the photo, Repenning is standing in front of his car, which has a POW license. (Photo courtesy C. Repenning.)

who owned a ranch at Summer Lake. On a trip to eastern Oregon in 1876, Whiteaker jokingly called the small depressions "fossil lakes, east and west," a name that stuck. Thomas Condon visited in that same year, as did the professional collector Charles Sternberg. Edward Cope himself appeared in 1879, camping at what he called the "bone yard" for several days. He produced extensive collections for the Philadelphia Academy of Sciences.

The next professionals at Fossil Lake were Annie Alexander, John Merriam's student from Berkeley, along with a field party. They were followed in 1923 and

1924 by Chester Stock and Eustace Furlong from the California Institute of Technology. Later work on the sandy playas was equally thorough. Ira Allison from Oregon State University listed not only the faunas but the environment and geology as well. Accompanied by a group of geologists from Oregon and California, Allison camped at Fossil and Summer lakes in 1939, examining all of the former lake basins in the vicinity. He proposed that the distribution of bones was the work of wind, but in 2005 Jim Martin and students concluded that the Fossil Lake paleoenvironment was the result of volcanic action.

In addition to the wealth of surface bones, Lake County has also recorded unusual finds such as bear tracks preserved in mudflats and the skull of a bighorn sheep in stream gravels. A huge bear, *Arctotherium*, walking in soft mud during the Ice Ages, left a series of alternating footprints across a distance of

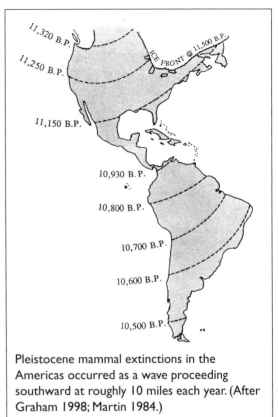

Pleistocene mammal extinctions in the Americas occurred as a wave proceeding southward at roughly 10 miles each year. (After Graham 1998; Martin 1984.)

15 feet before they became too faint to see. At the Drews Gap locale, professors Earl Packard and Ira Allison made casts and measured the prints, showing they corresponded closely to those of other bears of the same time period. Although the actual structure of the footpads of *Arctotherium* isn't known, Packard and Allison compared their findings to foot bones from the LaBrea tar pits in Los Angeles to reach their conclusions.

The skull of a prehistoric bighorn sheep *Ovis catclawensis*, unearthed by a bulldozer, is yet another uncommon fossil from near Adel. Once prevalent throughout the mountainous regions of the west, bighorn sheep were brought close to extinction by the 1900s because of hunting, disease, and destruction of their habitat. A worked pebble, possibly a human tool, was recovered with the skull. Today the specimen resides at the Lake County Museum in Lakeview.

Catastrophic widespread extinctions have long attracted the attention of paleontologists, and one of the most profound of these took place with animals of the late Pleistocene. Near the end of this epoch most of the bigger mammals disappeared from North and South America. Horses, camels, elephants, bison, and a variety of carnivores, which flourished throughout the later Tertiary and into the early Pleistocene, vanished here abruptly around 11,000 years ago. There are several theories as to why the extinctions occurred. The events may correspond to the appearance of humans who hunted out the creatures, mammal populations may have been at a low ebb, or disease may have played a role. Some suggestions seem improbable. One such is that the die-off might be related to fires from a comet shower torching grasslands and forests.

The theory currently favored by paleontologists is that dramatic fluctuations in the climate and environment during this time period were responsible for changes in the vegetation and subsequently in a reduction in the available plant food for herbivores. Once these larger mammals disappeared, something of a chain reaction was brought on, and scavengers and carnivores quickly followed suit. In the end it is doubtful that the demise of big mammals can be laid at the feet of one single cause.

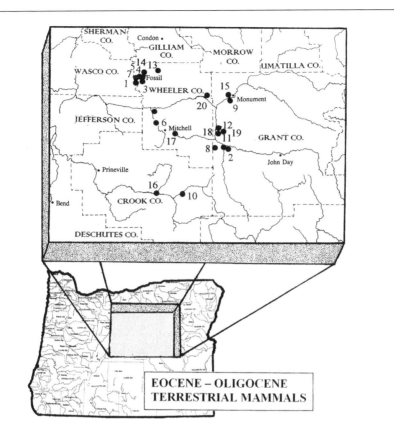

EOCENE – OLIGOCENE
TERRESTRIAL MAMMALS

EOCENE AND OLIGOCENE TERRESTRIAL MAMMAL LOCALITIES
The following are Eocene and Oligocene terrestrial mammal localities and a selection of authors who list faunas from the locales.

EOCENE
1. Clarno, Wheeler Co. Clarno Fm. Retallack 1981; Retallack, Bestland, and Fremd 1996, 2000; Stirton 1944.
2. Dayville, Grant Co. John Day Fm. Coleman 1949; Retallack, Bestland, and Fremd 2000.
3. Mammal Quarry, Wheeler Co. Hanson 1989, 1996; Mellett 1969; Pratt 1988; Retallack, Bestland, and Fremd 1996, 2000.
4. Nut Beds, Wheeler Co. Clarno Fm. Bestland et al. 1999; Hanson 1996; Retallack, Bestland, and Fremd 1996, 2000.
5. Whitecap Knoll, Wheeler Co. John Day Fm. Retallack 1991b; Retallack, Bestland, and Fremd, 2000.

OLIGOCENE
6. Bridge Creek valley, Wheeler Co. John Day Fm. Coombs et al. 2001; Fisher and Rensberger 1972; Leidy 1873; Merriam 1901; Merriam and Sinclair 1907; Rensberger 1971; Retallack, Bestland, and Fremd, 2000; Sinclair 1905; Stock 1946; Thorpe 1922b.
7. Clarno, Wheeler Co. John Day Fm. Matthew 1909; Retallack, Bestland, and Fremd 2000; Sinclair 1905; Thorpe 1921a, 1922b.
8. Cottonwood Creek, Grant Co. Matthew, 1899; Sinclair 1906; Thorpe 1921a, 1922b.
9. Courtrock, Grant Co. John Day Fm. Stirton and Rensberger 1964.

10. Crooked River valley, Crook Co. John Day Fm. Downs 1956; Leidy 1870, 1873; Maxson 1928.

11. Dayville, Grant Co. John Day Fm. Matthew 1899; Retallack, Bestland, and Fremd, 2000; Schultz and Falkenbach 1968.

12. Foree (4-E), Grant Co. John Day Fm. Fisher and Rensberger 1972; Rensberger 1976; Retallack, Bestland, and Fremd 2000.

13. Fossil (high school), Wheeler Co. John Day Fm. Brown 1959; Ferns, McClaughry, and Madin 2007.

14. Iron Mountain, Wheeler Co. John Day Fm. Retallack, Bestland, and Fremd 1996.

15. John Day River valley, Grant Co. John Day Fm. Cope 1879, 1880, 1882, 1883b; Eaton 1922; Matthew 1909; Merriam 1901, 1906; Rensberger 1971; Retallack, Bestland, and Fremd 2000; Schultz and Falkenbach 1968; Stirton 1940; Stock 1930, 1946; Toohey 1959.

16. Logan Butte, Crook Co. John Day Fm. Fisher and Rensberger 1972; Hanson 2000; Merriam 1906; Retallack, Bestland, and Fremd 2000; Stock and Furlong 1922; Thorpe 1921b, 1922b.

17. Painted Hills, Wheeler Co. John Day Fm. Retallack, Bestland, and Fremd 1996, 2000; Schultz and Falkenbach 1968.

18. Sheep Rock, Grant Co. John Day Fm. Fisher and Rensberger 1972; Rensberger 1976; Schultz and Falkenbach 1968.

19. Turtle Cove, Grant Co. John Day Fm. Black 1963; Eaton 1922; Fisher and Rensberger 1972; Lull 1921; Merriam and Sinclair 1907; Rensberger 1976; Retallack, Bestland, and Fremd 1996, 2000; Schultz and Falkenbach 1968; Sinclair 1905; Thorpe 1921a, 1922b.

20. Spray, Wheeler Co. John Day Fm. Sinclair 1901 .

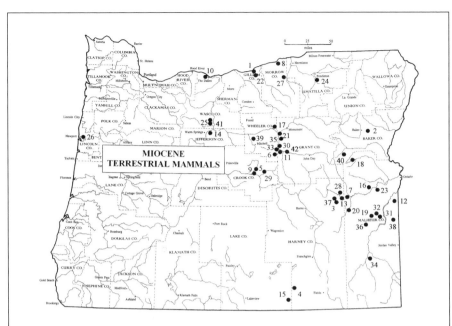

MIOCENE TERRESTRIAL MAMMAL LOCALITIES

The following are Miocene terrestrial mammal localities and a selection of authors who list faunas from the locales.

1. Arlington, Gilliam Co. The Dalles Fm. Black 1963; Fry 1973; Martin 1979, 1981, 1984; Shotwell 1958.
2. Baker City, Baker Co. Mascall? Fm. Downs 1952.
3. Bartlett Mtn., Harney Co. Drewsey Fm. Black 1963; Hall 1944; Hutchison 1966; Shotwell 1967, 1970; Shotwell and Russell 1963.
4. Beatys Butte, Harney Co. No Fm. Black 1963; Downs 1952; Hutchison 1968; Repenning 1967; Shotwell 1958; Stock 1930; Walker and Repenning 1965; Wallace 1946.
5. Beaver Creek valley, Crook Co. Mascall Fm. Downs 1956; Hay 1903, 1908.
6. Birch Creek valley, Wheeler Co. Mascall-Rattlesnake Fm. Downs 1956; Merriam, Stock, and Moody 1925.
7. Black Butte, Malheur Co. Juntura Fm. Black 1963; Hutchison 1966; O'Connor et al. 2003; Shotwell 1958, 1961, 1967, 1968.
8. Boardman, Morrow Co. No Fm. Shotwell 1958.
9. Camp Creek, Crook Co. Mascall Fm. Downs 1956; Hanson 2000; Merriam and Sinclair 1907; Rensberger 1971; Schultz and Falkenbach 1968.
10. Chenoweth Creek, Wasco Co. The Dalles Fm. Chaney 1944a; Condon 1902; Martin 1997.
11. Dayville, Grant Co. Mascall Fm. Downs 1956; Merriam 1901; Merriam, Stock, and Moody 1925.
12. Devils Gate, Malheur Co. Sucker Creek Fm. Downing 1992.
13. Drinkwater Pass, Harney Co. Drewsey Fm. Black 1963; Hutchison 1966; Shotwell 1967, 1968, 1970; Shotwell and Russell 1963.
14. Gateway, Jefferson Co. Mascall Fm. Downs 1956.
15. Guano Lake, Lake Co. Mascall Fm. Hutchison 1968; Repenning 1967.
16. Harper, Malheur Co. No Fm. Kaczmarska 1985; Furlong 1932.

17. Haystack Valley, Wheeler Co. John Day Fm. Fisher and Rensberger 1972; Hunt and Stepleton 2004; Lull 1921; Merriam and Sinclair 1907; Rensberger 1971, 1973, 1983; Schultz and Falkenbach 1968; Sinclair 1905; Thorpe 1921b, 1922a; Wang and Tedford 1992.

18. Ironside Mountain, Malheur Co. No Fm. Lowry 1943; Merriam 1916.

19. Juniper Creek valley, Malheur Co. Chalk Butte Fm. Shotwell 1967, 1970.

20. Juntura Basin, Malheur Co. Juntura Fm. Shotwell 1963.

21. Kimberly, Grant Co. John Day Fm. Coombs et al. 2001; Fisher and Rensberger 1972; Hunt and Stepleton 2004; Rensberger 1983; Schultz and Falkenbach 1968; Sinclair 1901.

22. Krebs Ranch, Gilliam Co. No Fm. Black 1963; Hutchison 1966, 1968; Martin 1980; Shotwell 1958, 1967.

23. Little Valley, Malheur Co. Chalk Butte Fm. Hutchison 1966; Shotwell 1967, 1970.

24. McKay Reservoir, Umatilla Co. McKay Fm. Black 1963; Downs 1956; Hutchison 1966; Martin 1979, 1981, 1983; Repenning 1968; Shotwell 1956, 1958, 1967.

25. Mutton Mountains, Wasco Co. John Day Formation. Fisher and Rensberger 1972; Woodburne and Robinson 1977.

26. Newport, Lincoln Co. Astoria Fm. Mitchell and Repenning 1963; Munthe and Coombs 1979.

27. Ordnance (Westend Blowout), Morrow Co. The Dalles Fm. Martin 1979, 1981, 1984; Shotwell 1958.

28. Otis Basin, Harney Co. Drewsey Fm. Black 1963; Hutchison 1966; Shotwell 1963, 1967.

29. Paulina Creek, Crook Co. Shotwell 1958.

30. Picture Gorge, Grant Co. John Day/Rattlesnake/Mascall Fms. Black 1963; Coombs et al. 2001; Colbert 1938; Downs 1956; Fisher and Rensberger 1972; Hay 1903, 1908; Martin 1983; Matthew 1909; McKenna 1990; Merriam and Sinclair 1907; Merriam, Stock, and Moody 1925; Rensberger 1971, 1973; Retallack 1991b; Rose and Rensberger 1983; Sinclair 1907; Thorpe 1921a, 1922b; Wilson 1938.

31. Quartz Basin, Malheur Co. Deer Butte Fm. Hutchison 1966, 1968; Shotwell 1963, 1967, 1968.

32. Red Basin, Malheur Co. Butte Creek Volcanic Sandstone. Hutchison 1966, 1968; Shotwell 1963, 1967, 1968.

33. Rock Creek, Grant Co. Mascall Fm. Downs 1956; Retallack 1991b.

34. Rome, Malheur Co. No Fm. Brattstrom and Sturn 1959; Colbert 1938; Furlong 1932; Hutchison 1966; Repenning 1967, 1968; Wilson 1934, 1938.

35. Rudio Creek, Grant Co. John Day Fm. Fisher and Rensberger 1972; Rensberger 1971, 1973; Sinclair 1905; Stirton and Rensberger 1964 .

36. Skull Springs, Malheur Co. Butte Creek Volcanic Sandstone. Black 1963; Downs 1952, 1956; Gazin 1932; Hutchison 1966; Shotwell 1958, 1968; Woodburne and Robinson 1977.

37. Stinking Water Creek, Harney Co. Juntura Fm. Hall 1944.

38. Succor Creek, Malheur Co. Sucker Creek Fm. Downing 1992; Scharf 1935.

39. Sutton Mountain, Wheeler County. John Day Fm. Coombs et al. 2001; Hunt and Stepleton 2004; Retallack, Bestland, and Fremd 1996.

40. Unity, Baker Co. No Fm. Lowry 1943; Retallack 2002.

41. Warm Springs, Wasco Co. John Day Fm. Woodburne and Robinson 1977.

42. White Hills, Grant Co. Mascall Fm. Downs 1952.

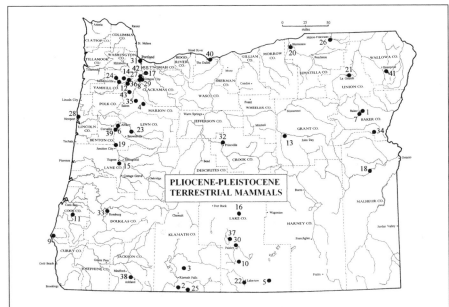

PLIOCENE AND PLEISTOCENE TERRESTRIAL MAMMAL LOCALITIES

The following are Pliocene and Pleistocene terrestrial mammal localities and a selection of authors who list faunas from the locales.

PLIOCENE

1. Always Welcome Inn (Baker City), Baker Co. No Fm. Van Tassell et al. 2007.
2. Enrico Ranch, Klamath Co. No Fm. Hutchison 1966.
3. Wilson's Quarry Pit, Klamath Co. Yonna Fm. Newcomb 1958.

PLEISTOCENE

4. Abiqua Creek, Marion Co. McCornack 1914.
5. Adel, Lake Co. Thoms and Smith 1973.
6. Albany, Linn Co. Hay 1927; McCornack 1914.
7. Baker City, Baker Co. Hay 1927; McCornack 1914.
8. Canby, Clackamas Co. McCornack 1914.
9. Cape Blanco, Curry Co. Elk River Beds. Leffler 1964.
10. Chewaucan Lake, Lake Co. Allison 1982.
11. Coquille River, Coos Co. Hay 1927; McCornack 1914.
12. Dayton, Yamhill Co. Hay 1927; McCornack 1914.
13. Dayville, Grant Co. Hay 1927.
14. Dundee, Yamhill Co. McCornack 1914.
15. Eugene, Lane Co. Hay 1927; McCornack 1914.
16. Fossil Lake, Lake Co. Cope 1889; Allison 1966; Elftman 1931; McCarville 2003.
17. Gladstone, Clackamas Co. McCornack 1914.
18. Harper, Malheur Co. Shotwell 1968.
19. Harrisburg, Linn Co. Hay 1927; McCornack 1914.
20. Hermiston, Umatilla Co. Hay 1927.
21. LaGrande, Union Co. Hay 1927; Quaintance 1969.
22. Lakeview, Lake Co. Packard and Allison 1980.
23. Lebanon, Linn Co. Hay 1927; McCornack 1914.

24. McMinnville, Yamhill Co. Hay 1927.
25. Merrill, Klamath Co. Hay 1927.
26. Milton-Freewater, Umatilla Co. Hay 1927.
27. Newberg, Yamhill Co. Hay 1927; McCornack 1914.
28. Nye Creek, Lincoln Co. Hay 1927.
29. Oregon City, Clackamas Co. Hay 1927.
30. Paisley, Lake Co. Hay 1927; McCornack 1914.
31. Portland, Washington Co. Hay 1927; McCornack 1914.
32. Prineville, Crook Co. Hay 1927; McCornack 1914.
33. Roseburg, Douglas Co. Hay 1927; McCornack 1914.
34. Rye valley, Baker Co. Hay 1927.
35. Silverton, Marion Co. Hansen and Packard 1949.
36. St. Paul, Marion Co. Hay 1927.
37. Summer Lake, Lake Co. Allison 1966; Cohen et al. 2004; Hay 1927.
38. Talent, Jackson Co. Hay 1927.
39. Tangent, Linn Co. McCornack 1914.
40. The Dalles, Wasco Co. Hay 1927; McCornack 1914.
41. Wallowa Lake, Wallowa Co. Spencer, Carson, and Orr 1985.
42. Wilsonville, Clackamas Co. McCornack 1914.
43. Woodburn, Marion Co. Hay 1927; Packard 1952.

REFERENCES

Abbott, W.H., 1970. Micropaleontology and paleoecology of Miocene non-marine diatoms from the Harper District, Malheur County, Oregon. Northwest Louisiana University, Master's thesis, 86p.

Aberra, G., and Retallack, G. 1991. Early Oligocene paleoenvironment of a paleosol from the lower part of the John Day Formation near Clarno, Oregon. Oregon Geology, v. 53, no. 6, pp. 131–136.

Addicott, W. O. 1964. A late Pleistocene invertebrate fauna from southwestern Oregon. Journal Paleontology, v. 38, no. 4, pp. 650–661.

———. 1966. New Tertiary marine mollusks of Oregon and Washington. Journal Paleontology, v. 40, no. 3, pp. 635–646.

———. 1972. Neogene molluscan paleontology along the West Coast of North America 1840–1969. Trends and status. Journal Paleontology, v. 46, no. 5, pp. 627–636.

———. 1980. Highlights in the 130-year history of marine Cenozoic stratigraphic paleontology on the Pacific Coast of North America. Geological Society America, Special Paper 184, pp. 3–15.

———. 1980. Miocene stratigraphy and fossils, Cape Blanco, Oregon. Oregon Geology, v. 42, no. 5, pp. 87–98.

———. 1981a. Brief history of Cenozoic marine biostratigraphy of the Pacific Northwest. Geological Society America, Special Paper 184, pp. 3–15.

———. 1981b. Significance of pectinids in Tertiary biochronology of the Pacific Northwest. Geological Society America, Special Paper 184, pp. 17–37.

———. 1983. Biostratigraphy of the marine Neogene sequence at Cape Blanco, southwestern Oregon. U.S. Geological Survey, Prof. Paper 774-G, Shorter Contributions to Geology, pp. G1–G17.

Adegoke, O.S. 1967. A probable pogonophoran from the early Oligocene of Oregon. Journal Paleontology, v. 41, no. 5, pp. 1090–1904.

Ahmad, R. 1986. Eocene geology of the Agness-Illahe area, southwest Oregon. Oregon Geology, v. 48, no. 2, pp. 15–31.

Allen, J., and Baldwin, E.M. 1944. Geology and coal resources of the Coos Bay Quadrangle, Oregon. Oregon Department of Geology and Mineral Industries, Bull. 27, 160p.

Allison, I. S. 1966. Fossil Lake, Oregon—its geology and fossil faunas. Oregon State Univ. Monogr., no. 9, 48p.

———. 1982. Geology of pluvial Lake Chewaucan, Lake County, Oregon. Oregon State University, Studies in Geology, no. 11, 78p.

Allison, I.S., and Bond, C.E. 1983. Identity and probable age of salmonids from surface deposits at Fossil Lake, Oregon. Copeia, v. 2, pp. 563–564.

Allmon, W.D. 2003. Boundaries, turnover, and the causes of evolutionary change. *In*: Prothero, D., Ivany, L., and Nesbitt, E., eds., From greenhouse to icehouse; the marine Eocene-Oligocene transition. New York, Columbia University Press, pp. 511–521.

Anderson, F.M., 1905. Cretaceous deposits of the Pacific Coast. California Academy Sciences, 3rd series, v. 2, 154p.

———. 1958. Upper Cretaceous of the Pacific Coast. Geological Soc. Amer., Memoir no. 71, 378p.

Applegate, S.P. 1968. A large fossil sand shark of the genus *Odontaspis* from Oregon. Ore Bin, v. 30, no. 2, pp. 32–36.

Armentrout, J.M. 1967. The Tarheel and Empire formations; geology and paleontology of the type sections, Coos Bay, Oregon. University of Oregon, Masters 155p.

———. 1975. Burrowing and boring molluscan faunas of the Pleistocene terraces of southwestern Oregon. American Association Petroleum Geologists, Abstract of Meeting., v. 2, pp. 1–2.

———. 1980. Field trip road log for the Cenozoic stratigraphy of Coos Bay and Cape Blanco, southwestern Oregon. *In*: Oles, K.F., et al., eds., Geologic field trips in western Oregon and southwestern Washington. Oregon Department Geology and Mineral Industries, Bull. 101, pp. 177–216.

Armentrout, J.M., ed. 1981. Pacific Northwest Cenozoic biostratigraphy. Geological Society America, Special Paper 184, 172p.

Armentrout, J.M., Cole, M.R., and Terbest, H., eds. 1979. Cenozoic paleogeography of the western United States. Society of Economic Paleontologists and Mineralogists, Pacific Coast Paleo. Symposium 3, 335p.

Armentrout, J.M., et al. 1983. Correlation of Cenozoic stratigraphic units of western Oregon and Washington. Oregon Department of Geology and Mineral Industries, Oil and Gas Investigation 7, 90p.

Arnold, C.A. 1937. Observations on the fossil flora of eastern and southeastern Oregon. University Michigan Museum Paleontology, Contributions, pt. l, v. 5, no. 8, pp. 79–102.

———. 1945. Silicified plant remains from the Mesozoic and Tertiary of western North America. Michigan Academy Sciences, Papers, v. 30, pp. 3–34.

———. 1952. Fossil Osmundaceae from the Eocene of Oregon. Palaeontographica B, v. 92, pp. 63–78.

———. 1953. Fossil plants of early Pennsylvanian type from central Oregon. Palaeontographica B., v. 93, pp. 61–68.

Arnold, C.A., and Daugherty, L.H. 1964. A fossil dennstaedtioid fern from the Eocene Clarno Formation of Oregon. University of Michigan, Contributions from the Museum of Paleontology, v. 19, no. 6, pp. 65–88.

Ash, S. R. 1991a. A new Jurassic flora from the Wallowa terrane in Hells Canyon, Oregon and Idaho. Oregon Geology, v. 53, no. 2, pp. 27–33.

———. 1991b. A new Jurassic *Phlebopteris* (Plantae, Filicales) from the Wallowa terrane in the Snake River Canyon, Oregon and Idaho. Jour. Paleo., v. 65, no. 2, pp. 332–329.

Ash, S. R., and Pigg, K.B. 1991. A new Jurassic *Isoetites* (Isoetales) from the Wallowa terrane in Hells Canyon, Oregon and Idaho. American Journal Botany, v. 78, no. 12, pp. 1636–1642.

Ash, S. R., and Read, C.B. 1976. North American species of *Tempskya* and their stratigraphic significance. U.S. Geological Survey, Prof. Paper 874.

Ashwill, M. 1983. Seven fossil floras in the rain shadow of the Cascade Mountains, Oregon. Oregon Geology, v. 45, no. 10, pp. 107–111.

———. 1987. Paleontology in Oregon: workers of the past. Oregon Geology, v. 49, no. 12, pp. 147–153.

———. 1996. The Pliocene Deschutes fossil flora of central Oregon: additions and taphonomic notes. Oregon Geology, v. 58, no. 6, pp. 131–141.

Axelrod, D.I. 1944. The Alvord Creek flora. Carnegie Institute, Washington, D.C., Publication 553, pp. 225–306.

———. 1966. A method for determining the altitudes of Tertiary floras. Paleobotanist, v. 14, no. 1–3, pp. 144–171.

Axelrod, D.I., and Bailey, H. 1969. Paleotemperature analysis of the Tertiary plants. Paleogeography, Paleoclimatology, Paleoecology, v. 6, pp. 163–195.

Bailey, I.W., and Sinnott, E.W. 1916. The climatic distribution of certain types of angiosperm leaves. American Journal Botany, v. 3, pp. 24–39.

Baldwin, E.M. 1945. Some revisions of the late Cenozoic stratigraphy of the southern Oregon coast. Journal Geology, v. 53, pp. 35–46.

———. 1950. Pleistocene history of the Newport, Oregon, region. Geol. Soc. Oregon Country Newsletter, v. 18, pp. 29–30.

———. 1964. Geology of the Dallas and Valsetz quadrangles, Oregon. Oregon Dept. Geology and Mineral Industries, Bull. 35, 52p.

———. 1973. Geology and mineral resources of Coos County, Oregon. Oregon Dept. Geology and Mineral Industries, Bull. 80, 82p.

———. 1974. Eocene stratigraphy of southwestern Oregon. Oregon Dept. of Geology and Mineral Industries, Bull. 83, 40p.

———. 1976. Geology of Oregon. Rev. ed. Dubuque, Ia.: Kendall-Hunt Publ. Co., 147p.

Baldwin, E.M., et al. 1955. Geology of the Sheridan and McMinnville Quadrangles, Oregon. U.S. Geological Survey, Oil and Gas Inv. Map, OM-155.

Bandy, O.L. 1941. Invertebrate paleontology of Cape Blanco. Oregon State University, Masters 137p.

———. 1944. Eocene foraminifera from Cape Blanco, Oregon. Journal Paleontology, v. 18, pp. 366–377.

———. 1950. Some later Cenozoic foraminifera from Cape Blanco, Oregon. Journal Paleontology, v. 24, pp. 269–281.

Barnes, L.G. 1976. Outline of eastern north Pacific fossil cetacean assemblages. Systematic Zoology, v. 25, no. 4, pp. 321–343.

———. 1987. An early Miocene pinnipeds of the genus *Desmatophoca* (Mammalia Otariidae) from Washington. Natural History Museum of Los Angeles County, Contributions in Science, no. 382, 20p.

———. 1989. A new enaliarctine pinniped from the Astoria Formation, Oregon, and a classification of the Otariidae (Mammalia: Carnivora). Los Angeles County Museum of Natural History, Contributions in Science, no. 403, pp. 1–26.

————. 1990. A new Miocene enaliarctine pinniped of the genus *Pteronarctos* (Mammalia: Otariidae) from the Astoria Formation, Oregon. Los Angeles County Museum of Natural History, Contributions in Science, no. 422, 20p.

Barnes, L.G., Domning, D.P., and C.E. Ray, C.E. 1985. Status of studies on fossil marine mammals. Society for Marine Mammalogy, Marine Mammal Science, v. 1, no. 1, pp. 15–53.

Barnes, L.G., and Mitchell, E.D. 1975. Late Cenozoic northeast Pacific Phocidae. Rappaports et Proces-verbaux des Reunions, Conseil International pour l'Exploration de la Mer, v. 169, pp. 34–42.

Barnes, L.G., et al. 1994. Classification and distribution of Oligocene Aetiocetidae (Mammalia; Cetacea; Mysticeti) from western North America and Japan. The Island Arc, v. 3, pp. 392–431.

Barnett, S.F., and Fisk, L.H. 1984. Palynology of the ?Miocene Alvord Creek Formation, southeastern Oregon. Palynology (Abstract). v. 8, p. 253,

Benton, M.J. 2006. Vertebrate paleontology. 3rd ed. Malden, Mass.: Blackwell, 455p.

Berggren, W.A., et al. 1985. Cenozoic geochronology. Geological Society America, Bull., v. 96, pp. 1407–1418.

Berglund, R.E., and Feldmann, R.M. 1989. A new crab, *Rogueus Orri* n. gen. and sp. (Decapod: Brachyura) from the Lookingglass Formation (Ulatisian Stage: Lower Middle Eocene) of southwestern Oregon. Journal Paleontology, v. 63, no. 1, pp. 69–73.

Berman, D.S. 1976. A new amphisbaenian (Reptilia: Amphisbaenia) from the Oligocene-Miocene John Day Formation, Oregon. Journal Paleontology, v. 50, no. 1, pp. 165–174.

Berta, A. 1991. New *Enaliarctos* (Pinnipedimorpha) from the Oligocene and Miocene of Oregon and the role of "Enaliarctids" in pinniped phylogeny. Smithsonian Contributions to Paleobiology, no. 69, 33p.

————. 1994a. New specimens of the pinnipediform *Pteronarctos* from the Miocene of Oregon. Smithsonian Contributions to Paleobiology, v. 78, 30p.

————. 1994b. A new species of phocoid pinniped *Pinnarctidon* from the early Miocene of Oregon. Journal of Vertebrate Paleontology, v. 14, pp. 405–413.

————. 1994c. Pinniped phylogeny. *In* Berta, A., and Demere, T.A., eds., Contributions in marine mammal paleontology honoring Frank C. Whitmore, Jr. San Diego Society of Natural History, Proceedings, May 1 1994; no. 29, pp. 33–56.

Berta, A., and Demere, T.A., eds. 1994. Contributions in marine mammal paleontology honoring Frank C. Whitmore, Jr. San Diego Society of Natural History, Proceedings, May 1 1994; no. 29, 268p.

Bestland, E.A. 1987. Volcanic stratigraphy of the Oligocene Colestin Formation in the Siskiyou Pass area of southern Oregon. Oregon Geology, v. 49, pp. 79–86.

————. 1997. Alluvial terraces and paleosols as indicators of early Oligocene climate change (John Day Formation, Oregon). Journal of Sedimentary Research, v. 67, pp. 840–855.

Bestland, E.A., et al. 1999. Geologic framework of the Clarno Unit, John Day Fossil Beds National Monument, central Oregon. Oregon Geology, v. 61, no. 1, pp. 3–18.

Bird, K. 1967. Biostratigraphy of the Tyee (Eocene) southwestern Oregon. University of Wisconsin, PhD. 209p.

Black, C.C. 1963. A review of the North American Tertiary Sciuridae. Harvard Univ., Museum Comparative Zoology, Bull. 130, pp. 113–248.

Blake, D.B., and Allison, R.C. 1970. A new west American Eocene species of the recent Australian ophiuroid *Ophiocrossota*. Journal Paleontology, v. 44, pp. 925–927.

Blake, M.C., et al. 1985. Technostratigraphic terranes in southwest Oregon. *In*: Howell, D.G., Technostratigraphic terranes of the circum-Pacific region. Earth Science Series, v. 1, CircumPacific Council for Energy and Mineral Resources, pp. 147–157.

Bland, A., Rose, R., and Currier, A. D. 2007. Jurassic crocodile discovered in Crook County, Oregon. Oregon Geology, v. 68, no. 1, pp. 24-26.

Blome, C.D. 1984. Upper Triassic radiolarian and radiolarian zonation from western North America. Bulletins of American Paleontology, v. 85, no. 318, 88p.

Blome, C.D., and Nestell, M.K. 1991. Evolution of a Permo-Triassic sedimentary mélange, Grindstone terrane, east-central Oregon. Geological Society America, Bull., v. 103, pp. 1280–1296.

———. 1992. Field guide to the geology and paleontology of pre-Tertiary volcanic arc and mélange rocks, Grindstone, Izee, and Baker terranes, east-central Oregon. Oregon Geology, v. 54, no. 6, pp. 123–141.

Blome, C.D., and Reed, K.M. 1992. Permian and early (?) Triassic radiolarian faunas from the Grindstone terrane, central Oregon. Journal Paleontology, v. 66, no. 3, pp. 351–383.

Blome, C.D., et al. 1986. Geologic implications of radiolarian-bearing Paleozoic and Mesozoic rocks from the Blue Mountains province, eastern Oregon. *In*: Vallier, T.L., and Brooks, H.C., eds., Geology of the Blue Mountains region of Oregon, Idaho, and Washington. U.S. Geological Survey, Professional Paper 1435, pp. 79–93.

Boggs, S., Orr, W.N., and Baldwin, E.M., 1973. Petrographic and paleontologic characteristics of the Rickreall limestone (Eocene) of southwestern Oregon. Journal Sedimentary Petrology, v. 43, pp. 644–654.

Bones, T.J. 1979. Atlas of fossil fruits and seeds from north central Oregon. Oregon Museum Science and Industry, Occasional Papers in Natural Science, no. 1, 6p.

Bostwick, D.A., and Koch, G.S. 1962. Permian and Triassic rocks of northeastern Oregon. Geological Society America, Bulletin, v. 73, pp. 410–422.

Bostwick, D.A., and Nestell, M.K. 1965. A new species of *Polydiexodina* from central Oregon. Journal Paleontology, v. 39, no. 4, pp. 611–614.

———. 1967. Permian tethyan fusulinid faunas of the northwestern United States. *In*: Adams, C.G., and Ager, D.V., eds., Aspects of tethyan biogeography. Systematics Association, Publication 7, pp. 93–102.

Bourgeois, J., and Leithold, E.L. 1984. Sedimentation, tectonics and sea-level change as reflected in four wave-dominated shelf sequences in Oregon and California. *In*: Larue, D.K., and Steel, R.J., eds., Cenozoic marine deposits of the Pacific margin. Society of Economic Paleontologists and Mineralogists, Pacific Section, pp. 1–16.

Brattstrom, B.H., and Sturn, A., 1959. A new species of fossil turtle from the Pliocene of Oregon with notes on other fossil *Clemmys* from western North America. Southern California Academy of Sciences, Bull. 58, pp. 65–71.

Brodkorb, P. 1958. Birds from the middle Pliocene of McKay, Oregon. Condor, v. 60, pp. 252–255.

———. 1961. Birds from the Pliocene of Juntura, Oregon. Florida Scientist, v. 24, no. 3, pp. 169–184.

Brouwers, E., et al. 1995. Paleogeography, paleoecology, and biostratigraphy of upper Paleocene to middle Eocene units of the Tyee basin, southwest Oregon. *In*: Fritsche, A.E., ed., Cenozoic paleogeography of the western United States – II: SEPM (Society for Sedimentary Geology, book 75, pp. 246–256.

Brown, R. W. 1936. Field identification of the fossil ferns called *Tempskya*. Washington Academy of Sciences, Journal, v. 26, no. 2. pp. 45–52.

———. 1937a. Additions to some fossil floras of the western United States. U.S. Geological Survey, Professional Paper 186-J, pp. 163–206.

———. 1937b. Further additions to some fossil floras of the western United States. Washington Academy Sciences, Journal, v. 27, no. 12, pp. 506–517.

———. 1940. New species and changes of name in some American fossil floras. Washington Academy Sciences, Journal, v. 30, pp. 344–356.

———. 1946. Alterations in some fossil and living floras. Washington Academy Sciences, Journal, v. 36, no. 10, pp. 344–355.

———. 1956. New items in Cretaceous and Tertiary floras of the western United States. Washington Academy of Sciences, Journal, v. 46, pp. 104–108.

———. 1959. A bat and some plants from the upper Oligocene of Oregon. Journal Paleontology, v. 33, no. 1, pp. 135–139.

Buffetaut, E. 1979. Jurassic marine crocodilians (Mesosuchia: Telosauridae) from central Oregon: first record in North America. Journal Paleontology, v. 54, no. 1, pp. 211–215.

Bukry, D., and Snavely, P. D. 1988. Coccolith zonation for Paleogene strata in the Oregon Coast range. *In*: Filewicz, M.V., and Squires, R.L., eds., Paleogene stratigraphy, West Coast of North America. Society of Economic Paleontologists and Mineralogists, Pacific Sect., Book 58, pp. 251–263.

Burns, C., and Mooi, R. 2003. An overview of Eocene-Oligocene echinoderm faunas of the Pacific Northwest. *In*: Prothero, D., Ivany, L., and Nesbitt, E., eds., From greenhouse to icehouse; the marine Eocene-Oligocene transition. New York, Columbia University Press, pp. 88–106.

Burns, C., Campbell, K.A., and Mooi R. 2005. Exceptional crinoid occurrences and associated carbonates of the Keasey Formation (early Oligocene) at Mist, Oregon, U.S.A. Paleogeography, Paleoclimatology, Paleoecology, v. 227, pp. 210–231.

Cameron, K. A., and Pringle, P. T. 1991. Prehistoric buried forests of Mount Hood. Oregon Geology, v. 53, no. 2, pp. 34–43.

Carroll, R.L. 1986. Vertebrate paleontology and evolution. New York, Freeman, 698p.

Cavender, T.M. 1968. Freshwater fish remains from the Clarno Formation Ochoco Mountains of north-central Oregon. Ore Bin, v. 30, no. 7, pp. 125–141.

————. 1969. An Oligocene mudminnow (Family Umbridae) from Oregon with remarks on relationships within the Esocoidei. Michigan University, Museum Zoology, Occasional Paper 660, 33p.

————. 1986. Review of the fossil history of North American freshwater fishes. *In*: Hocutt, C.H., and Wiley, E.O., eds., The zoogeography of North American freshwater fishes, pp. 699–724.

Cavender, T.M., Lundberg, J.G., and Wilson, R.L. 1970. Two new fossil records of the Genus *Esox* (Teleostei, Salmoniformes) in North America. Northwest Science, v. 44, no. 3, pp. 176–183.

Cavender, T.M., and Miller, R.R. 1972. *Smilodonichthys rastrosus*, a new Pliocene salmonid fish from Western United States. University of Oregon, Museum Natural History, Bull. 18, 44p.

Chaney, R.W. 1918. Ecological significance of the Eagle Creek flora of the Columbia River Gorge. Journal Geology, v. 26, pp. 577– 592.

————. 1920. Flora of the Eagle Creek Formation, Washington and Oregon. Univ. Chicago, Contrib. from the Walker Mus., v. 2, no. 5, pp. 115–151.

————. 1925a. A comparative study of the Bridge Creek flora and the modern redwood forest. Carnegie Institute Washington, Publication 349, pp. 1–22.

————. 1925b. The Mascall flora; its distribution and climatic relation. Carnegie Institute of Washington, Contributions to Paleontology no. 349, pp. 25–48.

————. 1927. Geology and paleontology of the Crooked River Basin with special reference to the Bridge Creek flora. Carnegie Institute Washington, Publ.346, pt.4, pp. 185–216.

————. 1938a. The Deschutes flora of eastern Oregon. Carnegie Institute, Washington, D.C., Contributions to Paleontology, v. 476, pp. 185–216.

————. 1938b. Miocene and Pliocene floras of western North America. Carnegie Institute, Washington, Contributions to Paleontology 476, 272p.

————. 1944a. The Dalles flora. Carnegie Institute, Washington, Publication 553, pp. 284–321.

————. 1944b. The Troutdale flora. Carnegie Institute, Washington, Publication 533, pp. 323–351.

————. 1956. The ancient forests of Oregon. Oregon State System of Higher Education, Condon Lectures, Eugene, Oregon. 56p.

Chaney, R.W., ed. 1944. Pliocene floras of California and Oregon. Carnegie Institute., Washington, Contributions to Paleontology 553, 407p.

Chaney, R.W., and Axelrod, D.I. 1959. Miocene floras of the Columbia River Plateau. Carnegie Institute Washington, Publication 617, 237p.

Chaney, R.W., and Sanborn, E.I. 1933. The Goshen flora of west central Oregon. Carnegie Institute, Washington, Publication 439 103p.

Clark, R. 1989. The odyssey of Thomas Condon. Portland: Oregon Historical Society Press, 569p.

Clifton, H.E., and Boggs, S. 1970. Concave-up pelecypod (*Psephidia*) shells in shallow marine sand, Elk River Beds, southwestern Oregon. Journal of Sedimentary Petrology, v. 40, no. 3, pp. 888–897.

Cockerell, T.D.A. 1927. Tertiary fossil insects from eastern Oregon. Carnegie Institute, Washington, Contribution to Paleontology, no. 346, pp. 64–65.

Cohen, A., et al. 2004. A paleoclimate record for the past 250,000 years from Summer Lake, Oregon, USA: II. Sedimentology, paleontology and geochemistry. Journal of Paleolimnology, v. 24, no. 2, pp. 151–182.

Colbath, G.K., and Steele, M.J. 1982. The geology of economically significant lower Pliocene diatomites in the Fort Fock basin near Christman Valley, Lake County, Oregon. Oregon Geology, v. 44, no. 10, pp. 111–118.

Colbath, S.L. 1985. Gastropod predation and depositional environments of two molluscan communities from the Miocene Astoria Formation at Beverly Beach State Park, Oregon. Journal of Paleontology v. 59, no. 4, pp. 849–869.

Colbert, E.H. 1938. Pliocene peccaries from the Pacific Coast region of North America. Carnegie Institute, Washington, Publication 487, pp. 243–269.

Coleman, R.G. 1949. A recently discovered entelodont from the John Day Basin. Oregon Academy Sciences, Proceedings, v. 2, p. 41.

Condon, T. 1902. The two Islands. Portland, Oregon: J.K. Gill Co., 211p.

———. 1906. A new fossil pinniped (*Desmatophoca oregonensis*) from the Miocene of the Oregon coast. University Oregon, Bull., Supplement, v. 3, pp. 1–14.

Coombs, M.C. 1978. Reevaluation of early Miocene North American *Moropus* (Perissodactyla, Chalicotheriidae, Schizotheriinae). Carnegie Museum of Natural History, Bull. 4, 62p.

———. 1998. Chalicotherioidea. *In*: Janis, C.M., Scott, K.M., and Jacobs, L.L., eds., Evolution of Tertiary mammals of North America. New York: Cambridge University Press, pp. 560–568.

Coombs, M.C., et al. 2001. Stratigraphy, chronology, biogeography, and taxonomy of early Miocene small chalicotheres in North America. Journal of Vertebrate Paleontology, v. 21, no. 3, pp. 607–620.

Cooper, G.A. 1957. Permian brachiopods from central Oregon. Smithsonian Miscellaneous Collections, v. 134, no. 12, 79p.

Cope, E.D. 1873. On the extinct Vertebrata of the Eocene of Wyoming, observed by the expedition of 1872, with notes on the geology. U.S. Geological and Geographical Survey of the Territories, Annual Report, v. 6, pp. 543–649.

———. 1878. Descriptions of new vertebrata from the upper Tertiary formations of the West. American Philosophical Society, Proceedings 17, pp. 219–231.

———. 1879. Observations on the faunae of the Miocene Territories of Oregon. U.S. Geographical and Geological Survey, Bull.5, pp. 55–59.

———. 1880. Extinct cats of North America. American Naturalist, v. 14, pp. 833–858.

———. 1882. On the Nimravidae and Canidae of the Miocene period. U.S. Geographical and Geological Survey, Bull.6, pp. 165–181.

———. 1883a. Extinct dogs of North America. American Naturalist, v. 17, pp. 235–249.

———. 1883b. Extinct Rodentia of North America. American Naturalist, v. 17, pp. 43–57 165–174, 370–381.

———. 1883c. A new Pliocene formation in the Snake River Valley. American Naturalist, v. 17, pp. 867–868.

————. 1883d. On the fishes of the Recent and Pliocene lakes of the western part of the Great Basin and of the Idaho Pliocene lake. Academy of Natural Sciences, Philadelphia, Proceedings, pp. 134–166.

————. 1883e. Vertebrata of the Tertiary formations of the West. U.S. Geographical and Geological Survey, Report, v. 3, pp. 762–1002.

————. 1886. Phylogeny of the Camelidae. American Naturalist, v. 20, pp. 611–624.

————. 1887. Perissodactyla. American Naturalist, v. 21, pp. 985–1007; 1060–1076.

————. 1888. On the Dictyolinae of the John Day Miocene of North America. American Philosophical Society, Proceedings, v. 25, pp. 62–79.

————. 1889a. The Edentata of North America. American Naturalist, v. 23, pp. 657–665.

————. 1889b. On a species of *Plioplarchus* from Oregon. American Naturalist, v. 12, pp. 970–982.

————. 1889c. The Silver Lake of Oregon and its region. American Naturalist, v. 23, pp. 970–982.

Cushman, J.A., Stewart, R.C., and Stewart, K.C. 1947. Five papers on foraminifera from the Tertiary of western Oregon. Oregon Dept. Geology and Mineral Industries, Bull. 36, pt. I–IV.

Dake, H.C. 1969. Oregon *Tempskya* locality. Gems and Minerals, no. 384, pp. 62–63.

Dall, W.H. 1909. Contributions to the Tertiary paleontology of the Pacific Coast. U.S. Geological Survey, Professional Paper 59, 216p.

Danner, W.R. 1968. Devonian of Washington, Oregon and western British Columbia. International Symposium, Devonian System, Calgary, 1967, v. 1, pp. 827–842.

David, L.R. 1956. Tertiary anacanthin fishes from California and the Pacific Northwest; their paleoecological significance. Journal Paleontology, v. 30, no. 3, pp. 568–607.

Demere, T.A., and Berta, A. 2001. Reevaluation of *Proneotherium repenningi* from the Miocene Astoria Formation of Oregon and its position as a basal odobenid (Pinnepedia: Mammalia). Journal Vertebrate Paleontology, v. 21, no. 2, pp. 279–310.

————. 2002. The Miocene pinniped *Desmatophoca oregonensis* Condon 1906 (Mammalia: Carnivora) from the Astoria Formation, Oregon. Smithsonian Contributions to Paleobiology, no. 93, pp. 113–147.

Detling, M.R. 1946. Foraminifera of the Coos Bay lower Tertiary, Coos County, Oregon. Journal Paleontology, v. 20, no. 4, pp. 348–361.

————. 1958. Some littoral foraminifera from Sunset Bay, Coos County, Oregon. Cushman Foundation Foraminiferal Research, Contribution v. IX, pt. 2, pp. 25–31.

Dickinson, W.R., and Thayer, T.P. 1978. Paleogeographic and paleotectonic implications of Mesozoic stratigraphy and structure in the John Day inlier of central Oregon. *In*: Howell, D.G., and McDougall, K.A., eds., Mesozoic paleogeography of the western United States (Pacific Coast Paleogeography Symposium 2), Soc. Economic Paleontologists and Mineralogists, Publ., 8, pp. 147–161.

Dickinson, W.R., and Vigrass, L.W. 1965. Geology of the Suplee-Izee area, Crook, Grant, and Harney counties, Oregon. Oregon Dept. Geology and Mineral Industries, Bull. 58 108 pp.

Diller, J.S. 1896. A geological reconnaissance in northwestern Oregon. U.S. Geological Survey 17th Annual Report, pt.1, pp. 441–520.

———. 1898. Description of the Roseburg Quadrangle. U. S. Geological Survey, Atlas, Roseburg Folio no. 49.

———. 1899. The Coos Bay coalfield, Oregon. U.S. Geological Survey 19th Annual Rept., pt. 3, pp. 309–370.

———. 1901. Description of the Coos Bay Quadrangle, Oregon. U.S. Geological Survey, Atlas, Folio 73.

———. 1902. Topographic development of the Klamath Mountains (California and Oregon). U.S. Geological Survey, Bull., 196, pp. 1–69.

———. 1903. Description of the Port Orford Quadrangle, Oregon. U.S. Geological Survey, Geology Atlas Folio 89, 6p.

———. 1907. The Mesozoic sediments of southwestern Oregon. American Journal Science, 4th ser., v. 23, pp. 401–421.

———. 1908. Strata containing the Jurassic flora of Oregon. Geological Society America, Bull.19, pp. 367–402.

Diller, J.S., and Kay, G.F. 1924. Description of the Riddle Quadrangle, Oregon. U.S. Geol. Survey, Atlas Folio no. 218.

Dixon, E. 1917. Bibliography of the geology of Oregon. University of Oregon, Geology series v. 1, no. 1, 125p.

Dodds, B.R. 1963. The relocation of geologic locales in Oregon. Ore Bin, v. 25, no. 7, pp. 113–128.

Domning, D.P. 1996. Bibliography and index of the Sirenia and Desmostylia. Smithsonian Contributions to Paleobiology, no. 80, 611p.

Domning, D.P., and Ray, C.E. 1986. The earliest Sirenian (Mammalia: Dugongidae) from the eastern Pacific Ocean. Marine Mammal Science, v. 2, no. 4, pp. 263–276.

Domning, D.P., Ray, C.E., and McKenna, M.C. 1986. Two new Oligocene Desmostylians and a discussion of Tethytherian systematics. Smithsonian Contributions to Paleobiology, no. 59, 56p.

Dott, R.H. 1966. Eocene deltaic sedimentation at Coos Bay, Oregon. Journal Geology, v. 74, no. 4, pp. 373–420.

———. 1971. Geology of the southwestern Oregon coast west of the 124th Meridian. Oregon, Department of Geology and Mineral Industries, Bull. 69, 63p.

Downing, K. 1992. Biostratigraphy, taphonomy, and paleoecology of vertebrates from the Sucker Creek Formation (Miocene) of southeastern Oregon. PhD diss., University of Arizona.

Downs, T. 1952a. A new mastodont from the Miocene of Oregon. University California, Publications in Geological Science, v. 29, no. 1, pp. 1–20.

———. 1956b. The Mascall fauna from the Miocene of Oregon. University of California, Publications in Geological Sciences, v. 31, pp. 199–354.

Durham, J.W. 1942. Coral faunas of Washington. Journal Paleontology, v. 16, no. 1, pp. 84–104.

————. 1944. Megafaunal zones of the Oligocene of northwestern Washington. University of California Publications in Geological Sciences, v. 27, no. 5, pp. 101–212.

Durham, J.W., Harper, H., and Wilder, B. 1942. Lower Miocene in the Willamette Valley, Oregon (Abstract). Geological Society America, Bull, v. 53, no. 4, pp. 425–452.

Dutro, T. 1985. Upper Mississippian brachiopods from Oregon and Washington—possible biogeographic affinities. Geological Society America, Abstracts with Program., v. 17, no. 6, pp. 352–353.

————. 1989. Paleotectonic significance of an early Namurian (Carboniferous) brachiopod fauna from northwestern North America. *In*: Yugan, J., and Chun, Li., eds., Onzieme Congres International de Stratigraphie et de Geologie du Carbonifere, Beijing, August 1987, Compte Rendu, Tome 3, pp. 327–335.

Eaton, G.T. 1922. The John Day Felidae in the Marsh Collection. American Journal Sci., ser. 5, v. 4, pp. 425–452.

Ehlen, J. 1967. Geology of state parks near Arago, Coos County, Oregon. Ore Bin, v. 29, no. 4, pp. 61–82.

Elftman, H.O. 1931. Pleistocene mammals of Fossil Lake, Oregon. American Museum Novitates, no. 481, 21p.

Emlong, D.R. 1966. A new archaic cetacean from the Oligocene of northwest Oregon. University of Oregon, Museum Natural History, Bull. 3, 51p.

Emry, R.J., ed. 2002. Cenozoic mammals of land and sea: tributes to the career of Clayton E. Ray. Washington, D.C.: Smithsonian Institution, Smithsonian Contributions to Paleobiology, no. 93, 372p.

Enochs, L.G., et al. 2002. Unraveling the mystery of the Oligocene flora at Sweet Home, Oregon. *In*: Moore, G.W., ed., Field guide to geologic processes in Cascadia; field trips to accompany the 98th Annual Meeting of the Cordilleran Section of the Geological Society of America, May 13–15 2002, Corvallis, Oregon. Portland, Oregon Department of Geology and Mineral Industries, Special Paper 36, pp. 155–166.

Erwin, D.M., and Schick, K.N. 2007. New Miocene oak galls (Cynipini) and their bearing on the history of cynipid wasps in western North America. Journal of Paleontology, v. 81, no. 3, pp. 568–580.

Eubanks, W. 1960. Fossil woods. Ore Bin, v. 22, no. 7, pp. 65–69.

————. 1962. The fossil flora of Thomas Creek. Ore Bin, v. 24, no. 2, pp. 26–27.

Evernden, J.F., and James, G.T. 1964. Potassium-argon dates of the Tertiary floras of North America. American Journal Science, v. 262, no. 8, pp. 945–974.

Evernden, J.F., et al. 1964. Potassium-argon dates and the Ceonzoic mammalian chronology of North America. American Journal Science, v. 262, p. 145–198.

Feldmann, R.M. 1974. *Hoploparia riddlensis*, a new species of lobster (Decapoda: Nephropidae) from the Days Creek Formation (Hauterivian, lower Cretaceous) of Oregon. Journal Paleontology, v. 48, no. 3, pp. 586–593.

————. 1989. *Lyreidus alseanus* Rathbun from the Paleogene of Washington and Oregon, U.S.A. Annals of the Carnegie Museum, v. 58, art. 2, pp. 61–70.

Ferns, M. L., McClaughry, J. D., and Madin, I. P. 2007. Preliminary assessment of the extent of the leaf fossil beds at Wheeler High School, Wheeler County, Oregon. Oregon Geology, v. 68, no. 1, pp. 34-41.

Fierstine, H.L. 2001. A new *Aglyptorhynchus* (Perciformes: Scombroidei: Blochiidae) from the late Oligocene of Oregon. Journal Paleontology, v. 21, no. 1, pp. 24–33.

Filewicz, M.V. , and Squires, R.L., eds. 1988. Paleogene stratigraphy, West Coast of North America. Society Economic Paleontologists and Mineralogists, Pacific Section, West Coast Paleogene Symposium, v. 58,

Fisher, R.V., and Rensberger, J.M. 1972. Physical stratigraphy of the John Day Formation, central Oregon. University Calif. Publications Geological Sciences, v. 101, pp. 1–45.

Flugel, E., Senowbari-Daryan, B., and Stanley, G.D. 1989. Late Triassic dasycladacean alga from northeastern Oregon: significance of first reported occurrence in western North America. Journal Paleontology, v. 63, no. 3, pp. 374–381.

Follo, M.F. 1992. Conglomerates as clues to the sedimentary and tectonic evolution of a suspect terrane: Wallowa Mountains, Oregon. Geological Society of American, Bulletin, v. 104, pp. 1561–1576.

———. 1994. Sedimentology and stratigraphy of the Martin Bridge limestone and Hurwal Formation (upper Triassic to lower Jurassic) from the Wallowa terrane, Oregon. U.S. Geological Survey, Professional Paper 1439, pp. 1–27.

Fontaine, W.M. 1905. Notes on some fossil plants from the Shasta Group of California and Oregon. U.S. Geological Survey, Monograph 48, pt. 1, pp. 221–273.

Fordyce, R.E. 2003. Cetacean evolution and Eocene-Oligocene oceans revisited. *In*: Prothero, D., Ivany, L., and Nesbitt, E., eds., From greenhouse to icehouse; the marine Eocene-Oligocene transition. New York: Columbia University Press, pp. 154–170.

Fouch, G.A. 1968. Geology of the northwest quarter of the Brogan Quadrangle, Malheur County, Oregon. Master's thesis, University of Oregon, 62p.

Fraser, N. M., Bottjer, D.J., and Fischer, A.G. 2004. Dissecting "*Lithiotis*" bivalves:; implications for the early Jurassic reef eclipse. Palaios, v. 19, pp. 51–67.

Fremd, T. 1988. Assemblages of fossil vertebrates in pre-ignimbrite deposits of the Turtle Cove Member, John Day Formation (Arikareean), from outcrops within the Sheep Rock Unit, John Day Fossil Beds National Monument. Journal Vertebrate Paleontology, Abstract, v. 8, p. A15.

Frey, R.W., and Cowles, J.G. 1972. The trace fossil *Tisoa* in Washington and Oregon. Ore Bin, v. 34, no. 7, pp. 113–119.

Fritsche, A.E., 1995. Cenozoic paleogeography of the western United States – II. Society for Sedimentary Geology (SEPM), Pacific Section, 309p.

Fry, W.E. 1973. Giant fossil tortoise of genus *Geochelone* from the late Miocene–early Pliocene of north central Oregon. Northwest Science v. 17, no. 4, pp. 239–249.

Fugier, L. 1891. World before the deluge. New York: Cassell, 518p.

Furlong, E.L. 1932. Distribution and description of skull remains of the Pliocene antelope *Sphenophalos* from the northern Great Basin province. Carnegie Institute, Washington, Publication 418, pp. 27–36.

Gabb, W.M. 1864. Triassic and Cretaceous fossils, California. Geological Survey California, Palaeontology, v. 1.

————. 1869. Cretaceous and Tertiary fossils. California, Geological Survey of California. Palaeontology, v. 2.

Gall, I. K. 1991. Cell wall structure of carbonized wood as related to ignimbrite deposition. Oregon Geology, v. 53, no. 5, pp. 109–112.

Gaona, M.T. 1984. Stratigraphy and sedimentology of the Osburger Gulch Sandstone member of the upper Cretaceous Hornbrook Formation . . . . *In*: Nilsen, T., ed., Geology of the upper Cretaceous Hornbrook Formation, Oregon and California. Pacific Sect., Society Economic Paleontologists and Mineralogists, v. 42, pp. 141–148.

Gaudry, A. 1867. *Animaux fossils et geologie de l'antique*. 2 vols. Paris, France.

Gazin, C.L. 1932. A Miocene mammalian fauna from southeastern Oregon. Carnegie Institute, Washington, Publication 418, pp. 39–86.

Getahun, A., and Retallack, G.J. 1991. Early Oligocene paleoenvironment of a paleosol from the lower part of the John Day Formation near Clarno, Oregon. Oregon Geology, v. 53, no. 6, pp. 131–136.

Gilluly, J. 1937. Geology and mineral resources of the Baker Quadrangle, Oregon. U.S. Geological Survey, Bull.879, 119p.

Gilmore, C.W. 1928. A new pterosaurian reptile from the marine Cretaceous of Oregon. U.S. National Museum, Proceedings, no. 2745, v. 73, art.24.

————. 1938. Fossil snakes of North America. Geological Society America, Paper no. 9, pt. 4, 37p.

Goedert, J.L. 1988. A new late Eocene species of Plotopteridae (Aves: Pelecaniformes) from northwestern Oregon. Calif. Academy Science, Proceedings, v. 45, no. 6, pp. 97–102.

Goldstrand, P.M. 1994. The Mesozoic geologic evolution of the northern Wallowa terrane, northeastern Oregon and western Idaho. *In*: Vallier, T.L., and Brooks, H.C., eds., Geology of the Blue Mountains region of Oregon, Idaho, and Washington: Stratigraphy, physiography. U.S. Geological Survey, Professional Paper 1439, pp. 29–53.

Goodwin, C.L. 1973. Stratigraphy and sedimentation of the Yaquina Formation, Lincoln County, Oregon. Corvallis, Oregon State University, Masters 121p.

Gordon, I.. 1985. The Paleocene Denning Spring flora of north-central Oregon. Oregon Geology, v. 47, no. 10, pp. 115–118.

Gottesfeld, A.S., Swanson, F.J., and Gottesfeld, L.M. 1981. A Pleistocene low-elevation subalpine forest in the Western Cascades, Oregon. Northwest Science, v. 55, no. 3, pp. 157–166.

Grabau, A.W., and Shimer, H.W. 1910. North American index fossils. 2 vols. New York: A.G. Seiler.

Graham, A.K. 1963. Systematic revision of the Sucker Creek and Trout Creek Miocene floras of southeastern Oregon. American Journal Botany, v. 50, no. 9, pp. 921–936.

———. 1965. The Sucker Creek and Trout Creek Miocene floras of southeastern Oregon. Kent State University, Research Series 9 (Bull. 53, no. 12), 147p.

Graham, R.W. 1998. The Pleistocene terrestrial mammal fauna of North America. *In*: Janis, C.M., Scott, K.M., and Jacobs, L.L., eds., Evolution of Tertiary mammals of North America, New York: Cambridge University Press, pp. 66–71.

Gregory, I. 1968. The fossil woods near Holley in the Sweet Home petrified forest, Linn County, Oregon. Ore Bin, v. 30, no. 4, pp. 57–76.

———. 1969a. Fossilized palm wood in Oregon. Ore Bin, v. 31, no. 5, pp. 93–110.

———. 1969b. Worm-bored poplar from the Eocene of Oregon. Ore Bin, v. 31, no. 9, pp. 184–185.

———. 1976. An extinct *Evodia* wood from Oregon. Ore Bin, v. 38, no. 9, pp. 135–139.

Gregory, W.K. 1933. Fish skulls: a study of the evolution of natural mechanisms. American Philosophical Society, Transactions, v. 23, no. 2, pp. 75–481.

Gregory, W.K., and Simpson, G.G. 1926. Cretaceous Mammal skulls from Mongolia. American Museum Novitates, no. 225, pp. 1–20.

Hahn, G., et al. 2000. Lower Permian trilobites from Oregon, USA. Geologica et Paleontologica, v. 34, pp. 125–135.

Hall, E.R. 1944. A new genus of American Pliocene badger, with remarks on the relationships of badgers of the Northern Hemisphere. Carnegie Institute, Washington, D.C., Contributions to Paleontology 551, pp. 9–23.

Hancock, A. 1962. The new mammal beds. Geological Society of the Oregon Country, Newsletter, v. 28, no. 6, pp. 39–40.

Hanger, R.A., and Strong, E.E. 1998. *Acteonina permiana*, a new species from the Permian Coyote Butte Formation, central Oregon . . . . Biological Society of Washington, Proceedings, v. 113, no. 4, pp. 795-798.

Hanger, R.A., Hoare, R.D., and Strong, E.E. 2000. Permian polyplacophora, rostroconchia, and problematica from Oregon. Journal Paleontology, v. 74, no. 2, pp. 192–198.

Hanna, G.D. 1920. Fossil mollusks from the John Day Basin in Oregon. University of Oregon, Publication, v. 1, no. 6, 8p.

———. 1922. Fossil freshwater mollusks from Oregon contained in the Condon Museum of the University of Oregon. University of Oregon Publication, v. 1, no. 12, 22p.

———. 1933. Freshwater diatoms from Oregon. Carnegie Institute, Washington, D.C., Contributions to Paleontology, no. 416, pp. 43–45.

———. 1963. Some Pleistocene and Pliocene freshwater mollusca from California and Oregon. California Academy Sciences, Occasional Paper no. 43, 20p.

Hansen, H.P. 1942a. The influence of volcanic eruptions upon post-Pleistocene forest succession in central Oregon. American Journal Botany, no. 29, pp. 214–219.

————. 1942b. A pollen study of peat profiles from lower Klamath Lake of Oregon and California. *In*: Cressman, L.S., Archaeological researches in northern Great Basin. Carnegie Institute, Washington, D.C., Publication 538, pp.103–114.

————. 1946. Postglacial forest succession and climate in the Cascades. American Journal Science, v. 244, pp. 710–734.

————. 1947. Postglacial forest succession, climate, and chronology in the Pacific Northwest. American Philosophical Society, Transactions, v. 37, pt.1, 130p.

Hansen, H.P., and Packard, E.L. 1949. Pollen analysis and the age of proboscidian bones near Silverton, Oregon. Ecology, v. 30, no. 4, pp. 461–468.

Hanson, C.B. 1989. *Teletaceras radinskyi*, a new primitive rhinocerotid from the late Eocene Clarno Formation, Oregon. *In*: Prothero, D.R., and Schoch, R.M., eds., The evolution of perissodactyls, New York: Oxford University Press, pp. 379–398.

————. 1996. Bridgerian-Duchesnean Clarno Formation, north-central Oregon. *In*: Prothero, D.R., and Emry, R.J., eds., The terrestrial Eocene-Oligocene transition in North America, New York, Cambridge University Press, pp. 206–239.

Hanson, D.A. 2000. Fossil mammals of the southern basin of the John Day Formation, Oregon. University of Oregon, Master's thesis, 199p.

Harper, R.E. 1946. Geology of the Molalla Quadrangle. Master's thesis, Oregon State University, 29p.

Hatcher, J.B. 1901. Some new and little known fossil vertebrates. Carnegie Institute, Washington, Museum, Annals, no. 1, pp. 128–144.

————. 1902. Oligocene Canidae. Carnegie Institution, Washington, Museum, v. 1, pp. 65–108.

Hay, O.P. 1903. Two new species of fossil turtles from Oregon. University of Calif., Publications in Geological Sciences, v. 3, no. 10, pp. 237–241.

————. 1908. The fossil turtles of North America. Carnegie Institute, Washington, D.C., Publ.75, pp. 1–568.

————. 1916. A contribution to the knowledge of the extinct sirenian *Desmostylus Hesperus*, Marsh. U.S. National Museum, Proceedings, v. 47, pp. 318–397.

————. 1927. The Pleistocene of the western region of North America and its vertebrated animals. Carnegie Institute of Washington, Publication 322B, 346p.

————. 1930. Second bibliography and catalogue of the fossil vertebrata of North America. 2 vols. Oxford: Clarendon Press.

Hedeen, C.D. 1999. Stratigraphy, paleontology, and paleoenvironment of the middle Tertiary Keasey Formation, northwest Oregon. Masters thesis, 180p.

Hergert, H.L. 1961. Plant fossils in the Clarno Formation, Oregon. Ore Bin, v. 23, no. 6, pp. 55–62.

Hickman, C.J.S. 1969. The Oligocene marine molluscan fauna of the Eugene Formation in Oregon. University of Oregon, Museum Natural History, Bull., no. 16, 112p.

————. 1972. Review of the bathyal gastropod genus *Phanerolepida* (Homalopomatinae) and description of a new species from the Oregon Oligocene. Veliger, v. 15, no. 2, pp. 89–112.

———— 1974. *Nehalemia hieroglyphica*, a new genus and species of Archaeogastropod (Turbinidae: Homalopomatinae) from the Eocene of Oregon. Veliger, v. 17, no. 2, pp. 89–91.

————. 1975. Bathyal gastropods of the family Turridae in the early Oligocene Keasey Formation in Oregon . . . . Bulletins of American Paleontology, v. 70, no. 292, 119p.

————. 1976. *Pleurotomaria* (Archaeogastropoda) in the Eocene of the northeastern Pacific; a review of the Cenozoic biogeography and ecology of the genus. Journal Paleontology, v. 50, no. 6, pp. 1090–1102.

————. 1980. Paleogene marine gastropods of the Keasey Formation in Oregon. Bulletins of American Paleontology, v. 78, no. 310, 111p.

————. 1984. Composition, structure, ecology, and evolution of six Cenozoic deep-water mollusk communities. Journal Paleontology, v. 58, no. 5, pp. 1215–1234.

————. 2003. Evidence for abrupt Eocene-Oligocene molluscan faunal change in the Pacific Northwest. *In*: Prothero, D., Ivany, L., and Nesbitt, E., eds., From greenhouse to icehouse; the marine Eocene-Oligocene transition. New York: Columbia University Press, pp. 71–87.

Holland, W.J., and Peterson, O.A. 1913. The osteology of the Chalicotheroidea with special reference to a mounted skeleton of *Moropus elatus Marsh* . . . . Memoirs of the Carnegie Museum, Pittsburgh, Carnegie Institute, v. III, no. 2, pp. 189–406.

Hoover, L. 1963. Geology of the Anlauf and Drain Quadrangles, Douglas and Lane Counties, Oregon. U.S. Geological Survey, Bull., v. 1122D, pp. D1–D62.

Hopkins, S. 2005. The evolution of fossoriality and the adaptive role of horns in the Mylagaulidae . . . . Proceedings of the Royal Society, v. 272, pp. 1705–1713.

————. 2006. Morphology of the skull in *Meniscomys* from the John Day Formation of central Oregon. PaleoBios 26 (1), pp. 1–9.

Hopkins, W.S. 1967. Palynology and its paleoecological applications in the Coos Bay area, Oregon. Ore Bin, v. 29, no. 9, pp. 161–183.

Howard, H. 1946. A review of the Pleistocene birds of Fossil Lake, Oregon. Carnegie Institute, Washington, D.C., Publication 551, pp. 141–195.

————. 1964. A new species of the 'Pigmy Goose", *Anabernicula*, from the Oregon Pleistocene, with a discussion of the genus. American Museum Natural History, Novitates, no. 2200, 14p.

Howard, J.K., and Dott, R.H. 1961. Geology of Cape Sebastian State Park and its regional relationships. Ore Bin, v. 23, no. 8, pp. 75–84.

Howe, H.V. 1922. Faunal and stratigraphic relationships of the Empire Formation, Coos Bay, Oregon. Calif. University, Department of Geological Sciences, Bull., v. 14, pp. 85–114.

Howell, D., and McDougall., K.A., eds. 1978. Mesozoic paleogeography of the Western United States: Pacific Section, Society of Economic Paleontologists and Mineralogists, Symposium 2, 573p.

Hoxie, L.R. 1965. The Sparta flora from Baker County, Oregon. Northwest Science, v. 39, no. 1, pp. 26–35.

Hu, H.H. 1948. How *Metasequoia*, the "Living Fossil" was discovered in China. New York Botanical Garden, Journal, v. 49, no. 585, pp. 201–207.

Hubbs, C.L., and Miller, R.R. 1948. The Great Basin, with emphasis on glacial and postglacial times, v. II: the zoological evidence. Utah University, Bulletin, v. 38, no. 20, pp. 17–166.

Hunt, R., and Stepleton, E. 2004. Geology and paleontology of the upper John Day beds, John Day River valley, Oregon; lithostratigraphic and biochronologic revision . . . . American Museum Natural History, Bull. no. 282, 90p.

Hunter, R.E. 1980. Depositional environments of some Pleistocene coastal terrace deposits, southwestern Oregon-case history of a progradational beach and dune sequence. Sedimentary Geology, v. 27, pp. 241–262.

Hutchison, J.H. 1966. Notes on some upper Miocene shrews from Oregon. University of Oregon, Museum of Natural History, Bull. 2, 23p.

———. 1968. Fossil Talpidae (Insectivora, Mammalia) from the later Tertiary of Oregon. University of Oregon, Museum of Natural History, Bull.11, 117p.

———. 1984. Cf. *Condylura* (Mammalia: Talpidae) from the late Tertiary of Oregon. Journal Vertebrate Paleontology, v. 4, no. 4, pp. 600–601.

———. 1992. Western North American reptile and amphibian record across the Eocene/Oligocene boundary and its climatic implications. *In*: Prothero, D.R., and Berggren, W.A., eds., Eocene-Oligocene climatic and biotic evolution. Princeton: Princeton University Press, pp. 451–463.

Imlay, R.W. 1967. The Mesozoic pelecypods *Otapira* Marwick and *Lupherella* Imlay, new genus in the United States. U.S. Geological Survey, Professional Paper 573-B, 11p.

———. 1968. Lower Jurassic (Pliensbachian and Toarcian) ammonites from eastern Oregon and California. U.S. Geol. Survey, Prof. Paper 593–C, pp. 1–51.

———. 1973. Middle Jurassic (Bajocian) ammonites from eastern Oregon. U.S. Geological Survey, Professional Paper 756, 100p.

———. 1980. Jurassic paleobiogeography of the conterminous United States in its continental setting. U.S. Geol. Survey, Professional Paper 1062, 134p.

———. 1981. Jurassic (Bathonian and Callovian) ammonites in eastern Oregon and western Idaho. U.S. Geological Survey, Professional Paper 1142, 24p.

———. 1986. Jurassic ammonites and biostratigraphy of eastern Oregon and western Idaho. *In*: Vallier, T.L., and Brooks, H.C., eds., Geology of the Blue Mountains region of Oregon, Idaho, and Washington. U.S. Geological Survey, Professional Paper 1435, pp. 53–57.

Imlay, R.W., and Jones, D.K. 1970. Ammonites from the *Buchia* zones in northwestern California and southwestern Oregon. U.S. Geological Survey, Professional Paper 647–B, 59p.

Imlay, R.W., et al. 1959. Relations of certain upper Jurassic and lower Cretaceous formations in southwestern Oregon. American Assoc. Petroleum Geologists, Bull.43, pp. 2770–2785.

Irwin, W.P. 1977. Review of Paleozoic rocks of the Klamath Mountains. *In*: Stuart, J.H., Stevens, C.H., and Fritsche, A.E., eds., Paleozoic paleogeography of the western United States. Society Economic Paleontologists and Mineralogists, Pacific Coast Paleogeography Symp. 1, pp. 441–454.

Irwin, W.P., and Galanis, S.P. 1976. Map showing limestone and selected fossil localities in the Klamath Mountains, California and Oregon. U.S. Geological Survey, Miscellaneous Field Studies Map MF-749.

Irwin, W.P., Jones, D.L., and Kaplan, T.A. 1978. Radiolarians from pre-Nevadan rocks of the Klamath Mountains, California and Oregon. *In*: Howell, D., and McDougall., K.A., Mesozoic paleogeography of the Western United States: Pacific Section, Society of Economic Paleontologists and Mineralogists, Symposium 2, pp. 303–310.

Irwin, W.P., Wardlaw, B.R., and Kaplan, T.A. 1983. Conodonts of the Western Paleozoic and Triassic Belt, Klamath Mountains, California and Oregon. Journal Paleontology, v. 57, no. 5, pp. 1030–1039.

Ivany, L.C., Nesbitt, E.A., and Prothero, D.R. 2003. The marine Eocene-Oligocene transition: a synthesis. *In*: Prothero, D., Ivany, L., and Nesbitt, E., eds., From greenhouse to icehouse; the marine Eocene-Oligocene transition. New York: Columbia University Press, pp. 525–534.

Janis, C.M., Scott, K.M., and Jacobs, L.L., eds. 1998. Evolution of Tertiary Mammals of North America. New York: Cambridge University Press, 691p.

Jehl, J.R. 1967. Pleistocene birds from Fossil Lake, Oregon. Condor, v. 69, no. 1, pp. 24–27.

Johnson, J.G., and Klapper, G. 1978. Devonian brachiopods and conodonts from central Oregon. Journal Paleontology, v. 52, no. 2, pp. 295– 299.

Johnson, W.R. 1965. Structure and stratigraphy of the southeastern quarter of the Roseburg 15<degree> Quadrangle, Douglas County, Oregon. Master's thesis, University of Oregon, 85p.

Jones, D.L. 1960. Lower Cretaceous (Albian) fossils from southwestern Oregon and their paleogeographic significance. Journal Paleontology, v. 34, no. 1, pp. 152–160.

Jones, D.L., et al. 1977. Wrangellia—a displaced terrane in northwestern North America. Canadian Journal Earth Sciences, v. 14, no. 11, pp. 2565–2577.

Jordan, D.S. 1907. The fossil fishes of California; with supplementary notes on other species of extinct fishes. Calif. Univ. , Dept. Geol. Sci., Bull.5, pp. 95–144.

Jordan, D.S., and Hannibal, H. 1923. Fossil sharks and rays of the Pacific slope of North America. Southern Calif. Academy Sciences, Bull., v. 22, pp. 27–68.

Kaczmarska, I. 1985. The diatom flora of Miocene lacustrine diatomites from the Harper Basin, Oregon, U.S.A. Acta Paleobotanica, v. 25, pp.33–99.

Keen, A.M., and Coan, E. 1974. Marine molluscan genera of Western North America; an illustrated key. 2nd ed. Stanford, Calif.: Stanford University Press, 208p.

Kennedy, G.L., and Lajoie, K.R. 1982. Aminostratigraphy and faunal correlations of late Quaternary marine terraces, Pacific Coast, USA. Nature, v. 299, October, pp. 545–547.

Kester, P. R. 2001. Paleoclimatic interpretation from the early Oligocene Willamette flora, Eugene, Oregon. University of Washington, Seattle, Master's thesis, 62p. (also: 2002, Geological Society of America, Cordilleran Section Meeting., Abstracts with Program, pp. A–9)

Kimmel, P. G. 1975. Fishes of the Miocene-Pliocene Deer Butte Formation, southeast Oregon. University of Michigan, Museum of Paleontology, Papers on Paleontology, no. 14, pp. 69–87.

———. 1982. Stratigraphy, age, and tectonic setting of the Miocene-Pliocene lacustrine sediments of the western Snake River plain, Oregon and Idaho. *In*: Bonnichsen, B., and Breckenridge, R.M., eds., Cenozoic geology of Idaho. Idaho Bureau of Mines and Geology, Bull. 26, pp. 550–578.

———. 1985. Fossil fish faunas and the biostratigraphy of sediments near Vale, Oregon. Geological Society America, Abstracts with Programs, v. 17, no. 4, p. 228.

Kleinhaus, L.C., Balcells-Baldwin, E.A., and Jones, R.E. 1984. A paleogeographic reinterpretation of some middle Cretaceous units, north-central Oregon: evidence for a submarine system. *In*: Nilsen, T. H., ed., Geology of the upper Cretaceous Hornbrook Formation, Oregon and California. Society Economic Paleontologists and Mineralogists, Pacific Section, v. 42, pp. 239–257.

Kleweno, W.P., and Jeffords, R.M. 1962. Devonian rocks in the Suplee-Izee area of central Oregon. Geological Society America, Special Paper 68, 34p.

Klucking, E.P. 1964. An Oligocene flora from the western Cascades. PhD diss., University of California, 372p.

Knowlton, F.H. 1900. Fossil plants associated with the lavas of the Cascade Range. U.S. Geological Survey 20th Annual Report, pt. 3, pp. 37–64.

———. 1902. Fossil flora of the John Day basin, Oregon. U.S. Geological Survey, Bull. 204, 127p.

———. 1910. Jurassic age of the "Jurassic flora of Oregon. " American Journal Science, 4th ser., v. 30, pp. 33–64.

Koch, J.G. 1966. Late Mesozoic stratigraphy and tectonic history, Port Orford-Gold Beach area, southwestern Oregon coast. American Assoc. Petroleum Geologists, Bull.50, no. 1, pp. 25–71.

Koch, J.G., and Camp, C.L. 1966. Late Jurassic ichthyosaur from Sisters Rocks coastal southwestern Oregon. Ore Bin, v. 28, no. 3, pp. 65–68.

Kooser, M.A., and Orr, W.N. 1973. Two new decapod species from Oregon. Journal Paleontology, v. 47, no. 6, pp. 1044–1046.

Krause, W.F. 1999. The Miocene Metasequoia Creek flora on the Columbia River in northwestern Oregon. Oregon Geology, v. 61, no. 5, pp. 111–114.

Lakhanpal, R.N. 1958. The Rujada flora of west central Oregon. University California, Publications in Geological Sciences, v. 35, no. 1, pp. 1–66.

LaMaskin, T.A. 2008. Late Triassic (Carnian-Norian) mixed carbonate-volcaniclastic facies of the Olds Ferry terrane, eastern Oregon and western Idaho. Geological Society America, Special Paper 442, pp. 251–267.

LaMaskin, T.A., et al. 2008. Tectonic controls on mudrock geochemistry, Mesozoic rocks of eastern Oregon and Western Idaho, U.S.A.: implications for cordilleran tectonics. Journal Sedimentary Research, v. 78, pp. 787–805.

Larue, D.K., and Steel, R.J., eds. 1983. Cenozoic marine deposits of the Pacific margin. Society of Economic Paleontologists and Mineralogists, Pacific Section, 239p.

LeConte, J. 1892. Elements of geology. Rev. and enl. New York: Appleton, 640p.

Leffler, S.R. 1964. Fossil mammals from the Elk River Formation, Cape Blanco, Oregon. Journal Mammalogy, v. 45, pp. 53–61.

Leidy, J. 1870. Remarks on a collection of fossils from . . . Thomas Condon. Academy Natural Sciences, Philadelphia, Proceedings, v. 22, pp. 111–113.

————. 1873. Contributions to the extinct vertebrate fauna of the western territories. U.S. Geol. Survey, Report of Territories (Hayden), v. 1, 358p.

Lesquereux, L. 1878. Fossil plants of the auriferous gravel deposits of the Sierra Nevadas. Cambridge, Museum Comparative Zoology, Memoirs, v. 6, no. 2, 62p.

————. 1879. Atlas of the coal flora of Pennsylvania and the Carboniferous formation throughout the United States. Harrisburg, Second Geological Survey, 87 plates.

————. 1888. Recent determinations of fossil plants from Kentucky, Louisiana and Oregon. U.S. Natl. Mus., Proc., v. 11, pp. 11–38.

Lewis, R.Q. 1950. The geology of the southern Coburg Hills including the Springfield-Goshen area. University of Oregon, Master's thesis, 78p.

Linder, R.A. 1986. Mid-Tertiary echinoids and Oligocene shallow marine environments in the Oregon central Western Cascades. Master's thesis, University of Oregon, 196p.

Linder, R.A., and Orr, W.N. 1983. Mid-Tertiary echinoids from the Oregon western Cascades. Oregon Academy Sciences, 41st Annual Conference, Proceedings, v. 19, p. 104.

Linder, R.A., Durham, J.W., and Orr, W.N. 1988. New late Oligocene echinoids from the central Western Cascades of Oregon. Journal Paleontology, v. 62, no. 6, pp. 945–958.

Lindberg, D.R., and Hickman, C.S. 1986. A new anomalous giant limpet from the Oregon Eocene (Mollusca: Patellida). Journal Paleontology, v. 60, no. 3, pp. 661–668.

Lockley, M.G. 1990. How volcanism affects the biostratigraphic record. In: Lockley, M.G., and Rice, A., eds., Volcanism and fossil biotas. Geological Society America, Special Paper 244, pp. 1–12.

Lohman, K.E. 1936. Diatoms in the Mascall Formation from Tipton and Austin, Oregon. Carnegie Institute, Washington, D.C., Publication 455, pp. 9–11.

————. 1937. [no title] In: Moore, B.N. 1937. Non-metallic mineral resources of eastern Oregon. U.S. Geological Survey, Bull.875, various pagings.

Lowry, W.D. 1940. The geology of the Bear Creek area, Crook and Deschutes counties, Oregon. Master's thesis, Oregon State University, 78p.

————. 1943. The geology of the northeast quarter of the Ironside Mountain Quadrangle, Baker and Malheur counties, Oregon. University of Rochester, PhD diss.

Lowther, J.S. 1967. A new Cretaceous flora from southwestern Oregon (Abstract). Northwest Science, v. 41, no. 1, p. 54.

Lucas, F. A. 1902. Animals of the past. New York, McClure, Phillips, 258p.

Lull, R.S. 1921. New camels in the Marsh Collections. American Journal Science, ser. 5, v. 1, no. 5, pp. 392–404.

Lupher, R.I. 1930. Stratigraphy and correlation of the marine Jurassic deposits of central Oregon. PhD diss., California Institute of Technology.

———. 1941. Jurassic stratigraphy of central Oregon. Geological Society America, Bull., v. 52, pt.1, pp. 219–269.

Lupher, R.I., and Packard, E.L. 1930. The Jurassic and Cretaceous rudistids of Oregon. Univ. of Oregon, Publications in Geology, v. 1, no. 3, pp. 203–212.

Lutz, F.E. 1935. Fieldbook of insects. 3rd ed. New York: Putman, 562p.

MacGinitie, H.D. 1933. The Trout Creek flora of southeastern Oregon. Carnegie Institute, Washington, D.C., Publ. 416, pp. 21–68.

Mahood, A.D. 1981. *Stephanodiscus rhombus*, a new diatom species from Pliocene deposits at Chiloquin, Oregon. Micropaleontology, v. 27, no. 4, pp. 379–383.

Mamay, S.H., and Read, C.B. 1956. Additions to the flora of the Spotted Ridge Formation in central Oregon. U.S. Geological Survey, Professional Paper 274-I, pp. 211–226.

Manchester, S.R. 1979. *Triplochitoxylon* (Sterculiaceae): a new genus of wood from the Eocene of Oregon and its bearing on xylem evolution in the extant genus *Triplochiton.* American Journal Botany, v. 66, no. 6, pp. 699–708.

———. 1980. *Chattawaya* (Sterculiaceae): a new genus of wood from the Eocene of Oregon and its implications for xylem evolution of the extant genus *Pterospermum.* American Journal Botany, v. 67, no. 1, pp. 59–67.

———. 1981. Fossil plants of the Eocene Clarno Nut Beds. Oregon Geology, v. 43, no. 6, pp. 75–81.

———. 1983. Fossil wood of the Engelhardiaceae (Juglandaceae) from the Eocene of North America: *Engelhardioxylon* gen. nov. Botanical Gazette, v. 144, no. 1, pp. 157–163.

———. 1986. Vegetative and reproductive morphology of an extinct plane tree (Platanaceae) from the Eocene of Western North America. Botanical Gazette, v. 147, no. 2, pp. 200– 226.

———. 1987a. Extinct ulmaceous fruits from the Tertiary of Europe and Western North America. Review of Palaeobotany and Palynology, v. 52, pp. 119–129.

———. 1987b. The fossil history of the Juglandaceae. Missouri Botanical Garden, Monographs in Systematic Botany, v. 21, 137p.

———. 1988. Fruits and seeds of *Tapiscia* (Staphyleaceae) from the middle Eocene of Oregon, USA. Tertiary Research, v. 9, no. 1–4, pp. 59–66.

———. 1991. *Cruciptera,* a new juglandaceous winged fruit from the Eocene and Oligocene of western North America. Systematic Botany, v. 16, pp. 715–725.

———. 1994. Fruits and seeds of the middle Eocene Nut Beds flora, Clarno Formation, Oregon. Palaeontographica Americana, no. 58, 205p.

———. 1995. Yes, we had bananas. Oregon Geology, v. 57, no. 2, pp. 41–43.

———. 2000. Late Eocene fossil plants of the John Day Formation, Wheeler County, Oregon. Oregon Geology, v. 62, no. 3, pp. 51–64.

Manchester, S.R., and Crane, S.R. 1987. A new genus of Betulaceae from the Oligocene of western North America. Botanical Gazette, v. 148, pp. 263–273.

Manchester, S.R., and Kress, W.J. 1993. Fossil bananas (Musaceae) *Ensete oregonensis* sp. nov from the Eocene of western North America and its phytogeographic significance. American Journal of Botany, v. 80, pp. 1264–1272.

Manchester, S.R. and Meyer, H.W. 1987. Oligocene fossil plants of the John Day Formation, Fossil Oregon. Oregon Geology, v. 49, no. 10, pp. 115–117.

Marsh, O.C. 1871. Notice of some new fossil mammals from the Tertiary formation. American Journal Science, ser.3, v. 2, pp. 41–42.

———. 1873. Notice of new Tertiary mammals. American Journal Science, ser.3, v. 5, pp. 407–410; 485–488.

———. 1874. Notice of new equine mammals from the Tertiary formations. American Journal Science, ser.3, v. 7, pp. 147–158.

———. 1877. Notice of some new vertebrate fossils. American Journal Science, ser.3, v. 14, no. 81, pp. 249–258.

———. 1894. Eastern division of the *Miohippus* beds, with notes on some of the characteristic fossils. American Journal Science, ser.3, v. 48, pp. 91–94.

———. 1895. Reptilia of the *Baptanodon* beds. American Journal Science, v. 50, no. 3, pp. 405–406.

Martin, J. 1979. Hemphillian rodents from northern Oregon and their relationships to other rodent faunas in North America. Ph.D. diss., University of Washington, 265p.

———. 1981. Contents of coprolites from Hemphillian sediments in northern Oregon . . . . South Dakota Academy Sciences, Proceedings, v. 60, pp. 105–115.

———. 1983. Additions to the early Hemphillian (Miocene) Rattlesnake fauna from central Oregon. South Dakota Academy Sciences, Proceedings, v. 62, pp. 23–33.

———. 1984. A survey of Tertiary species of *Perognathus* (Perognathinae) and a description of a new genus of Heteromyinae. *In*: Mengel, R.M., ed., Papers in vertebrate paleontology honoring Robert Warren Wilson. Carnegie Museum of Natural History, Special Publ., no. 9, pp. 90–121.

———. 1997. Progression of borophagines (Carnivora: Canidae) in the Pacific Northwest. Journal Vertebrate Paleontology (Abstracts), Suppl. v. 17, no. 3, p. 62A.

Martin, J., et al. 2005. Lithostratigraphy, tephrochronology, and rare earth element geochemistry of fossils at the classical Pleistocene Fossil Lake area, south central Oregon. Journal of Geology, v. 113, pp. 139–155.

Martin, P. S., and Klein, R.G., eds. 1984. Quaternary extinctions, a prehistoric revolution. Tucson: University of Arizona Press, 892p.

Matthew, W.D. 1899. A provisional classification of the fresh-water Tertiary of the West. American Museum Natural History, Bull.12, pp. 19–75.

———. 1909. Faunal lists of the Tertiary mammalia of the west. U.S. Geological Survey, Bull.361, pp. 91–138.

Maxson, J.H. 1928. *Merychippus isonesus* (Cope) from the later Tertiary of the Crooked River basin, Oregon. Carnegie Institute, Washington, D.C., Contributions to Paleontology 393, pp. 55–58.

McAfee, R. 2003. Confirmation of the sloth genus *Megalonyx* (Xenarthra: Mammalia) from the John Day region and its implication. Journal of Vertebrate Paleontology, Supplement (Abstracts), v. 23, no. 3, pp. 77A.

McCarville, K. 2003. Avian taphonomy of Fossil Lake, Oregon. Geol. Society America, Abstracts with Programs, v. 35, no. 6, p. 497.

————. 2003. A new interpretation for a classical locality: Fossil Lake, Oregon. Journal Vertebrate Paleontology, Abstracts, Suppl., v. 23, no. 3, p. 77A.

McClammer, J.U. 1978. Paleobotany and stratigraphy of the Yaquina flora (latest Oligocene–earliest Miocene) of western Oregon. Master's thesis, University of Maryland, 174p.

McCornack, E.C. 1914. A study of Oregon Pleistocene. University of Oregon, Bull., n. s., v. 12, no. 1, 16p.

————. 1928. Thomas Condon, pioneer geologist of Oregon. Eugene: University of Oregon Press, 355p.

McDougall, K., 1975. The microfauna of the type section of the Keasey Formation of northwestern Oregon. *In*: Weaver, D.W., et al., eds., Paleogene symposium and selected technical papers, pp. 343–359.

McFadden, B.J. 1984. Systematica and phylogeny of *Hipparion, Neohipparion, Nannippus*, and *Cormohipparion* (Mammalia, Equidae) from the Miocene and Pliocene of the New World. American Museum Natural History, Bull., v. 179, art. 1, 196p.

McFadden, J. J. 1986. Fossil flora near Gray Butte, Jefferson County, Oregon. Oregon Geology, v. 48, no. 5, pp. 51– 55.

McInelly, G.W., and Kelsey, H.M. 1990. Late Quaternary tectonic deformation of the Cape Arago-Bandon region of coastal Oregon as deduced from wave-cut platforms. Journal Geophysical Research, v. 95, no. B5, pp. 6699–6713.

McIntosh, W.C., Manchester, S.R., and Meyer, H.W. 1997. Age of the plant-bearing tuffs of the John Day Formation at Fossil Oregon, based on 40/Ar/39/Ar single-crystal dating. Oregon Geology, v. 59, no. 1, pp. 3–5.

McKee, T.M. ?1970. Preliminary report on the fossil fruits and seeds from in the mammal quarry of the lower Tertiary Clarno Formation, Oregon. Oregon Museum Science and Industry, Student Research Center, 17p.

McKeel, D.R., and Lipps, J.H. 1972. Calcareous plankton from the Tertiary of Oregon. Paleogeography, Paleoclimatology, Paleoecology, v. 12, no. 1–2, pp. 75–93.

————. 1975. Eocene and Oligocene planktonic foraminifera from the central and southern Oregon Coast Range. Journal Foraminiferal Research, v. 5, no. 4, pp. 249–269.

McKenna, M.C. 1990. Plagiomenids (Mammalia: ?Dermoptera) from the Oligocene of Oregon, Montana, and South Dakota … *In*: Bown, T.M., and Rose, K.D., eds., Dawn of the age of mammals in the northern Rocky Mountain interior, North America. Geological Society America, Special Paper 243, pp. 211–234.

McKillip, C.J. 1992. Paleoenvironment and biostratigraphy of the Eugene Formation near Salem, Oregon. Master's thesis, University of Oregon, 109p.

McKnight, B.K. 1984. Stratigraphy and sedimentation of the Payne Cliffs Formation, southwestern Oregon. *In*: Nilsen, T.H., ed., Geology of the upper Cretaceous Hornbrook Formation, Oregon and California. Society of Economic Paleontologists and Mineralogists, Pacific Section, v. 42, pp. 187–194.

McRoberts, C.A. 1990. Systematic paleontology, stratigraphic occurrence, and paleoecology of halobiid bivalves from the Martin Bridge Formation (Upper Triassic) Wallowa terrane, Oregon. Master's thesis, University of Montana, 156p.

————. 1993. Systematics and biostratigraphy of halobiid bivalves from the Martin Bridge Formation (upper Triassic), northeast Oregon. Journal of Paleontology, v. 67, pp. 198–210.

McWilliams, R.G. 1968. Paleogene stratigraphy and biostratigraphy of central-western Oregon. PhD diss., University of Washington, 167p.

Meek, F.B. 1864. California, Geological Survey of California, Palaeontology, v. 1.

Mehringer, P.J., 1985. Late-Quaternary pollen records from the interior Pacific Northwest and northern Great Basin of the United States. *In*: Bryant, V.M., and Holloway, R.G., eds. Pollen records of Late-Quaternary North American sediments. American Association of Stratigraphic Palynologists, Dallas, Texas, pp. 167–189.

Mellett, J.S. 1969. A skull of *Hemipsalodon* (Mammalia, order Deltatheridia) from the Clarno Formation of Oregon. American Museum Novitates, no. 2387, 19p.

Merriam, C.W. 1941. Fossil Turritellas from the Pacific coast region of North America. University of California, Publications in Geological Sciences, v. 26, no. 1, 214p.

————.1942. Carboniferous and Permian corals from central Oregon. Journal of Paleontology, v. 16, pp. 372–381.

Merriam, C.W., and Berthiaume, S.A. 1943. Late Paleozoic formations of central Oregon. Geological Society America, Bull., v. 54, pt. 1, pp. 145–171.

Merriam, J.C. 1901. Contribution to the geology of the John Day basin. University California, Publications in Geological Sciences, v. 9, pp. 269–314.

————. 1906. Carnivora from the Tertiary formations of the John Day region. University California, Publications in Geological Sciences, v. 5, pp. 1–64.

————. 1913. Tapir remains from the late Cenozoic beds of the Pacific Coast region. University California, Publications in Geological Sciences, v. 7, no. 9, pp. 169–175.

————. 1916. Mammalian remains from a late Tertiary formation at Ironside, Oregon. University California, Publications in Geological Sciences, v. 10, no. 9, pp. 129–135.

————. 1930. *Allocyon*, a new canid genus from the John Day beds of Oregon. University California, Publications in Geological Sciences, v. 19, no. 9, pp. 229–244.

Merriam, J.C., and Gilmore, C.W. 1928. An ichthyosaurian reptile from marine Cretaceous of Oregon. Carnegie Institute, Washington, D.C., Contributions to Paleontology, no. 393, pp. 1–4.

Merriam, J.C., and Sinclair, W.J. 1907. Tertiary faunas of the John Day region. University of California, Publications in Geological Sciences, v. 5, no. 11, pp. 171–205.

Merriam, J.C., and Stock, C. 1927. A hyaenarctid bear from the later Tertiary of the John Day basin, Oregon. Carnegie Institute, Washington, D.C., Publication 346, pp. 39–44.

Merriam, J.C., Stock, C., and Moody, C.L. 1925. The Pliocene Rattlesnake Formation and fauna of eastern Oregon, with notes on the geology of the Rattlesnake and Mascall deposits. Carnegie Institute, Washington, D.C., Publication 347, pp. 43–92.

Merrill, G.P. 1969. The first one hundred years of American geology. New York: Hafner, 773p.

Meyer, H.W. 1973. The Oligocene Lyons flora of northwestern Oregon. Ore Bin, v. 35, no. 3, pp. 37–51.

Meyer, H.W., and Manchester, S. 1997. The Oligocene Bridge Creek flora of the John Day Formation, Oregon. University of California, Publications in Geological Sciences, v. 141, 195p.

Miles, G.A. 1977. Planktonic foraminifera of the lower Tertiary Roseburg, Lookingglass, and Flournoy formations (Umpqua Group), southwest Oregon. PhD diss., University of Oregon, 360p.

Miller, A.H. 1911. Additions to the avifauna of the Pleistocene deposits at Fossil Lake, Oregon. University Calif. Publications in Geological Sciences, v. 7, no. 5, pp. 61–115.

———. 1931. An auklet from the Eocene of Oregon. University Calif. Publications in Geological Sciences, v. 20, pp. 23–26.

———. 1944. Some Pliocene birds from Oregon and Idaho. Condor, v. 46, pp. 25–32.

Miller, C.N. 1992. Structurally preserved cones of *Pinus* from the Neogene of Idaho and Oregon. International Journal of Plant Science, v. 153, no. 1, pp. 147–154.

Miller, L. 1899. Journal of first trip of University of California to John Day beds of eastern Oregon. University of Oregon, Museum of Natural History, Bull. 19, 21p.

Miller, M., and Wright, J.E. 1987. Paleogeographic implications of Permian tethyan fossils from the Klamath Mountains, California. Geology, v. 15, pp. 266–269.

Miller, P.R., and Orr, W.N. 1986. The Scotts Mills Formation: Mid-Tertiary geologic history and paleogeography of the central Western Cascade Range, Oregon. Oregon Geology, v. 48, no. 12, pp. 139–151.

———. 1988. Mid-Tertiary transgressive rocky coast sedimentation: central Western Cascade Range, Oregon. Journal Sedimentary Petrology, v. 58, no. 6, pp. 959–968.

Miller, R.R. 1965. Quaternary freshwater fishes of North America. *In*: Wright, J. E., and Frey, D. G., eds., Quaternary history of the United States. Princeton, N.J.: Princeton University Press, pp. 569–581.

Miller, W., ed. 2007. Trace fossils: concepts, problems, prospects. Boston: Elsevier, 611p.

Mitchell, E. 1966. Faunal succession of extinct North Pacific marine mammals. Norsk Hvalfangst-Tidende, no. 3, pp. 47–60.

———. 1966. The Miocene pinniped *Allodesmus*. University of California, Publications in Geological Sciences, v. 61, 46p.

———. 1975. Parallelism and convergence in the evolution of Otariidae and Phocidae. Rapports et Process-Verbaux des Reunions. Conseil International pour l'Exploration de la Mer, v. 169, pp. 12–26.

Mitchell, E., and Repenning, C.A. 1963. The chronologic and geographic range of Desmostylians. Los Angeles County Museum of Natural History, Contributions to Science, v. 78, pp. 1–20.

Mitchell, E., and Tedford, R.H. 1973. The Enaliarctinae, a new group of extinct aquatic Carnivora and a consideration of the origin of the Otariidae. American Museum Natural History, Bull.151, pp. 201–284.

Mobley, B.J. 1956. Geology of the southwest quarter of the Bates Quadrangle, Oregon. University of Oregon, Master's thesis, 66p.

Moore, B.N. 1937. Non-metallic mineral resources of eastern Oregon. U.S. Geological Survey, Bull. 875, pp. 17–118.

Moore, E.J. 1963. Miocene marine mollusks from the Astoria Formation in Oregon. U.S. Geological Survey, Professional Paper 419, 190p.

———. 1971. Fossil mollusks of coastal Oregon. Oregon State University, Monograph no. 10, 64p.

———. 1976. Oligocene marine mollusks from the Pittsburg Bluff Formation in Oregon. U.S. Geological Survey, Professional Paper 922, pp. 1– 66.

———. 1984. Middle Tertiary molluscan zones of the Pacific Northwest. Journal Paleontology, v. 58, no. 3, pp. 718–737.

———. 1994. Fossil shells from Oregon beach cliffs. Corvallis: Chintimini Press, 88p.

———. 2000. Fossil shells from western Oregon. Corvallis: Chintimini Press, 131p.

Moore, E.J., and Addicott, W.O. 1987. The Miocene Pillarian and Newportian (Molluscan) Stages of Washington and Oregon and their usefulness in correlations from Alaska to California. U.S. Geological Survey, Bull. 1664, Contributions to Paleontology and Stratigraphy, pp. A1–A13.

Moore, E.J., and Moore, G.W. 2002. Miocene molluscan fossils and stratigraphy, Newport, Oregon. *In*: Moore, G.W., ed., Field guide to geologic processes in Cascadia; field trips to accompany the 98th Annual Meeting of the Cordilleran Section of the Geological Society of America, May 13–15 2002, Corvallis, Oregon. Portland, Oregon, Department of Geology and Mineral Industries, Special Paper 36, pp. 155–166.

Moore, R.C., ed. 1969. Treatise on invertebrate paleontology. Boulder Colorado, University of Kansas and the Geological Society of America, various volumes.

Moore, R.C., and Vokes, H.E. 1953. Lower Tertiary crinoids from northwestern Oregon. U.S. Geological Survey, Professional Paper 233–E, pp. 113–148.

Morris, E.M., and Wardlaw, B.R. 1986. Conodont ages for limestones of eastern Oregon and their implication for pre-Tertiary mélange terranes. *In*: Vallier, T.L., and Brooks, H.C., eds., Geology of the Blue Mountain region of Oregon, Idaho, and Washington. U.S. Geological Survey, Professional Paper 1435, pp. 59–63.

Morrison, R.F. 1964. Upper Jurassic mudstone unit named in Snake River canyon, Oregon-Idaho boundary. Northwest Science, v. 38, no. 3, pp. 83–87.

Muhs, D.R., et al. 1990. Age estimates and uplift rates for late Pleistocene marine terraces: southern Oregon portion of the Cascadia forearc. Journal Geophysical Research, v. 95, no. B5, pp. 6685–6698.

Mumford, D.G. 1989. Geology of Elsie-lower Nehalem River area . . . northwestern Oregon. Master's thesis, Oregon State University, 392p.

Munthe, J., and Coombs, M.C. 1979. Miocene dome-skulled chalicotheres (Mammalia, Perissodactyla) from the western United States: a preliminary discussion of a bizarre structure. Journal Paleontology, v. 53, no. 1, pp. 77–91.

Murchey, B.L., and Jones, D.L. 1994. The environmental and textonic significance of two coeval Permian radiolarian-sponge associations in eastern Oregon. *In*: Vallier, T., and Brooks, H., eds., Geology of the Blue Mountains region of Oregon, Idaho, and Washington. U.S. Geological Survey, Professional Paper 1439, pp. 183–198.

Myers, J.A. 2003. Terrestrial Eocene-Oligocene vegetation and climate in the Pacific Northwest. *In*: Prothero, D.R., et al., eds., From greenhouse to icehouse. New York: Columbia University Press, pp. 171–185.

Myers, J.A., Kester, P.R., and Retallack, G.J. 2002. Paleobotanical record of Eocene-Oligocene climate and vegetational change near Eugene, Oregon. *In*: Moore, G.W., ed., Field guide to geologic processes in Cascadia; field trips to accompany the 98th Annual Meeting of the Cordilleran Section of the Geological Society of America, May 13–15 2002, Corvallis, Oregon. Portland, Oregon, Department of Geology and Mineral Industries, Special Paper 36, pp. 145–154.

Nations, J.D. 1970. The family Cancridae and its fossil record on the West Coast of North America. PhD diss., University of California.

Nauss, A.L., and Smith, P. L. 1988. *Lithiotis* (bivalvia) bioherms in the lower Jurassic of east-central Oregon, U.S.A. Paleogeography, Paleoclimatology, Paleoecology, v. 65, pp. 253–268.

Naylor, B.G. 1979. A new species of *Taricha* (Caudata: Salamandridae), from the Oligocene John Day Formation of Oregon. Canadian Journal of Earth Sciences, v. 16, pp. 970–973.

Nesbitt, E.A. 2003. Changes in shallow-marine faunas from the northeastern Pacific margin across the Eocene/Oligocene boundary. *In*: Prothero, D., Ivany, L., and Nesbitt, E., eds., From greenhouse to icehouse; the marine Eocene-Oligocene transition. New York: Columbia University Press, pp. 57–70.

Newberry, J.S. 1882. Brief descriptions of plants, chiefly Tertiary from western North America. U.S. National Museum, Proceedings 5, pp. 502–514.

Newcomb, R.C. 1958. Yonna Formation of the Klamath River Basin, Oregon. Northwest Science, v. 32, no. 2, pp. 41–48.

Newton, C.R. 1983. Paleozoogeographic affinities of Norian bivalves from the Wrangellian, Peninsular, and Alexander terranes, western North America. *In*: Stevens, C.H., ed., Pre-Jurassic rocks in western North American suspect terranes. Society Economic Paleontologists and Mineralogists, Pacific Sect., pp. 37–48.

———. 1986. Late Triassic bivalves of the Martin Bridge Limestone, Hells Canyon, Oregon: taphonomy, paleoecology, paleozoogeography. *In*: Vallier, T.L., and Brooks, H.C., eds., Geology of the Blue Mountains region of Oregon, Idaho, and Washington. U.S. Geological Survey, Professional Paper 1435, pp. 7–17.

———. 1987. Biogeographic complexity in Triassic bivalves of the Wallowa terrane, northwestern United States: Oceanic islands, not continents, provide the best analogues. Geology, v. 15, pp. 1126–1129.

Newton, C.R., et al. 1987. Systematics and paleoecology of Norian (late Triassic) bivalves from a tropical island arc: Wallowa terrane, Oregon. Journal Paleontology, v. 61, no. 4, Suppl. 4, 83p.

Nicholson, N. A. 1897. The ancient life-history of the earth. New York: Appleton, 406 pp.

Niem, A.R., and Niem, W.A. 1985. Oil and gas investigation of the Astoria basin, Clatsop and northernmost Tillamook counties, northwest Oregon. Oregon Department of Geology and Mineral Industries, Oil and Gas Investigation 14.

Niem, A.R., et al. 1994. Sedimentary, volcanic, and textonic framework of forearc basins and the Mist gas field, northwest Oregon. *In*: Swanson, D.A., and Haugerud, R.A., eds., Geologic field trips in the Pacific Northwest. Geological Society of America, Annual Meeting 1994, pp. IF1–IF42.

Niklas, K.J., and Giannasi, D.E. 1978. Angiosperm paleobiochemistry of the Succor Creek flora (Miocene) Oregon, USA. American Journal Botany, v. 65, no. 9, pp. 943– 952.

Nilsen, T.H 1984. Stratigraphy, sedimentology, and tectonic framework of the upper Cretaceous Hornbrook Formation, Oregon and California. *In*: Nilsen, T., ed. Geology of the upper Cretaceous Hornbrook Formation, Oregon and California. Pacific Sect., Society Economic Paleont. and Mineralogists, v. 42, pp. 51–88.

Nyborg, B.O. 2002. Fossil decapod crustaceans from the Astoria Formation, Washington and Oregon, U.S.A.: paleobiogeographic implications. Geological Society America, Abstracts with Programs, v. 34, no. 5, p. 41.

Nyborg, T.G. 2002. Fossil decapod crustaceans from the Astoria Formation, Washington and Oregon, U.S.A.: depositional environment implications. Geological Society America, Abstracts with Programs, v. 34, no. 5, p. 40.

O'Connor, J., et al. 2003. The mustelid *Sthenictis* in Mongolia: immigrant from North America. Geological Society America, Abstracts with Programs, v. 35, no. 6, p. 498.

Oliver, E. 1934. A Miocene flora from the Blue Mountains, Oregon. Carnegie Institute, Washington, D.C., Publ. 455, pt. 1, pp. 1–27.

Orr, E.L., and Orr, W.N. 1984. Bibliography of Oregon paleontology, 1792–1983. Oregon Department Geology and Mineral Industries, Special Paper 17, 82p.

———. 1999a. Geology of Oregon. Dubuque, Iowa: Kendall-Hunt, 254p.

———. 1999b. Oregon fossils. Dubuque, Iowa: Kendall-Hunt, 381p.

Orr, E.L., Orr, W.N., and Baldwin, E.M. 1992. Geology of Oregon. 4th ed., Dubuque, Iowa; Kendall-Hunt, 254p.

Orr, W.N. 1986. A Norian (late Triassic) ichthyosaur from the Martin Bridge Limestone, Wallowa Mountains, Oregon. *In*: Vallier, T.L., and Brooks, H.C., eds., Geology of the Blue Mountains Region of Oregon, Idaho, and Washington. U.S. Geological Survey, Professional Paper 1435, pp. 41–47.

Orr, W.N., and Faulhaber, J. 1975. A middle Tertiary cetacean from Oregon. Northwest Science, v. 49, pp. 174–181.

Orr, W.N., and Katsura, K.T. 1985. Oregon's oldest vertebrates (Ichthyosauria [Reptilia]). Oregon Geology, v. 47, no. 7, pp. 75–77.

Orr, W.N., and Kooser, M.A. 1971. Oregon Eocene decapods Crustacea. Ore Bin, v. 33, pp. 119–129.

Orr, W.N., and Miller, P.R. 1982. Mid Tertiary stratigraphy of the Oregon western Cascades. Geological Society America, Abstracts with Programs, v. 14, no. 4, p. 222.

Orr, W.N., and Miller, P.R. 1983. Fossil cetacea (whales) in the Oregon Western Cascades. Oregon Geology, v. 45, no. 9, pp. 95– 98.

———. 1984. The trace fossil *Cylindrichnus* in the Oregon Oligocene. Oregon Geology, v. 46, no. 5, pp. 51– 52.

Orr, W.N., and Orr, E.L. 1981. Handbook of Oregon plant and animal fossils. Eugene, Oregon, 285p.

———. 2002. Geology of the Pacific Northwest. 2nd ed. Long Grove, Ill.: Waveland Press, 337p.

Orr, W.N., Ehlen, J., and Zaitzeff, J.B. 1971. A late Tertiary diatom flora from Oregon. California Academy Sciences, Proceedings, 4th Ser., v. 37, no. 16, pp. 489–500.

Osborn, H.F. 1910. Age of mammals. New York: Macmillan, 635p.

Packard, E.L. 1921. The Trigoniae from the Pacific Coast of North America. Oregon University, Publications, v. 1, no. 9, pp. 1–40.

———. 1923. An aberrant oyster from the Oregon Eocene. Oregon University, Publications, v. 2, no. 4, pp. 1–6.

———. 1935. Additional cetacean material from Astoria Formation. Mineralogist, v. 3, no. 6, pp. 9–10

———. 1940. A new turtle from the marine Miocene of Oregon. Oregon State University, Studies in Geology, no. 2, 31p.

———. 1942. Additional information on the carapace of *Psephophorus? oregonensis* Packard. Geological Society Oregon Country, Newsletter, v. 10, pp. 99–100

———. 1947a. Fossil baleen from the Pliocene of Cape Blanco, Oregon. Oregon State University, Studies in Geology, v. 5, pp. 1–12.

———. 1947b. A fossil sea lion from Cape Blanco, Oregon. Oregon State University, Studies in Geology, v. 6, pp. 13–22.

———. 1947c. A pinniped humerus from the Astoria Miocene of Oregon. Oregon State University, Studies in Geology, v. 7, pp. 23–32.

———. 1952. Fossil edentates of Oregon. Oregon State University, Studies in Geology, v. 8, 15p.

Packard, E.L., and Allison, I.S. 1980. Fossil bear tracks in Lake County, Oregon. Oregon Geology, v. 42, no. 4, pp. 71–72.

Packard, E.L., and Jones, D.L. 1962. A new species of *Anisoceras merriami* from Oregon. Journal Paleontology, v. 36, no. 5, pp. 1047–1050.

———. 1965. Cretaceous pelecypods of the genus *Pinna* from the Pacific Coast region of North America. Journal of Paleontology, v. 39, no. 5, pp. 901–905.

Packard, E.L., and Kellogg, A.R. 1934. A new cetothere from the Miocene Astoria Formation of Newport, Oregon. Carnegie Institute, Washington, D.C., Publ. 477, pp. 1–62.

Pearl, C.A. 1999. Holocene environmental history of the Willamette Valley, Oregon . . . University of Oregon, Master's thesis, 150p.

Pearson, H.S. 1927. On the skulls of early Tertiary Suidae . . . . Royal Society of London, Philosophical Transactions, B215, pp. 389–460.

Peck, D.L., Imlay, R.W., and Popenoe, W.P. 1956. Upper Cretaceous rocks of parts of southwestern Oregon and northern California. American Association of Petroleum Geologists, Bull., v. 40, no. 8, pp. 1968–1984.

Peck, D.L., et al. 1964. Geology of the central and northern parts of the Western Cascade Range in Oregon. U.S. Geological Survey, Professional Paper 449, 56p.

Pessagno, E.A., and Blome, C.D. 1980. Upper Triassic and Jurassic Pantanellinae from California, Oregon, and British Columbia. Micropaleontology, v. 26, no. 3, pp. 225–273.

————. 1982. Bizarre Nassellarina (Radiolaria) from the middle and upper Jurassic of North America. Micropaleontology, v. 28, no. 3, pp. 289–318.

————. 1986. Faunal affinities and tectonogenesis of Mesozoic rocks in the Blue Mountains province of eastern Oregon and western Idaho. *In*: Vallier, T.L., and Brooks, H.C., eds. Geology of the Blue Mountains region of Oregon, Idaho, and Washington. U.S. Geological Survey, Professional Paper 1435, pp. 65–78.

————.1990. Implications of new Jurassic stratigraphic, geochronometrics, and paleolatitudinal data from the western Klamath terrane (Smith River and Rogue Valley subterranes). Geology, v. 18, no. 7, pp. 665–668.

Pessagno, E.A., and Whalen, P. A. 1982. Lower and middle Jurassic Radiolaria (multicyrtid Nassellariina) from California, east-central Oregon, and the Queen Charlotte Islands, B.C. Micropaleontology, v. 28, no. 2, pp. 111–169.

Pessagno, E.A., Whalen, P.A., and Yeh, K. 1986. Jurassic Nassellariiana (Radiolaria) from North American geologic terranes. Bulletins of American Paleontology, v. 91, no. 326, 75p.

Pessagno, E.A., et al. 1993. Jurassic radiolaria from the Josephine ophiolite and overlying strata, Smith River subterrane (Klamath Mountains), northwestern California and southwestern Oregon. Micropaleontology, v. 39, no. 2, pp. 93–166.

Peterson, J.V. 1964. Plant fossils of the Clarno Formation. Earth Science, v. 17, no. 1, pp. 11–15.

Peterson, O.A. 1905. Description of new rodents and discussion of the origin of *Doemonelix*. Carnegie Institute, Washington, D.C., Museum, Memoir 2, pp. 139–191.

————. 1909. A revision of the Entelodontidae. Carnegie Institute, Pittsburgh, Museum Memoir 4, pp. 42–158.

————. 1910. Description of new carnivores from the Miocene of western Nebraska. Carnegie Institute, Pittsburgh, Museum Memoir 4, pp. 205–278.

————. 1920. The American diceratheres. Carnegie Institute, Pittsburgh, Museum Memoir 7, no. 7, pp. 339–456.

Pigg, J.S. 1961. Lower Tertiary sedimentary rocks in the Pilot Rock and Heppner areas, Oregon. University of Oregon, Master's thesis, 67p.

Popenoe, W.P., Imlay, R.W., and Murphy, M.A. 1960. Correlation of the Cretaceous formations of the Pacific Coast (United States and northwestern Mexico). Geological Society America, Bull., v. 71, pp. 1491–1540.

Pouchet, F.A. 1882. The universe; or, the wonders of creation. Portland, Maine: Hallett, 761p.

Pratt, J.A. 1988. Paleoenvironment of the Eocene/Oligocene Hancock Mammal Quarry, upper Clarno Formation, Oregon. Master's thesis, University of Oregon, 104p.

Prothero, D.R. 1998. The chronological, climatic, and paleogeographic background to North American mammalian evolution. *In*: Janis, C.M., Scott, K.M., and Jacobs, L.L., eds. 1998. Evolution of Tertiary mammals of North America. New York: Cambridge University Press, pp. 9–36.

———. 2000. Magnetic stratigraphy and tectonic rotation of the Eocene-Oligocene Keasey Formation, northwest Oregon. Journal of Geophysical Research, v. 105, no. B7, pp. 16,473–16,480.

———. 2003. Pacific Coast Eocene-Oligocene marine chronostratigraphy: a review and an update. *In*: Prothero, D., Ivany, L., and Nesbitt, E., eds., From greenhouse to icehouse; the marine Eocene-Oligocene transition. New York: Columbia University Press, pp. 1–13.

———. 2005. The evolution of North American rhinoceroses. New York: Cambridge University Press, 218p.

Prothero, D.R., ed. 2001. Magnetic stratigraphy of the Pacific Coast Cenozoic. SEPM (Society for Sedimentary Geology), Pacific Section, Proceedings 1997; and AAPG-SEPM, Cordilleran Section, 394p.

Prothero, D.R., and Berggren, W.A., eds. 1992. Eocene-Oligocene climatic and biotic evolution. Princeton: Princeton University Press, 568p.

Prothero, D.R., and Foss, S. 2007. The evolution of artiodactyls. Baltimore: Johns Hopkins University Press, 367p.

Prothero, D.R., and Schoch, R.M., eds. 1989. The evolution of Perissodactyls. New York: Oxford University Press, 537p.

———. 2002. Horns, tusks, and flippers: the evolution of hoofed mammals. Baltimore: Johns Hopkins University Press, 311p.

Prothero, D.R., Ivany, L.C., and Nesbitt, E.A., eds. 2003. From greenhouse to icehouse; the marine Eocene-Oligocene transition. New York: Columbia University Press, 541p.

Prothero, D.R., et al. 2001a. Magnetic stratigraphy and tectonic rotation of the Oligocene Alsea, Yaquina, and Nye formations, Lincoln County, Oregon. *In*: Prothero, D.R., ed., Magnetic stratigraphy of the Pacific Coast Cenozoic. SEPM (Society for Sedimentary Geology), Pacific Section, no. 91, pp. 184–194.

———. 2001b. Magnetic stratigraphy and tectonic rotation of the upper middle Eocene Cowlitz and Hamlet formations, western Oregon and Washington. *In*: Prothero, D.R., ed., Magnetic stratigraphy of the Pacific Coast Cenozoic. SEPM (Society for Sedimentary Geology), Pacific Section, no. 91, pp. 75–95.

Quaintance, C.W. 1969. *Mylodon*, furthest north in Pacific Northwest. American Midland Naturalist, v. 81, no. 2, pp. 593–594.

Ramp, L. 1969. Dothan (?) fossils discovered. Ore Bin, v. 31, no. 12, pp. 245–246.

Rathbun, M.J. 1926. The fossil stalk-eyed Crustacea of the Pacific slope of North America. U.S. National Museum, Bulletin 138, 155p.

Rau, W. 1981. Pacific northwest Tertiary benthonic foraminiferal biostratigraphic framework; an overview. Geological Society America, Special Paper 184, pp. 67–84.

Ray, C.E. 1976. Fossil marine mammals of Oregon. Systematic Zoology, v. 25, no. 4, pp. 420–436.

Ray, C.E., Domning, D.P. , and McKenna, M.C. 1994. A new specimen of *Behemotops proteus* (Order Desmostylia) from the marine Oligocene of Washington. *In*: Berta, A., and Demere, T.A., eds. 1994. Contributions in marine mammal paleontology honoring Frank C. Whitmore, Jr. San Diego Society of Natural History, Proceedings, v. 29, pp. 205–222.

Read, C.B., and Brown, R.W. 1937. American Cretaceous ferns of the genus *Tempskya*. U.S. Geological Survey, Professional Paper 186–F, pp. 127–128.

Read, C.B., and Merriam, C.W. 1940. A Pennsylvanian flora from central Oregon. American Journal Science, v. 138, no. 2, pp. 107–111.

Rensberger, J.M. 1971. Entophychine pocket gophers (Mammalia, Geomyoidea) of the early Miocene John Day Formation, Oregon. University of California, Publications in Geological Sciences., v. 90, 209p.

———. 1973. Pleurolicine rodents (Geomyoidea) of the John Day Formation, Oregon, and their relationships to taxa from the early and middle Miocene, South Dakota. University of California, Publications in Geological Sciences, v. 102, 95p.

———. 1976. John Day Fossil Beds National Monument. Report to the National Park Service, 61p.

———. 1979. Hemphillian rodents from northern Oregon and their relationships to other rodent faunas in North America. Ph.D. diss., University of Washington.

———. 1983. Successions of Meniscomyine and Allomyine rodents (Aplodontidae) in the Oligo-Miocene John Day Formation, Oregon. University of California Publications in Geological Sciences, v. 124, 157p.

Repenning, C.A. 1967. Subfamilies and genera of the Soricidae. U.S. Geological Survey, Professional Paper 565, 74p.

———. 1968. Mandibular musculature and the origin of the subfamily Arvicolinae (Rodentia). Acta Zoologica, Cracoviensia, v. 13, no. 3, pp. 29–72.

Repenning, C.A., and Mitchell, E.D. 1966. The Miocene pinniped *Allodesmus*. Journal of Paleontology, v. 61, no. 3, pp. 1–46.

Repenning, C.A., and Tedford, R. 1977. Otaroid seals of the Neogene. U.S. Geological Survey, Professional Paper 992, pp. 1–93.

Repenning, C.A., Weasma, T.R., and Scott, G.R. 1995. The early Pleistocene (latest Blancan–earliest Irvingtonian) Froman Ferry fauna . . . . U.S. Geological Survey, Bulletin 2105, 86p.

Retallack, G.J. 1981. Preliminary observations on fossil soils in the Clarno Formation (Eocene to early Oligocene) near Clarno, Oregon. Oregon Geology, v. 43, no. 11, pp. 147–150.

———. 1987. Maladaptation of Oligocene mammalian faunas of North America. Geological Society. America, Abstracts with Programs, v. 19, no. 6, p. 443.

———. 1991a. A field guide to mid-Tertiary paleosols and paleoclimatic changes in the high desert of central Oregon—Part 1. Oregon Geology, v. 53, no. 3, pp. 51–59.

———. 1991b. A field guide to mid-Tertiary paleosols and paleoclimatic changes in the high desert of central Oregon—Part 2. Oregon Geology, v. 53, no. 4, pp. 75–80.

———. 2002. Late Miocene (Clarendonian) fossil plants and animals from Unity, Baker County, Oregon. Geological Society American, Abstracts with Programs, v. 34, no. 5, p. 10.

———. 2004. Late Oligocene bunch grassland and early Miocene sod grassland paleosols from central Oregon, USA. Paleogeography, Paleoclimatology, Paleoecology, v. 207, pp. 203–237.

Retallack, G.J., Bestland, E.A., and Fremd, T.J. 1996. Reconstructions of Eocene and Oligocene plants and animals of central Oregon. Oregon Geology, v. 58, no. 3, pp. 51–69.

———. 2000. Eocene and Oligocene paleosols of central Oregon. Geological Society of America, Special Paper 344, 192p.

Retallack, G.J., et al. 2004. Eocene-Oligocene extinction and paleoclimatic change near Eugene, Oregon. Geological Society of America, Bull., v. 116, no. 7/8, pp. 817–839.

Richardson, H.E. 1950. The geology of the Sweet Home petrified forest. University of Oregon, Master's thesis, 44p.

Ries, J.E. 1989. Undescribed Eocene floras from the Coast Range of northwestern Oregon: paleogeographic and climatic implications. Geological Society America, Abstracts with Programs, p. 134.

Roberts, M.C., and Whitehead, D.R. 1984. The palynology of a nonmarine Neogene deposit in the Willamette Valley, Oregon. Review of Palaeobotany and Palynology, v. 41, pp. 1–12.

Robinson, P. T., Brem, G.F., and McKee, E.H. 1984. John Day Formation of Oregon: a distal record of early Cascade volcanism. Geology, v. 12, pp. 229–232.

Romer, A.S. 1966. Vertebrate paleontology. Chicago: University of Chicago Press, 468p.

Rooth, G.H. 1974. Biostratigraphy and paleoecology of the Coaledo and Bastendorff formations, southwestern Oregon. Ph.D. diss., Oregon State University, 270p.

———. 1987. Miocene *Gyrolithes* (lebensspurn) from the Astoria Formation, Lincoln County, Oregon. Oregon Geology, v. 49, no. 4, pp. 47–48.

Rose, K.D., and Rensberger, J.M. 1983. Upper dentition of *Ekgmowechashala* (Omomyid primate) from the John Day Formation, Oligo-Miocene of Oregon. Folia Primatologica, v. 41, p. 102–111.

Ross, C.P. 1938. The geology of part of the Wallowa Mountains. Oregon Department of Geology and Mineral Industries, Bull. 3, 74p.

Roth, B. 1979. Late Cenozoic marine invertebrates from northwest California and southwest Oregon. Ph.D. diss., University of California, Berkeley, 2 vols.

Rudy, P. , and Rudy, L. 1983. Oregon estuarine invertebrates. Contract No. 79-111, U.S. Fish and Wildlife Service, Washington, D.C., 224p.

Sada, K., and Danner, W.R. 1973. Late lower Carboniferous *Eostaffella* and *Hexaphyllia* from central Oregon, U.S.A. Paleontological Society Japan, Transactions and Proceedings, no. 91, pp. 151–160.

Sanborn, E.I. 1937. The Comstock flora of west central Oregon. Carnegie Institute, Wash., Publ. 465, pp. 1–28.

———. 1947. The Scio flora of western Oregon. Oregon State Univ., Studies in Geology, v. 4, pp. 1–47.

Saul, L.R. 1988. New late Cretaceous and early Tertiary Perissityidae (Gastropoda) from the Pacific slope of North America. Los Angeles County Museum, Natural History Museum, Contributions in Science, no. 400, pp. 1–25.

Savage, N. M., and Amundson, C.T. 1979. Middle Devonian (Givetian) conodonts from central Oregon. Journal Paleontology, v. 53, pp. 1395–1400.

Scharf, D.W. 1935. A Miocene mammalian fauna from Succor Creek, southeastern Oregon. Carnegie Institute, Washington, D.C., Publication 453, pp. 97–118.

Schenck, H.G. 1923. A preliminary report of the geology of the Eugene Quadrangle, Lane and Linn counties. Master's thesis, University of Oregon, 104p.

———.1927. Marine Oligocene of Oregon. University of California, Publications in Geological Sciences, v. 16, no. 12, 449–460.

———. 1928. Stratigraphic relations of western Oregon Oligocene formations. University of California, Publications in Geological Sciences, v. 18, 50p.

———. 1931. Cephalopods of the genus *Aturia* from western North America. University of California, Publications in Geological Sciences, v. 19, pp. 435–490.

———. 1936. Nuculid bivalves of the genus *Acila*. Geological Society America, Special Paper 4, 149p.

Schenck, H.G., and Kleinpell, R.M. 1936. Refugian stage of Pacific Coast Tertiary. American Association of Petroleum Geologists, Bull., v. 20, no. 2, pp. 215–225.

Schoch, R.M. 1989. A review of the tapiroids. *In*: Prothero, D.R., and Schoch, R.M., eds., The evolution of Perissodactyls, New York: Oxford University Press, pp. 298–320.

Schultz, C., and Falkenbach, C.H. 1968. The phylogeny of the oreodonts. American Museum Natural History, Bull., v. 139, 498p.

Scott, R.A. 1954. Fossil fruits and seeds from the Eocene Clarno Formation of Oregon. Palaeontographica B., v. 96, pp. 66– 97.

Scott, W.B. 1913. History of land mammals in the western hemisphere. New York: Macmillan.

Sea, D.S. and Whitlock, C. 1995. Postglacial vegetation and climate of the Cascade Range, central Oregon. Quaternary Research, v. 43, pp. 370–381.

Sellick, J.T.C. 1994. Phasmida (stick insect) eggs from the Eocene of Oregon. Paleontology, v. 37, no. 4, pp. 913–921.

Senowbari-Daryan, B., and Stanley, G.D. 1988. Triassic sponges (Sphinctozoa) from Hells Canyon, Oregon. Journal Paleontology, v. 62, no. 3, pp. 419–423.

Seward, A.C. 1898. Fossil plants. 2 vols. Cambridge: University Press.

Shimer, H.W. 1914. An introduction to the study of fossils. New York: Macmillan, 450p.

Shotwell, J.A. 1951. A fossil sea-lion from Fossil Point, Oregon (abstract). Geological Society America, Bull., v. 2, p. 97.

———. 1956. Hemphillian mammalian assemblage from northeastern Oregon. Geological Society of America, Bull., v. 67, pp. 717–738.

———. 1958. Inter-community relationships in Hemphillian (Mid-Pliocene) mammals. Ecology, v. 39, no. 2, pp. 271–282.

———. 1961. Late Tertiary biogeography of horses in the northern Great Basin. Journal Paleontology, v. 35, no. 1, pp. 203–217.

———. 1963. The Juntura basin: studies in earth history and paleoecology. American Philosophical Society, Transactions, v. 53, 77p.

———. 1964. Community succession in mammals of the late Tertiary. *In*: Imbrie, J., and Newell, N., eds., Approaches to paleoecology. New York: Wiley, pp. 135–150.

———. 1967. Late Tertiary geomyoid rodents of Oregon. University of Oregon, Museum Natl. History, Bull. 9, 51p.

———. 1968. Miocene mammals of southeastern Oregon. University of Oregon, Museum Natl. History, Bull. 14, 67p.

———. 1970. Pliocene mammals of southeast Oregon and adjacent Idaho. University of Oregon, Museum Natural History, Bull. 17, 103p.

Shotwell, J.A., and Russell, D.E. 1963. Mammalian fauna of the upper Juntura Formation, the Black Butte local fauna. *In*: Shotwell, J.A., The Juntura Basin, American Philosophical Society, Transactions, v. 53, 77p.

Shroba, C.S. 1992. Paleoecology and taphonomy of middle Tertiary and Recent sediments from Oregon and Washington, and biogeographic affinities of the Wallowa terrane, eastern Oregon. Ph.D. diss., University of Oregon, 181p.

Shroba, C.S., and Orr, W.N. 1995. Tertiary facies and paleoenvironments along the Coast Range and Western Cascade margin, western Oregon. *In*: Fritsche, A.E., ed., Cenozoic paleogeography of the western United States—II: Pacific Section, SEPM, book 75, pp. 257–273.

Shufeldt, R.W. 1891. On a collection of fossil birds from the *Equus* beds of Oregon. American Naturalist, v. 25, pp. 259–262.

———. 1912. Prehistoric birds of Oregon. Overland Monthly, v. 60, pp. 536–642.

———. 1913. Review of the fossil fauna of the desert region of Oregon, with a description of additional material collected there. American Museum Natural History, Bull.25, pp. 259–262.

———. 1915. Fossil birds in the Marsh Collection of Yale University. Connecticut Academy Arts Science, Transactions, v. 19, pp. 1–110.

Sinclair, W.J. 1901. Discovery of a new fossil tapir in Oregon. Journal Geology, v. 9, pp. 702–707.

———. 1905. New or imperfectly known rodents and ungulates from the John Day series. University of California, Publications in Geological Sciences, v. 4, no. 6, pp. 125–143.

———. 1906. Some edentate-like remains from the Mascall beds of Oregon. University of California, Publications in Geological Sciences, v. 5, no. 2, pp. 65–66.

Skinner, J.W., and Wilde, G.L. 1966. Permian fusulinids from Pacific Northwest and Alaska. Part 2: Permian fusulinids from Suplee area . . . . University of Kansas, Paleontological Contributions, v. 4, pp. 11–15.

Sliter, W.V. , Jones, D.L., and Throckmorton, C.K. 1984. Age and correlation of the Cretaceous Hornbrook Formation, California and Oregon. *In*: Nilsen, T.H., ed., Geology of the upper Cretaceous Hornbrook Formation, Oregon and California. Society Economic Paleontologists and Mineralogists, Pacific Section, v. 42, pp. 89–98.

Smith, G.A., et al. 1998. Late Eocene-early Oligocene tectonism, volcanism, and floristic change near Gray Butte, Central Oregon. Geological Society America, Bull., v. 110, no. 6, pp. 759–778.

Smith, G.R. 1975. Fishes of the Pliocene Glenns Ferry Formation, southwest Idaho. Museum of Paleontology, University of Michigan, Claude W. Hibbard Memorial Volume 5, pp. 1–68.

———. 1981. Late Cenozoic freshwater fishes of North America. Annual Review Ecology and Systematics, v. 12, pp. 163–193.

Smith, G.R., et al. 1982. Fish biostratigraphy of late Miocene to Pleistocene sediments of the western Snake River Plain, Idaho. *In*: Bonnichsen, B., and Breckenridge, R.M., eds., Cenozoic geology of Idaho. Idaho Bureau of Mines and Geology, Bull.26, pp. 519–541.

Smith, G.S. 1988. Paleoenvironmental reconstruction of Eocene fossil soils from the Clarno Formation of eastern Oregon. Master's thesis, University of Oregon, 167p.

Smith, H.V. 1932. The fossil flora of Rockville, Oregon. Master's thesis, University of Oregon, 41p.

———. 1938. Additions to the fossil flora of Sucker Creek, Oregon. Michigan Academy of Sciences, Arts and Letters, Papers, v. 24, pp. 107–121.

———. 1940. Notes of the systematic and ecological implications of the Miocene flora of Sucker Creek, Oregon and Idaho. American Midland Naturalist, v. 24, no. 2, pp. 437–443.

Smith, J.P. 1912. Occurrence of coral reefs in the Triassic of North America. American Journal Science, 4th ser., v. 33, pp. 92–96.

Smith, P. L. 1980. Correlation of the members of the Jurassic Snowshoe Formation in the Izee Basin of east-central Oregon. Canadian Journal Earth Science, v. 17, no. 12, pp. 1603–1608.

Smith, P. L., et al. 1988. An ammonite zonation for the lower Jurassic of Canada and the United States: the Pliensbachian. Canadian Journal Earth Science, v. 25, pp. 1503–1523.

Smith, W.D., and Allen, J.E. 1941. Geology and physiology of the northern Wallowa Mountains, Oregon. Oregon Department of Geology and Mineral Industries, Bull. 12, 64p.

Snavely, P. D., and Baldwin, E.M. 1948. Siletz River volcanic series, northwestern Oregon. American Assoc. Petroleum Geologists., Bull., v. 32, pp. 806–812.

Snavely, P.D., and Vokes, H.E. 1949. Geology of the coastal area between Cape Kiwanda and Cape Foulweather, Oregon. U.S. Geological Survey, Oil and Gas Investigations, Preliminary map 97.

Snavely, P. D., and Wagner, H.C. 1963. Miocene geologic history of western Oregon and Washington. Washington (State) Div. of Mines and Geology, Report of Investigations, no. 22, 25p.

Snavely, P. D., MacLeod, N. S., and Rau, W.W. 1969. Geology of the Newport area, Oregon. Ore Bin, v. 31, no. 2, pp. 25–47.

Snavely, P. D., Rau, W.W., and Wagner, H.C. 1964. Miocene stratigraphy of the Yaquina Bay area, Newport, Oregon. Ore Bin, v. 26, no. 8, pp. 133–151.

Snavely, P. D., et al. 1975. Alsea Formation; an Oligocene marine sedimentary sequence in the Oregon Coast Range. U.S. Geological Survey, Bull.1395–F, 21p.

Snoke, A.W., and Barnes, C.G, eds. 2006. Geological studies in the Klamath Mountains province, California and Oregon. Geological Society America, Special Paper 410, 505p.

Sorauf, J.E. 1972. Middle Devonian coral faunas (rugose) from Washington and Oregon. Journal Paleontology, v. 46, no. 3, pp. 426–439.

Spencer, P. K., Carson, R., and Orr, W.N. 1985. Pleistocene vertebrates and sediments near Wallowa Lake, Oregon. Geological Society of America, Abstracts with Programs, v. 17, no. 6, pp. 410.

Squires, R.L. 1989. Pteropods (Molluscs: Gastropoda) from Tertiary formations of Washington and Oregon. Journal Paleontology, v. 63, no. 4, pp. 443–448.

———. 2003. Turnovers in marine gastropod faunas during the Eocene-Oligocene transition, West Coast of the United States. *In*: Prothero, D., Ivany, L., and Nesbitt, E., eds., From greenhouse to icehouse; the marine Eocene-Oligocene transition. New York: Columbia University Press, pp. 14–35.

Squires, R.L., and Saul, L.R. 2002. New early late Cretaceous (Cenomanian) mollusks from Oregon. Journal Paleontology, v. 76, no. 1, pp. 43–51.

Squires, R.L., and Saul, L.R. 2004. The pseudomelaniid gastropod *Paosia* from the marine Cretaceous of the Pacific Slope. Journal Paleontology, v. 78, no. 3, pp. 484–500.

Stanley, G.D. 1986. Late Triassic coelenterate faunas of western Idaho and northeastern Oregon: implications for biostratigraphy and paleogeography. *In*: Vallier, T.L., and Brooks, H.C., eds., Geology of the Blue Mountains region of Oregon, Idaho, and Washington. U.S. Geological Survey, Professional Paper 1435, pp. 23–35.

———. 1987. Travels of an ancient reef. Natural History, v. 11, pp. 35–42.

———. 1988. The history of early Mesozoic reef communities: a three-step process. Palaios, v. 3, pp. 170–183.

Stanley, G.D., and Beauvais, L. 1990. Middle Jurassic corals from the Wallowa terrane, west-central Idaho. Journal Paleontology, v. 64, no. 3, pp. 352–362.

Stanley, G.D., and Senowbari-Daryan, B. 1986. Upper Triassic, Dachstein-type, reef limestone from the Wallowa Mountains, Oregon: first reported occurrence in the United States. Palaios, v. 1, pp. 172–177.

Stanley, G.D., and Whalen, M.T. 1989. Triassic corals and spongimorphs from Hells Canyon, Wallowa terrane, Oregon. Journal Paleontology, v. 63, no. 6, pp. 800–819.

Stearns, R.E.C. 1902. Fossil shells of the John Day region. Science, n. s., v. 15, pp. 153–154.

———. 1906. Fossil mollusca from the John Day and Mascall beds of Oregon. University of California, Publications in Geological Sciences., v. 5, no. 3, pp. 67–70.

Steele, J.D., and Jenks, J.W. 1887. Popular zoology. New York: Chautauqua Press, 307p.

Steere, M. L. 1954. Fossil localities of Lincoln County beaches, Oregon. Ore Bin, v. 16, no. 4, pp. 21–26.

———. 1955. Fossil localities in the Coos Bay area, Oregon. Ore Bin, v. 17, no. 6, pp. 39–43.

———. 1957. Fossil localities of the Sunset Highway area, Oregon. Ore Bin, v. 19, no. 5, pp. 37–44.

——— 1958. Fossil localities of the Eugene area, Oregon. Ore Bin, v. 20, no. 6, pp. 51–59.

———. 1959. Fossil localities of the Salem-Dallas area, Oregon. Ore Bin, v. 21, no. 6, pp. 51–58.

Steere, M.L., ed. 1977. Fossils in Oregon; a collection of reprints from the Ore Bin. Oregon Department of Geology and Mineral Industries, Bull. 92, 227p.

Stevens, C.H., and Rycerski, B.A. 1983. Permian colonial rugose corals in the western Americas—aids in positioning of suspect terranes. In: Stevens, C.H., ed., Pre-Jurassic rocks in western North American suspect terranes. Society Economic Paleontologists and Mineralogists, Pacific Section, pp. 23–33.

Stevens, C.H., Miller, M.M., and Nestell, M. 1987. A new Permian Waagenophyllid coral from the Klamath Mountains, California. Journal Paleontology, v. 61, no. 4, pp. 690–699.

Stewart, R.E. 1956. Stratigraphic implications of some Cenozoic foraminifera from western Oregon. Ore Bin, v. 18, nos.1, 7; v. 19, no. 2.

Stirton, R.A. 1940. Phylogeny of North American Equidae. University of California, Department of Geological Sciences, Bull., 25, no. 4, pp. 165–198.

———. 1944. A rhinoceros tooth from the Clarno Eocene of Oregon. Journal Paleontology, v. 18, no. 3, pp. 65–67.

Stirton, R.A., and Rensberger, J.M. 1964. Occurrence of the insectivore genus Micropternodus in the John Day Formation of central Oregon. Southern California Academy of Sciences, Bull., v. 63, pt.2, pp. 57–80.

Stock, C. 1925. Cenozoic gravigrade edentates of western North America . . . . Carnegie Institute, Washington, D.C., Publication 331, 206p.

———. 1930. Carnivora new to the Mascall Miocene fauna of eastern Oregon. Carnegie Institute, Washington, D.C., Contributions to Paleontology, Publ.404, pp. 43–48.

———. 1946. Oregon's wonderland of the past, the John Day. Science Monthly, v. 63, pp. 57–65.

Stock, C., and Furlong, E.L. 1922. A marsupial from the John Day Oligocene of Logan Butte, eastern Oregon. University California, Department Geological Sciences, Bull., v. 13, no. 8, pp. 311–317.

Stokesbary, W.A. 1933. A faunal study of the Dallas area in an attempt to determine the faunal horizon . . . . Master's thesis, Oregon State University, 73p.

Storer, R.W. 1989. The Pleistocene western grebe Aechmophorus (Aves, Podicipedidae) from Fossil Lake, Oregon: a comparison with recent material. University of Michigan, Contributions from the Museum of Paleontology, v. 27, pp. 321–326.

Stricker, L., and Taylor, D.G. 1989. A new marine crocodile (Mesosuchia: Metriorhynchidae) from the Snowshoe Formation (Jurassic) Oregon. Journal Vertebrate Paleontology, v. 9, Suppl. to No. 3, p. 40A.

Stucky, R.K. 1992. Mammalian faunas in North America of Bridgerian to early Arikareean "Ages" (Eocene and Oligocene. *In*: Prothero, D.R., and Berggren, W.A., eds., Eocene-Oligocene climatic and biotic evolution. Princeton: Princeton University Press, pp. 464–493.

Sudworth, G.B. 1908. Forest trees of the Pacific slope. U.S. Dept. of Agriculture, 441p.

Taggart, R. 1990a. Palynologic evidence of Miocene plant succession induced by fluvial emplacement of volcaniclastic sediments, Oregon (abstract). Palynology, v. 14, p. 218.

———. 1990b. The vanished forests of Succor Creek. Natural Science, v. IV, no. 1, pp. 15–18.

Taggart, R., and Cross, A.T. 1974. History of vegetation and paleoecology of upper Miocene Sucker Creek beds of eastern Oregon. Birbal Sahni Institute Paleobotany, Special Publ., no. 3, pp. 125–132.

———.1990. Plant successions and interpretations in Miocene volcanic deposits, Pacific Northwest. Geological Society America, Special Paper 244, pp. 57–68.

Taylor, D.G. 1982. Jurassic shallow marine invertebrate depth zones, with exemplification from the Snowshoe Formation, Oregon. Oregon Geology, v. 44, no. 5, pp. 51–56.

———. 1988. Middle Jurassic (late Aalenian and early Bajocian) ammonite biochronology of the Snowshoe Formation, Oregon. Oregon Geology, v. 50, no. 11–12, pp. 123–138.

Taylor, D.G., and Guex, J. 2002. The Triassic/Jurassic system boundary in the John Day inlier, east-central Oregon. Oregon Geology, v. 64, pp. 3–28.

Taylor, D.W. 1960. Distribution of the freshwater clam *Pisidium ultramonanum*; a zoogeographic inquiry. American Journal Science, Bradley Volume, v. 258–A, pp. 325–334.

———. 1963. Mollusks of the Black Butte local fauna. American Philosophical Society, Transactions, v. 53, pt.1, pp. 35–40.

Tedford, R.H., Barnes, L.G., and Ray, C.E. 1994. The early Miocene littoral ursoid *Kolponomos*: systematics and mode of life. *In*: Berta, A., and Demere, T.A., eds., Contributions in marine mammal paleontology honoring Frank C. Whitmore, Jr. San Diego Society of Natural History, Proceedings, v. 29, pp. 11–32.

Terry, J.S. 1968. *Mediargo*, a new Tertiary genus in the family Cymatidae. Veliger, v. 11, no. 1, pp. 42–44.

Thompson, G.C., Yett, J.R., and Green, K.E. 1984. Subsurface stratigraphy of the Ochoco Basin, Oregon. Oregon Department Geology and Mineral Industries, Oil and Gas Investigations 8, 22p.

Thoms, R.E. 1965. Biostratigraphy of the Umpqua Formation, southwest Oregon. Ph.D. diss., University of California, 215p.

Thoms, R.E., and Smith, H.C. 1973. Fossil bighorn sheep from Lake County, Oregon. Ore Bin, v. 35, no. 8, pp. 125–134.

Thorpe, M.R. 1921a. John Day Eporeodons, with description of new genera and species. American Journal of Science., ser.5, v. 2, pp. 93–111.

———. 1921b. John Day Promerycochoeri with description of five new species and one new subgenus. American Journal Science, ser.5, v. 1, pp. 215–144.

———. 1921c. Two new forms of Argiochoerus. American Journal Science, ser.5, v. 2, pp. 111–119.

———. 1921d. Two new fossil carnivora. American Journal Science, ser.5, v. 1, pp. 477–483.

———. 1922a. *Araeocyon*, a probable old world migrant. American Journal Science, ser.5, v. 3, pp. 371–377.

———. 1922b. Oregon Tertiary Canidae with descriptions of new forms. American Journal Science, ser.5, v. 3, pp. 162–176.

———. 1925. A new species of extinct peccary from Oregon. American Journal Science, ser.5, v. 7, pp. 393–397.

Tidwell, W.D. 1975. Common fossil plants of western North America. Provo, Utah: Brigham Young University Press, 197p.

Tidwell, W.D., Parker, L.R., and Vaughn, K.F. 1986. *Pinuxylon woolardii* sp. nov., a new petrified taxon of Pinaceae from the Miocene basalts of eastern Oregon. American Journal Botany, v. 73, no. 11, pp. 1517–1524.

Toohey, L. 1959. The species of *Nimravus* (Carnivora, Felidae). American Museum of Natural History, Bull., v. 118, art.2, pp. 75–112.

Treasher, R.C., and Hodge, E.T. 1936. Bibliography of the geology and mineral resources of Oregon. Portland, State Planning Board, 224p.

Trimble, D.E., 1963. Geology of Portland, Oregon, and adjacent areas. U.S. Geological Survey, Bull. 1119, 119p.

Turner, F.E. 1938. Stratigraphy and mollusca of the Eocene of western Oregon. Geological Society America, Special Paper 10, 130p.

Uyeno, T., and Miller, R.R. 1963. Summary of late Cenozoic freshwater fish records for North America. University of Michigan, Museum Zoology, Occasional Papers, no. 631, 34p.

Vallier, T.L. 1967. The geology of part of the Snake River Canyon and adjacent areas in northeastern Oregon and western Idaho. Ph.D. diss., Oregon State University, 267p.

Vallier, T.L., and Brooks, H.C. 1970. Geology and copper deposits of the Homestead area, Oregon and Idaho. Ore Bin, v. 32, no. 3, pp. 37–57.

Vallier, T.L., and Brooks, H.C., eds. 1986. Geology of the Blue Mountains region of Oregon, Idaho, and Washington: geologic implications of Paleozoic and Mesozoic paleontology and biostratigraphy . . . . U.S. Geological Survey, Professional Paper 1435, 93p.

———. 1994. Geology of the Blue Mountains region of Oregon, Idaho, and Washington; stratigraphy, physiography, and mineral resources . . . . U.S. Geological Survey, Professional Paper 1439, 198p.

Vanderhoof, V. L. 1937. A study of the Miocene sirenian *Desmostylus*. University California, Publications in Geological Sciences, v. 42, pp. 169–262.

Vanderhoof, V. L., and Gregory, J.T. 1940. A review of the genus *Aleurodon*. University of California, Publications in Geological Sciences., v. 25, no. 3, pp. 143–164.

Van Frank, R. 1955. *Palaeotaricha oligocenica*, new genus and species; an Oligocene salamander from Oregon. Brevoria, no. 45, p. 1–12.

VanLandingham, S.L. 1990. Observations on the biostratigraphy of Pliocene and Pleistocene diatomites from the Terrebonne district, Deschutes County, Oregon. Micropaleontology, v. 36, no. 2, pp. 182–196.

Van Tassell, J., McConnell, V. , and Smith, G.R. 2001. The mid-Pliocene Imbler fish fossils, Grande Ronde Valley, Union County, Oregon, and the connection between Lake Idaho and the Columbia River. Oregon Geology, v. 63, no. 3, pp. 77–84, 89–96.

Van Tassell, J., et al. 2007. Early Pliocene (Blancan) Always Welcome Inn local fauna, Baker City, Oregon. Oregon Geology, v. 68, no. 1, pp. 3–23.

Vokes, H.E., Norbisrath, H., and Snavely, P. D. 1949. Geology of the Newport-Waldport area, Lincoln County, Oregon. U.S. Geol. Survey, Oil and Gas Investigation Map, OM-88.

Vokes, H.E., Snavely, P.D., and Myers, D.A. 1951. Geology of the southern and southwestern border areas, Willamette Valley, Oregon. U.S. Geological Survey, Oil and Gas Investigation Map, OM-110.

Wagner, N.S., et al. 1963. Marine Jurassic exposures in Juniper Mountain area of eastern Oregon. American Association of Petroleum Geologists, Bull. 47, no. 4, pp. 687, 701.

Walker, G.W., ed. 1990. Geology of the Blue Mountains region of Oregon, Idaho, and Washington: Cenozoic geology of the Blue Mountains region. U.S. Geological Survey, Professional Paper 1437, 135p.

Walker, G.W., and Repenning, C.A. 1965. Reconnaissance geologic map of the Adel Quadrangle, Lake, Harney, and Malheur counties, Oregon. U.S. Geological Survey, Miscellaneous Geologic Investigation Map I-457.

Walker, G.W., and Robinson, P. T. 1990. Cenozoic tectonism and volcanism of the Blue Mountains region. *In*: Walker, G.W., ed., Geology of the Blue Mountains region of Oregon, Idaho, and Washington. U.S. Geological Survey, Professional Paper 1437, pp. 119–134.

Walker, G.W., and Robinson, P. T. 1990. Paleocene(?), Eocene, and Oligocene (?) rocks of the Blue Mountains region. *In*: Walker, G.W., ed., Geology of the Blue Mountains region of Oregon, Idaho, and Washington: Cenozoic geology of the Blue Mountains region. U.S. Geological Survey, Professional Paper 1437, pp. 13–27.

Wallace, R.E. 1946. A Miocene mammalian fauna from Beatty (Beatys) Buttes, Oregon. Carnegie Institute, Washington, D.C., Publication 551, pp. 113–134.

Wang, X., and Tedford, R.H. 1992. The status of genus *Nothocyon* Matthew 1899 (Carnivora): an arctoid not a canid. Journal of Vertebrate Paleontology, v. 12, no. 2, pp. 223–229.

Ward, L. 1905. Status of the Mesozoic floras of the United States. U.S. Geological Survey Monograph XLVIII, pt. 1 and pt. 2.

Ward, P. D., and Westerman, G.E.G. 1977. First occurrence, systematics, and functional morphology of *Nipponites* (Cretaceous Lytoceratina) from the Americas. Journal of Paleontology, v. 51, no. 2, pp. 267–273.

Wardlaw, B.R., and Jones, D.L. 1980. Triassic conodonts from eugeoclinal rocks of western North America and their tectonic significance. Rivista Italiana di Paleontologia e Stratigrafia, v. 85, no. 3–4, pp. 895–908.

Wardlaw, B.R., Nestell, M., and Dutro, J.T. 1982. Biostratigraphy and structural setting of the Permian Coyote Butte Formation of central Oregon. Geology, v. 10, pp. 13–16.

Warren, W.C., and Norbisrath, H. 1946. Stratigraphy of upper Nehalem River basin, northwestern Oregon. American Association of Petroleum Geologists, Bull., v. 30, no. 2, pp. 213–237.

Warren, W.C., Norbisrath, H., and Grivetti, R.M. 1945. Geology of northwestern Oregon, west of the Willamette River and north of latitude 45<degrees> 15<minutes>. U.S. Geological Survey, Oil and Gas Investigation Map, OM 42.

Washburne, C.W. 1903. Notes on the marine sediments of eastern Oregon. Journal Geology, v. 11, pp. 224–229.

———. 1914. Reconnaissance of the geology and oil prospects of northwestern Oregon. U.S. Geological Survey, Bull. 590, 111p.

Weaver, C.E. 1942. Paleontology of the marine Tertiary formations of Oregon and Washington. University of Washington, Publications in Geology, v. 5, pt. I, II, III, 790p.

———. 1945. Stratigraphy and paleontology of the Tertiary formations at Coos Bay, Oregon. University of Washington, Publications in Geology, v. 6, no. 2, pp. 31–62.

Weaver, C.E., et al. 1944. Correlation of the marine Cenozoic formations of western North America. Geological Society America, Bull., v. 55, pp. 569–598.

Welton, B.J. 1972. Fossil sharks in Oregon. Ore Bin, v. 34, no. 10, pp. 161–170.

———. 1974. *Heptranchias howellii* (Reed 1946) (Selachii, Hexanchidae) in the Eocene of the United States and British Columbia. PaleoBios, no. 17, 15p.

———. 1979. Late Cretaceous and Cenozoic Squalomorphii of the northwest Pacific Ocean. Ph.D. diss., University of California, Berkeley, 553p.

Welton, B.J., and Farish, R.F. 1993. The collector's guide to fossil sharks and rays from the Cretaceous of Texas. Lewisville, Texas: Before Time Publ. 204p.

Whalen, M.T. 1988. Depositional history of an upper Triassic drowned carbonate platform sequence: Wallowa terrane, Oregon and Idaho. Geological Society America, Bull., v. 100, pp. 1097–1110.

Wheeler, G. 1982. Problems of the regional stratigraphy of the Strawberry Volcanics. Oregon Geology, v. 44, no. 1, pp. 3–7.

White, C.A. 1885. On invertebrate fossils from the Pacific Coast. U.S. Geological Survey, Bull. 51, pp. 28–32.

White, J.D.L. 1994. Intra-arc basin deposits within the Wallowa terrane, Pittsburg Landing area, Oregon and Idaho. *In*: Vallier, T.L., and Brooks, H.C., eds., Geology of the Blue Mountains region of Oregon, Idaho, and Washington; stratigraphy, physiography, and mineral resources . . . . U.S. Geological Survey, Professional Paper 1439, pp. 75–89.

White, J.D.L., and Vallier, T.L. 1994. Geologic evolution of the Pittsburg Landing area, Snake River Canyon, Oregon and Idaho. *In*: Vallier, T.L., and Brooks, H.C., eds., Geology of the Blue Mountains region of Oregon, Idaho, and Washington: stratigraphy, physiography, and mineral resources . . . . U.S. Geological Survey, Professional Paper 1439, pp. 55–74.

White, J.D.L., et al. 1992. Middle Jurassic strata link Wallowa, Olds Ferry, and Izee terranes in the accreted Blue Mountains island arc, northeastern Oregon. Geology, v. 20, no. 8, pp. 729–732.

Whiting, M.C., and Schrader, H. 1985. Late Miocene to early Pliocene marine diatom and silicoflagellate floras from the Oregon coast and continental shelf. Micropaleontology, v. 31, no. 3, pp. 249–270.

Whitlock, C. 1992. Vegetational and climatic history of the Pacific Northwest during the last 20,000 years . . . . The Northwest Environmental Journal, v. 8, pp. 5–28.

Whitmore, F.C., and Sanders, A.E. 1976. Review of the Oligocene cetacea. Systematic Zoology, v. 25, no. 4, pp. 304–320.

Wilkinson, W.D., 1959. Field guidebook; College Teachers Conference in Geology. Oregon Department of Geology and Mineral Industries, Bull. 50, 148p.

Wilkinson, W.D., and Oles, K.S. 1968. Stratigraphy and paleoenvironments of Cretaceous rocks, Mitchell Quadrangle, Oregon. American Association Petroleum Geologists, Bull., v. 52, no. l, pp. 120–161.

Wilson, R.W. 1934. A new species of *Dipoides* from the Pliocene of eastern Oregon. Carnegie Institute, Washington, D.C., Publication 453, pp.19–23.

———. 1938. Pliocene rodents of western North America. Carnegie Institute, Washington, D.C., Publication 487, pp. 21–73.

Wing, S.L. 1998. Tertiary vegetation of North America as a context for mammalian evolution. *In*: Janis, C.M., Scott, K.M., and Jacobs, L.L., eds. 1998. Evolution of Tertiary mammals of North America. New York: Cambridge University Press, pp. 37–60.

Wolfe, J.A. 1954. The Collawash flora of the upper Clackamas River basin, Oregon. Geological Society of the Oregon Country, Newsletter, v. 20, np. 10, pp. 89–94.

———. 1960. Early Miocene floras of northwest Oregon. Ph.D. diss., University of California.

———. 1962. A Miocene pollen sequence from the Cascade Range of northern Oregon. U.S. Geological Survey, Professional Paper 450-C, pp. 81–84.

———. 1969. Neogene floristic and vegetational history of the Pacific Northwest. Madrono, v. 20, pp. 83–110.

———. 1981. Paleoclimatic significance of the Oligocene and Neogene floras of the northwestern United States. *In*: Niklas, K.J., Paleobotany, paleoecology, and evolution, New York: Praeger, v. 2, pp. 79–101.

———. 1992. Climatic, floristic, and vegetational changes near the Eocene/Oligocene boundary in North America. *In*: Prothero, D.R., and Berggren, W.A., eds., Eocene-Oligocene climatic and biotic evolution. Princeton, N.J.: Princeton University Press, pp. 421–436.

———. 1993. A method of obtaining climatic parameters from leaf assemblages. U.S. Geological Survey, Bull. v. 2040, pp. 1–71.

————. 1994. Tertiary climatic changes at middle latitudes of western North America. Paleogeography, Paleoclimatology, Paleoecology, v. 108, pp. 195–205.

Wolfe, J.A., and Hopkins, D.M. 1967. Climatic changes recorded by Tertiary land floras in northwestern North America. *In*: Hatai, K., ed., Tertiary correlations and climatic changes in the Pacific. Pacific Science Congress ll, Tokyo 1966, Symposium 25, pp. 67–76.

Wolfe, J.A., and Tanai, T. 1987. Systematics, phylogeny, and distribution of *Acer* (maples) in the Cenozoic of western North America. Journal of the Faculty of Science, Hokkaido University, Ser. IV, v. 22, no. l, 246p.

Wolfe, J.A., Forest, C.E., and Molnar, P. 1998. Paleobotanical evidence of Eocene and Oligocene paleolatitudes in midlatitude western North America. Geological Society America, Bull., v. 110, no. 5, pp. 664–678.

Wood, A.E. 1935. Evolution and relationship of the heteromyid rodents. Carnegie Museum, Annals, v. 24, pp. 73–262.

————. 1957. What, if anything, is a rabbit? Evolution, v. 11, pp. 417–425.

Woodburne, M.O., and Robinson, P. T. 1977. A new late Hemingfordian mammal fauna from the John Day Formation, Oregon, and its stratigraphic implications. Journal Paleontology, v. 51, no. 4, pp. 740–757.

Worona, M.A., and Whitlock, C. 1995. Late Quaternary vegetation and climate history near Little Lake, central Coast Range, Oregon. Geological Society America, Bull., v. 107, no. 7, pp. 867–876.

Wortman, J.L., and Matthew, W.D. 1899. The ancestry of certain members of the Canidae, the Viverridae, and Procyonidae. American Museum Natural History, Bull.12, pp. 109–138.

Wright, J.E., and Frey, D.G., eds. 1965. The Quaternary history of the United States. Princeton, N.J.: Princeton University Press, 723p.

Yochelson, E.L. 1961. Occurrences of the Permian gastropod *Omphalotrochus* in northwestern United States. U.S. Geological Survey, Professional Paper 424-B, pp. 237–239.

Zittel, K.A., von 1913. Textbook of palaeontology. 3 vols. London: Macmillan.

Zullo, V. A. 1964. The echinoid genus *Salenia* in the eastern Pacific. Paleontology, v. 7, pt.2, pp. 331–349.

————. 1969. A late Pleistocene marine invertebrate fauna from Bandon, Oregon. California Academy of Sciences, Proceedings, 4th ser., v. 39, no. 12, pp. 346–361.

Zullo, V. A., and Chivers, D.D. 1969. Pleistocene symbiosis, pinnotherid crabs in pelecypods from Cape Blanco, Oregon. Veliger, v. 12, no. 1, pp. 72–73.

INDEX